Applications of Petroleum Tools
for Field Geologists

2ND EDITION
Applications *of* Petroleum Tools *for* Field Geologists

Key Topics on Petroleum Engineering Methods
A Guide to Lithostratigraphically Correlation Charts (Model Type-North Africa)
Disclosed several Petroleum, Paleontological and Structural Geological Issues

Fakhry A. Assaad

Mill City Press

Mill City Press, Inc.
2301 Lucien Way #415
Maitland, FL 32751
407.339.4217
www.millcitypress.net

© 2019 by Fakhry A. Assaad

All rights reserved. No part of this publication may be reproduced, stored in a retrieval system, or transmitted, in any form or by any means, electronic, mechanical, photocopying, recording, or otherwise, without the prior written permission of the author.

Printed in the United States of America

ISBN-13: 978-1-54566-243-4

Acknowledgement

put here
The acknowledgement of Dr. Chris Haynes.

Preface

The author worked for more than a decade, in several consecutive contracts as a petroleum geologist engineer and as a District Geologist, of the Exploration Directorate of the Algerian State Petroleum Company "Sonatrach/"/ (Societé Nationale de Transport et de Hydrocarbures); started at a virgin area in NE Algeria (District I), "normally given for the fresh appointees, a smart "move by senior officials"

During the second year of his first contract with "Sonatrach", the author participated in "The Eighth Arab Petroleum Congress, No 84 "B3", 1972"; organized by the Secretariat General of the league of Arab States; and submitted several lectures at the French Institute Algerienne de Petroliér (IAP), Institute Nationale de Hydrocarbures" (INH); as well as to the graduate students of the Geology Dept at the University of Alabama, besides the undergraduate students at Shelton State Community College, in Tuscaloosa, Alabama.

The author has been engaged in several internal reports of Sonatrach, submitted by "The Bulletin of "La Division de Hydrocarbures- Analytique Petroliér" of the Exploration Directorate, together with two scentific papers at the Journal of Petroleum Geology, (JPG, UK).

The author also participated in several scientific papers, at the Bulletin of the Fac. Sc., Cairo University, and as a co-author with a visitor professor, Dr. Hans Peter Jordan: {Der Einfluß "Saurgen Regens" auf Büden und Grundwasserleiter-cin Überliek, (1990)}.

The author personally worked with Late Dr. Philip E. LaMoreaus, the previous State Geologist of Alabama, in different scientific activities in USA, as an Author, co-Author, Editor, and Senior Editor, also, editing new scientific books submitted by the Environmental Geology magazine (EGM), Heidelberg, Germany; updated the AGI (normally once every four years), and reviewed a PHD Thesis (in four weeks. Period) on behalf of Dr.Phil E.LaMoreaux for a Graduate student at the University of South Africa.

It is interesting to mention that the previous boss of the author at the northern High Plateaux, Mr. Ait Hammouda had been transferred to together with the author to the main Dept of the Triassic Province of the southern Saharan Platform "District IV", and later updated several petroleum , and Staratighraphical isuues, which led to the discovery of the Oil Ring structure at the Base of the Hassi R'Mel Gas Field, **prepareed 15 copies of his discovery**, to be distributed to the Algerian Ministry of Petroleum and Energy together with the Head Quarters of top officials of the State Company " Sonatrach!!".

However, it was a fruitful period for the author when given a free hand as a senior geologist to continue studying on his own, different issues of Stratigraphy, structural geology, updated the old petroleum terms of the Triassic reservoirs, that had been previously used by the French companies and other delegates, each working separately, and caused a real confusion to define the proper tops of key reservoirs.

Later, the author, took the responsibility of the Canadian International Petrodata Incorporation (IPI CO) that dealt with the new Data Base Retrieval System, **while actually following up the drilling progress of the produtive wells of his own discovery**.

Finally, it may worth mentioning that the author started his career as a manager of the Darhib Talc Mine Southeast of the Eastern Sahara of Egypt and discovered the main Trunk of the rich Copper deposits of

Chalcopyrite or "Fools Gold" (70% copper). It was then nationalized, as a stratigic mineral, by the military Egyptian authority during Nasser Time. Ironically, the author faced an intended preplanned plot against him by the head of the Tribe "Sheikh Ali El-Barun., **at** the working site, while detonating the dinamete for collecting the rich copper ore,

It is worthy mentioning that the author, while starting his career at Derhib Talc Mine of the eastern Desert of Egypt, found several hills of iron and copper ore deposits over the location of the mine, and lately, while preparing a mining map, after defining its location on a topographic map, he discovered in a neglected draft deep under the ground the "Chalcopyrite ore, known as fools gold", which is 70% rich of copper" , (Assaad, 1960); while preparing the mining map; though, after reviewing the government contract with the owner of the mine, it had been nationalized by the military Egyptian authority during Nasser time, as a strategic mineral.

The author was very lucky to avoid the intended explosion by the chief of the tribe at a nearby working site of the Talc mine, just before going down to collect more specimens of the new discovered rich copper ore.

Author

*On the back cover of the book,
together with the pct. of the author*

Part (I) constitutes of five chapters on Key Topics of Petroleum engineering methods-
Part (II) A guide to lithostratigraphic corelation charts (A Model type- North Africa)-
Part (III) - Potenial hydrocarbon in North Africa (Journal Articles/Google Scholar "https://scholar.google.com"): (New Discoveries of on-shore & Off-shore oil and Gas fields in both Libya and Egypt)
Part (IV) Application of different petroleum tools for Field Geologists, started with citing a key well location at Nador Sud area (NAS-1) for drilling at a Virgin area in Northeast High Plateau of Algeria; carried out a regional electric log correlation with the nearest old wells at the southern Saharan Platform (ONR-4) inspite of the very complicated stratigraphy of the Atlas Trough in between, actiually due to the presence of a huge "SINK HOLE", several hudreds of thousands years ago, that followed by much younger deposits; which led to a denial of such corelation by the French scientists and other delegates of the Eastern Block, working for the very CO. of "Sonatrach".

It is interesting to mention that Mr. Ait Hammouda, the Algerian previous boss of the author at the northern High Plateaux was transferred together with the author to the main Dept of the Triassic Province of the southern Saharan Platform (DIstrict IV),

The author prepared 15 copies of the new discovery to be distributed among the Algerian Ministry of Petroleum and Energy together with the Head Quarters of top officials of the State Company " Sonatrach!!".

However, it was a fruitful period for the author when he was given a free hand as a senior geologist to update the old petroleum terms of the Triassic reservoirs, that had been previouly used by the French companies and other Romanian, Polish, and Russian deligates, each workuing separately, and caused a real confusion to define the proper tops of key reservoirs.

Later, the author, took the responsibility of the Canadian International Petrodata Incorporation (IPI CO) that dealt with the new Data Base Retrieval System, while been *actually checking the drilling progress of the oil produtive wells of his own discovery of the Oil Ring structure*.

List of Contents

PART I
Petroleum Exploration Methods

Chapter (1) Introduction –Petroleum Hydrocarbons 4
- 1.1 Historical Aspect 4
- 1.2 Petroleum Occurrence and Chemical Composition 4
- 1.3 Properties of Crude Oils 5
- 1.4 Natural Gas - Definition 5
- 1.5 Petroleum Hydrocarbon-Non-Reservoir Rocks 6
- 1.6 Petroleum Reservoir Rocks 6
- 1.7 Petroleum Migration and Accumulation 7
  - 1.7.1 Introduction 7
  - 1.7.2 Oil Migration 7
  - 1.7.3 The Rôle of Connate Water 7
  - 1.7.4 Differential Entrapment of Petroleum Hydrocarbons- Gussow Theory 8
  - 1.7.5 Petroleum Accumulation 9
    - 1.7.5.1 Other factors affecting oil accumulation 9
- 1.8 The Capacity of Oil and Gas Traps 10
  - 1.8.1 General Review 10

References 11
Cited References 11

Chapter (2) Plate Tectonics and Sedimentary Basins 12
- 2.1 Scope 12
- 2.2 Plate Tectonics and Plate Interactions 13
  - 2.2.1 Introduction 13
  - 2.2.2 Plate Tectonic Features 14
  - 2.2.3 Plate Interactions 14
- 2.3 Continental Shelves 17
  - 2.3.1 Development of Sedimentary basins 18
    - 2.3.1.1 Geodynamic Settings of Sedimentary Basins 19
    - 2.3.1.2 Sedimentary Basins and Hydrocarbon Occurrences 20
  - 2.3.2 Rifted Margins and Sedimentary Prism 21
  - 2.3.3 Foreland Basins 22

    2.4   Historical Aspect of the sub-seafloor fluid 23
    2.5   Subduction and Oil Migration 23
    2.6.   Convection Currents and Subduction Zones 24

References 24
Cite References 25

Chapter (3) Surface Geophysical Petroleum Methods, Remote Sensing, and Satellite Images in Salt Dome Exploration 26

    3.1   Introduction 26
    3.2   Magnetic Survey 26
    3.3   Gravimetric Survey 26
    3.4   Seismic Exploration Survey 27
         3.4.1   General 27
         3.4.2   Seismic Refraction Methods 27
         3.4.3   Seismic Reflection Method 27
               3.4.3.1   Multiple Reflections and Marine Exploration 27
               3.4.3.2   Seismic Profiling of Diapiric Structures 28
    3.5.   Remote Sensing and Land-Satellite Images in Salt Dome Exploration 28

References 28

Chapter (4) Drilling Technology in Petroleum Geology 29

    4.1   Introduction 29
    4.2   Petroleum Drilling Operations 31
         4.2.1   Discussion 31
         4.2.2   Types of Drilling Operations 33
    4.3   Drilling Fluids 34
         4.3.1   Properties of Drilling Muds 34
         4.3.2   Composition and Nature of the Drilling Fluids 34
         4.3.3   Different kinds of Drilling Mud Fluids 35
         4.3.4   Mud Systems in Salt structures 35
         4.3.5   Salt Dome Drilling 36
    4.4   Drilling Hazards 36
    4.5   Drill Stem Testing (DST) 36
         4.5.1   Scope 36
         4.5.2   The Halliburton Formation Testing Procedure 38
    4.6   General Remarks 39
         4.6.1   The DST-Procedure 39
         4.6.2   Required Conditions and Reasons for Carrying Out a DST in a Petroleum Reservoir Formation 40

Appendix 4.A – Selected Graphs for Drill Stem Testing in different Reservoir Rocks and Interpretation **(Figs 4.7-4.39)**. 43

References 50

Chapter (5) Geophysical Well Logging Methods of Oil and Gas Reservoirs and Well Log Interpretations 52

    5.1   Introduction 52
    5.2   Borehole Parameters and Rock Properties 52
    5.3   Resistivity Measurements by Well Electric Logs 55
         5.3.1   Definition 55

		5.3.2 Annulus and Resistivity Profiles – Hydrocarbon Zone	56
5.4	Formation Temperature (T_f)		57
5.5	Specific Log Types		57
	5.5.1	Spontaneous Potential Logs (SP)	57
	5.5.2	Resistivity Logs (R)	58
		5.5.2.1 Induction Electric Devices	59
		5.5.2.2 Electrode Resistivity Logs	60
		5.5.2.3 Micrologs (ML)	62
	5.5.3	Porosity Logs	62
		5.5.3.1 Sonic Logs	62
		5.5.3.2 Density Logs	63
		5.5.3.3 Neutron Logs	63
		5.5.3.4 Combination Neutron-Density Logs	64
		5.5.3.4.1. Examples and Interpretations	64
	5.5.4	Gamma Ray Logs	68
5.6	Well Design and Well Type Completions		70
	5.6.1	Scope	70
	5.6.2	Open Hole Completions	71
	5.6.3	Perforated Completion	71
	5.6.4	Screening Techniques	72
		5.6.4.1 Screen Liners	72
		5.6.4.2 Gravel Packing	72
	5.6.5	Other Well Type Completion	73

References 73

PART II
A Guide to Compurterized Lithostratigraphic Correlation Charts - Petroleum Resources -A Model Type /North Africa

Chapter (6) Mineral and Petroleum Resources- Petroleum Provinces, and Computerized Lithostratigraphic Correlation Charts of the Sedimentary Section in North Africa 76

6.1	Introduction		76
6.2	Mineral and Petroleum Resources in North Africa		77
	6.2.1	Scope	77
	6.2.2	The Algerian Territory	77
	6.2.3	Thge Libyan Territory	78
	6.2.4	The Egyptian Territory	80
6.3	Structural Geology of Algeria		80
	6.3.1	The Algerian Alpine	81
	6.3.2	The Algerian Saharan Platform	81
6.4	Petroleum Classification of Provinces		84
	6.4.1	Petroleum Province-Definition	84
	6.4.2	Total Petroleum System	84
	6.4.3	Geological Provinces of the Algerian Sahara	84
6.5	Regions of the Saharan Platform		85
	6.5.1	The Western Saharan Region	85
	6.5.2	Northeast Triassic Region	86
	6.5.3	The Eastern Region	86
6.6	An Approach to a Computerized Lithostratigraphic Charts in North Africa		86

		6.6.1	Discussion	86
		6.6.2	A Numeric Coding of Different Elements of the Geologic Provinces of the Algerian Sahara	89
		6.6.3	Structural Settings of the Algerian Sahara	90
		6.6.4	Stratigraphy of the Algerian Sahara	92
	6.7	A Geological Study of an Exploratory well (RY -1), at the Eastern Border of the Algerian Sahara		97
	6.8	Petroleum Geology of the Libyan Sahara		98
		6.8.1	Scope	98
		6.8.2	Stratigraphy of Sebha Area, West of Libya	98
		6.8.3	A General Lithostratigraphic Comparison of Sedimentary Section, West of Libya and Northeast Algerian Sahara	101
	6.9	The Egyptian Territory (Northeast Africa)		102
		6.9.1	The Western Desert of Egypt	102

- 6.9.1.1 The Qattara -Siwa Depression ... 104
- 6.9.1.2 The Bahariya Oases ... 105
  - 6.9.1.2.1 The Sedimentary section ... 106
  - 6.9.1.2.2 El Bahariya Petroleum Exploration Well #51 ... 106
- 6.9.1.3 Geological Results of Assiut-Kharga well ... 108
  - 6.9.1.3.1 Discussiosn ... 110
  - 6.9.1.3.2 Local Correlation ... 110

- 6.10 The Nile Delta of Egypt ... 111
- 6.11 The Gulf of Suez, Egypt ... 113
- 6.12 The Miocene Reservoir of El-Morgan off-Shore Oil ... 115

References ... 117
Web Sites ... 118
Cited References ... 118
Appendix 6A-1 - Geological Time Scale ... 119

PART III
Several Article Papers of Chapter 7-ABC on petroleum activiies in North Africa
Part III -7ABC - Several Article papers on the present Petroleum Activities in North Africa and comprise two Chapters at the northern High Plateaus of Algeria (District #1):

Chapter (7) ABC – Several Articles Scientific Papers on the Petroleum Activities in Northeast Africa ... 122

- 7A Potential Hydrocarbon in Northwest Africa ... 122
  - 7A.1 Introduction ... 122
  - 7A.2 Hyrocarbon Resrves in North Africa ... 123
  - 7A.3 Stratigraphy ... 124
  - 7A.4 Geological History and the Mesozoic Era of North Africa ... 125

References ... 126
Cited Refrence ... 126

- 7B Petroleum Geology and Potential Hydrocabnon Plays of Heavy Oil (Off-Shore Mediterranian Conference/ Ganub El Wadi Petrolrum Holding CO.) ... 127
  - 7B.1 Abstract ... 127
  - 7B.2 Introduction ... 127
  - 7B.3 Stratigraphy and Structural Events ... 129
  - 7B.4 Examples from Egypt: Abu Dubra Oil Area (Geographic setting) ... 130

	7B.5	Conclusion and Recommendations	131
Cited References **131**
7C	The Sirte Basin Province of Libya-0Sirte-Zelten-Total Petroleum System	133
	7C.1 Foreword	133
	7C.2 Abstract	133
	7C.3 Introduction	134
	7C.4 Geology & Boundary	136
	7C.5 Geographic Settings	137

PART IV
A Long Term Petroleum Project "A Precise Research Study on a Virgin Location in NW Africa, correlated with Regional structures which led to the discovery of an Oil-Ring within an important Gas Field" -for the first time in the Middle East

Chapter (8) A Regional Petoleum Study of Algeria-A General Review. 140

8.1	Historical Aspect	140
8.2	Location and morphology	141
8.3	Geological Framework of the Algerian Alpine	141
	8.3.1 The Northern Off-shore Area	141
	8.3.2 The Tellian Atlas	141
	8.3.3 The Hodna Basin	142
	8.3.4 The High Plateaux	142
	8.3.5 The Saharan Atlas elongated Trough	142
8.4	Geodynamic Evolution	144
8.5	Tectonic Outline	145
8.6	Updated Study on the Main Geological Regions of Algeria	145
	8.6.1 The Scope	145
	8.6.2 The Western Saharan Region	146
	8.6.2.1 The Tanezzuft -Timimoun Total Petroleum System (205801) of the Grand Erg/Ahnet Province	146
	8.6.3 The Northeast Central Triassic Region	149
	8.6.4 The Eastern Geologic Provinces	150
	8.6.4.1 The Trias/Ghdames Geologic Province	151
	8.6.4.2 The Illizi Province	153
8.7	A Geological Aspect- The Allochthonous Triassic Evaporitic Region of the Betic–Maghrebean Domain	156
	8.7.1 Introduction	156
	8.7.2 Discussion	156
	8.7.3 Western Betic Cordillera (Spain)	157
	8.7.4 The Central Betic cordillera (South¤Spain)	157
	8.7.5 The Perifaine Nappe Rif Cordillera (Morocco)	157
	8.7.6 "Zone des Domes" of Tell Cordillera (Algeria and Tunisia)	158
8.8	Comparison with the Triassic Evaporites of the Maghrebian Domain and Betic Cordillera (Spain)	158
	8.8.1 Historical Aspect	158

References **158**
Coted References **159**

Chapter (9) Discovery of a Triassic basin at a Virgin Area in northeast Algeria, Defined its trend and Boundary – Discussed the Lithology, Electric well Log, and Palynological correlation, with that of the Triassic Province of the Saharan Platform, and determine its extention–Stratigraphy and Sedimentation, Stratighraphic Evolution 160

- 9.1 Introduction 160
- 9.2 Triassic Deposits and Triassic Outcrops 161
- 9.3 An extended Regional Triassic Province of the Saharan Platform 162
  - 9.3.1 Historical Aspect 162
  - 9.3.2 A Discovery of a new Triassic Basin, NE of the High Plateaux 162
  - 9.3.3 The Triassic section to the NE of the High Plateau is a mirror image of that of the Saharan Platform 165
  - 9.3.4 A Lithotartigraphic Correlation of the northern Triassic basin with that of the Triassic Province of the Saharan Platform 167
    - 9.3.4.1 The Lower series (SI) 169
    - 9.3.4.2 The Triassic Argilo-sandstone series (T1+T2) 170
    - 9.3.4.3 The Lower Shales 171
    - 9.3.4.4 The Evaporite Deposits 171
- 9.4 Stratigraphy and Sedimentology 172
  - 9.4.1 Scope 172
  - 9.4.2 Discussion 173
  - 9.4.3 Stratigraphic Evolution 175

References 179

Chapter (10) The Petroleum Reservoir Characteristics of the Triassic Province of the Algerian Saharan Platform- Emphasizing the Dome-like structure and the Discovery of an Oil Ring structure at the base of the Hassi R'Mel Gas field – Application of the Gussow Theory-Tectonic settings of the related structures-The Evolution of Hydrocarbon Migration-Methodology 180

- 10.1 The Triassic Province and the Oil and Gas Fields of the Algerian Saharan Platform 180
- 10.2 Petroleum Reservoir Rocks of the Algerian Saharan Platform 181
  - 10.2.1 Introduction 181
  - 10.2.2 Petroleum Characteristics 182
    - 10.2.2.1 The Paleozoic Reservoirs 182
    - 10.2.2.2 The Mesozoic Reservoirs 182
- 10.3 The Main Productive Structures 184
  - 10.3.1 The Tinrhemt High 184
  - 10.3.2 The Oued M'ya Depression 184
  - 10.3.3 The Rhadames (Ghadames) Depression 187
- 10.4 Petrophysical Aspects, and Lithostratigraphy of Hassi R'Mel, Oued Noumer, and Haoud Berkaoui oil and gas fields 188
  - 10.4.1 Scope 188
  - 10.4.2 The Hassi R' Mel Gas Field 189
    - 10.4.2.1 Location 189
    - 10.4.2.2 Structural Geology 190
    - 10.4.2.3 Lithostratigraphy 191
    - 10.4.2.4 Hydrocarbon Perspective of Hassi R'Mel Gas Field 195
  - 10.4.3 Oued Noumer and Ait Kheir Fields 199
    - 10.4.3.1 Topography 199
    - 10.4.3.2 General Geology 200

		10.4.3.3 Classification of the Triassic Sandstone Formation at the Oued Noumer Area	200
		10.4.3.4 Characteristics of the Reservoir Rocks	201
	10.4.4	Haoud Berkaoui and Haniet El-Mokta Oil Fields	205
10.5	Application of Gussow Theory on the Triassic Reservoirs. The Efolution of the Hydrocarbon Migration -Methodology		205
	10.5.1 Discussion		205

References 206

Chapter (11) - An Approach to Halokinematics and Interplate Tectonics- Development of The Triassic Salt domes in North Algeria –its Stratigraphic Relation with that of the Gulf of Mexico - The Economic Aspects of Salt structures 207

11.1	Introduction	207
11.2	Paleostructural History	208
11.3	Evaporite Rocks of the Eastern and Western Hemispheres	210
	11.3.1 Scope	210
	11.3.2 Development of Salt structures and Salt Basins	210
	11.3.3 Types of Salt Movements and the controlling factors-North Algeria	211
	11.3.3.1 Halokinesis and the Triassic/Liassic Salt Dome Structures,	211
	11.3.3.2 Salt Domes and Piercing Diapirs	213
	11.3.3.3 Disclosure of several Issues related to the Discovery of a new Triassic basin in NE Algeria	216
	a) The Mis-interpretation of the disappeance of salt deposits at Cedraia area (CED-1 well)	216
	b) The Salt Diapirs at "Rocher de Sel" of Djelfa, south of Cedraia area	216
11.4	Halotectonics and Salt Outcrops of the High Plateaus	216
11.5	Halotectonics and the salt outcrops at the Saharan Atlas	218
11.6	Plate Tectonics and Halokinematics	218
	11.6.1 Salt movements and plate kinematics	219
11.7	Eruptive Rocks, and Interplate Tectonics	220
11.8	Evaporite Deposits in NW Africa and the Gulf of Mexico	222
	11.8.1 The Triassic/Liassic Salt Dome Structures of the Gulf of Mexico and its Correlation with that of the Arabian Maghreb, NW Africa	222
	11.8.2 Salt Dome structures and the Alpine in North Algeria	224
	11.8.3 Development of Salt Basins in the Gulf of Mexico	224
	11.8.4 Configuration and Composition of Salt Dome Structures in the Gulf of Mexico	224
11.9	Separation of Gulf of Mexico and Africa	225
11.10	Is Oil Migration related to the partial subduction along the Tellian Flexure at the Western Coastal Zone of Algeria	225
11.11	Economical Aspects of Salt Structures	226

References 226
Appendix 11A.1- More about Evaporites –Special Readings 227

Chapter (12) - An estimate of Petroleum Reserves of the Triassic reservoirs at the Hassi R'Mel-M'Zab High and Oued M'ya Basin, NE of the Algerian Saharan Platform 228

12.1	Scope	228
12.2	Introduction	228
12.3	Petroleum Hydrocarbon Reserves of Oued Noumer Area	229
	12.3.1 Discussion	229

   12.3.2 Reservoir Characteristics 231
   12.3.3 Hydrocarbon Potenialities 232
   12.3.4 Parameters used for Measurement of Reserves 234
   12.3.5 An Approximate Estimate of Reserves of Oued Noumer and Ait Kheir Fields 235
 12.4 Petroleum Hydrocarbon reserves of Hassi R'Mel Gas field 235
   12.4.1 Historical and Petrographical Aspect 235
   12.4.2 Sedimentology 236
   12.4.3 Reservoir Characteristics 237
   12.4.4 An Approximate Estimate of Gas Reserves 239

References 241
Appendix 12.A1 –English Translation of Table 12.5. 242
Appendix 12. A2- A possible Gravimetric Method of Hydrocarbon Prospects 243
Appendix 12.A3 243

List of Figures

Part I
Petroleum Exploration Methods

Fig 1.1abc – Paraffins Series, Cycloparaffins and Benzine ring (Gatlin, C., 1960) 5
Fig 1.2 – Diagrammatic sketch showing the path water takes in flowing from intake area
   A to outlet area B as it passes across synclines and anticlines Levorsen, (1954) 8
Fig 1.3 – Differential Entrapment of petroleum hydrocarbons – Gussow Theory 8
Fig 2.1 – Position of Main Tectonic plates 13
Fig 2.2 – Development of foreland basin and suture belt from continental collision between
   arc-trench system and rifted-margin sediment prism 15
Fig 2.3 – A diagram illustrating subduction of an oceanic plate, with evolution of volcanic
   island arc, marginal basin and trench. 16
Fig 2.4 – Principal kinds of plate interactions at Divergent and convergent Plate Junctures. 17
Fig 2.5 – World coninental Shelves (Petrol.Press Services (London, 1951) 18
Fig. 2.6 – A cross section of sedimentary basin "geosyncline Trough" 18
Fig 2.7 – Schematic Diagrams to illustrate sedimentary basins associated with intraoceanic &
   continental margin magmatic arcs 19
Fig 2.8 – Schematic diagrams to illustrate sedimentary basins associated with crustal collision
   to form intercontinental suture belt with collison orogen. 20
Fig 2.9 – Growth of rifted margin sediment prism.(stippled) 22
Fig 2.10 – Tectonic Position (Schematic) of Foreland Fold -Thrust Belt " Petroleum,
   Press Service, London, 1951". 23
Fig 2.11 – Schematic cross section depicting various types of sub-sea floor fluid flow
Fig 4.1 – Standard Cable Tool Drilling System 30
Fig 4.2 – Basic Components of a rotary drilling rig 31
Fig 4.3 – Specific Applications of Directional Drilling 32
Fig 4.4 – Three Diagrams showing major types of Directional Drilling Methods on Salt Domes ... 33
Fig 4.5 – Typical Conventional Drill-Stem Test Tools 37
Fig 4.6a – A detailed illustration of the Drill Stem Testing 38
Fig 4.6b – A record of Fluid passage Diagram for a conventional Bottom Section / DST. ....
Figs 4.7- 4-14 – Different illustrations of Drill Stem Testing graphs 39
Appendix 4A (Fig 4.15-4.39) – Selected graphs of Drill Stem Testing in different reservoir rocks 43
Drill Stem Testing – Author's example – Explain two flow periods & two shut-in- periods ....
Fig 5.1 – A Diagram shows the invasion of fluids through the surrounding rock 54
Fig 5.2 – Resitivity Profile - Hydrocarbon Zone 56

Fig 5.3 – SP deflection with resistivity of the mud filtrate much greater than formation water 58
Fig 5.4 – Dual Induction Focused log curves through a hydrocarbon-bearing zone 60
Fig 5.5 – Dual Laterolog –Microspherically Focused Log curves through hydrocarbon
bearing zone. 61
Fig 5.6 – A Combnation of Neutron.Density Log with Gamma Ray log and Caliper ....
Fig 5.6a – Chart for correcting Neutron -Density Log porosities for lithology where
freshwater-based drilling mud is used ....
Fig 5.6b – Chart for correcting Neutron-Density Log porosities for lithology where
saltwater-based mud is used ....
Fig 5.7 – A generalized lithology logging with combination Gamma Ray-Neutron-Density log 67
Fig 5.8 – Schematic illustration of neutron-density responses in gas-bearing sandstones
(modified after Truman et al, 1972)... 68
Fig 5.9 – Gamma Ray/Density log shiows curves and scales. 70
Fig 5-10 – A sketch of open hole complletion in oil -productive well 71
Fig 5.11 – Method of plottiing screen analysis data... 72
Fig 5.12 – A simplefied method of gravel packing cvommenly used in Gulf Coast 73

Part II
An Approach to a Computerized Lithostratigraphic Correlation Chart - (Model Type North Africa)

Fig 6.1 – North Africa – Geology and oil and gas fields. 77
Fig 6.2a – The mineral resources of the Sahara 78
Fig 6.2b – Fort Polignac oil field 79
Fig 6.2c – The East Libyan Oil Field... 79
Fig 6.2d – The Mineral Resources of Egypt 80
Fig.6.3 – Natural boundaries of Algeria and the main morphological zones 82
Fig 6.4 – North Central Africa 85
Fig 6.4a – A postulated Slatch Diagram showing the Tectonic structures in Algeria 91
Fig 6.5 – Regional structural map of the Algerian Sahara 91
Fig 6.6 – Old and recent classifications of Lower Liassic/Triassic formations of Algiers 93
Fig 6.7 – Location map of RY-1 well at the western border of Algeria 100
Fig 6.8 – Cross section in the Berkine Basin.. 101
Fig 6.9 – Location map of Sebha area, Libya... 101
Fig 6.10 – Location map of El-Razak and Abu Gharadig Oil and Gas Fields, NW
of the Western Desert of Egypt.. 104
Fig 6.11 – A Geological map of Bahariya Oases 105
Fig 6.12 – A geological cross section from Mersa Matruh, northwest coast of Egypt to
Kharga-Assiut well on the eastern escarpment of Kharga Oases. 108
Fig 6. 13 – Location map of Kharga-Assiut well, Egypt.. 108
Fig 6.14 – A geological cross section passing by Gebel Um-El Ghanaiem, Um El Kosour
and Assiut- Kharga wells 111
Fig 6.15 – Location map of Northern Nile Delta, Egypt.. 112
Fig 6.16 – Generalized lithostratigraphic column of the Nile Delta.. 113
Fig 6.17 – Location map of Gulf of Suez showing the Morgan oil field and the location of
CC' cross section. 114
Fig 6.18 – A Structural cross section CC' in the Gulf of Suez area. 114
Fig 6.19 – A simplified Lithostratigraphic log of the Morgan oil field, Egypt..................................... 116

List of Charts

Chart 6.1 – A Lithostratigraphic Correlation Chart of sediments in Northeast Center of the Algerian Sahara 83
Chart 6.1a – A Lithostratigraphic correlation chart of the Cenozoic Formation, NE of the Algerian Saharan Platform, Algeria 88
Chart 6.1b – A Computer application of a Lithostratigraphic Chart of The Mesozoic Formation, NE of the Algerian Saharan Platform, Algeria 95
Chart 6.1c – A Lithostratigraphic correlation Chart of The Paleozoic section of the Algerian, Saharan Platform 96
Chart 6.2 – Bahariya Oases-well # 1 (old #51) 107
Chart 6.3 – Assiut-Kharga composite well log 109
Appendix. 6A.1 – A Geological Time Scale 119

Part III
Chapter (7-ABC) - Potential Hydrocarbon in N. Africa-

Fig 7.A1 – Hydrocarbon resources in North Africa 123
Fig 7.A.2 – Paleogeographic Map of Godwana land 125
Fig 7.B.1 – Location Mapof the sudy areasin both Egypt and Libya 128
Fig 7B.2 – Stratigraphic column of the study areas 129
Fig 7B.3 – Abu Durba Area (3-figs) 130
Fig 7.C-1 – Sirte Zellen Total petroleum system 130
Fig 7C-2 – Structural elements of Sirte Basin 135
Fig 7C-3 – Cetral Mediterranean Sea 136

Part IV
Petroleum Geology of NW Africa

Fig 8.1a – Tectonic map of the Saharan basin nothwest Afriuca 144
Fg 8.1b – A postulated sketch Diagram showing Tectonic Structures 144
Fig 8.2 – Distribution of proven Hydrocarbon Resources in North Africa 146
Fig 8.3 – North Central Africa showing USGSdefined geologic provinces and major stuctures 148
Fig 8.4a – Stratigraphic cross sectionAÁ through Grand/Ahnet/Ahnet and neighboring providences . 148
Fig 8.4b – Stratigraphic cross section BB' North to South through Oued M'ya and Mouydir 149
Fig 8.5 – Columner section; stratigraphic Nomenclature and Correlation of the eastern basins. 150
Fig 8.6 – North Central Africa showing USGS defined geologic provinces and major structures 151
Fig 8.7a – Stratigraphic cross sections AA', through Trias/Ghadames and Illizi provinces 152
Fig 8.7b – Stratigraphic cross sections BB'through Oued M'ya and Ghadames Basins 152
Fig 8.8 – Schematic cross section of the Triassic Illizi petroleum Province 154
Fig 8.9 – Illizi Provnce showing 2-stratigraphic locations AA" & CC' together with geologic provinces and Major structures 155
Fig 8.10 – Stratigraphic Cross sections CC' through Mouydir &Ilizi basins 155
Fig 8.11a – Landsat image of the Betic- Maghrebian Domain, and the location of the cross section (white strip, Fig. 2). 157
Fig 8.11b – A cross section through the Western Betic Cordillera (Spain), based on well -log data, Lanaja, (1987) 157
Fig 9.1 – A Geologic Map in NE Central Algeria - locations and cross sections 161
Fig 9.2 – Stratigraphic Correlation among wells in NE Central Algerian High Plateaus ... 163
Fig 9.3a – A Seismic section at Nador Sud "noisy area" 163
Fig 9.3b – An Isochrone Map on top of Triassic Deposit at Nador Sud area 163

Fig 9.4 – A Lithology Well Log of NAS-1 164
Fig 9.5 – A Geologic Cross section among wells in NE Central of the High Plateaus 165
Fig 9.6 – A Geological cross section of wells at the main three Structures of the
High Plateaus, Saharan Atlas, and the Saharan Platfrom 166
Fig 9.7ab – Two Lithostratigraphic cross sections of Triassic beds among wells of the
High Plateaus, and the Saharan Platform (D2-Datum plane) 166
Fig 9.8 – Electric Log Correlation of the Triassic Formation of wells in the High Plateaus
and the Saharan PlatformFig 9.9a – A facies Map of Trias/Lias deposits in
NE Central Algeria 173
Fig 9.9b – A Postulated Regional; Facies Map of the Triassic Province of Algeria 174
Fig 9.10 – A NE/SW Schematic Structural Cross section showing the Evolution of
Sedimentary Formation among the main structures in Algeria 177
Fig 9.11a – Lithostratigraphic Diagrams for the encountered wells in the studied area 178
Fig 9.11.b – A NW/SE Regional structural cross section among wells in East of Algeria 179
Fig 10.1 – A Locaion Map showing different Oil and Gas Fields of Algeria 185
Fig 10.2 – NE/SW structural cross section in Haoud Berkaoui oil field 185
Fig 10.3a – Isobaths mapS in Haoud Beraoui and Ben Kahla oil fields, showing the location
of three sections I-I, II-II and III-III 186
Fig 10.3b – Three structural cross sections IV, V & VI at both Haoud Berkaoui/Guellala
and Ben Kahla oil fields 186
Fig 10.4 – A Subcrop map of the Discordance Hercynian, shows Limits of deposition, of the
Triassic sediments, NE center of Algiers 188
Fig 10.5 – A geographic Location Map of Hassi R'Mel Gas Field. 189
Fig 10.6 – Old Isopaches map of the Triassic Argilo-sandstone "B" series at Hassi R'Mel
Field as an open structure (SN Repal, Feb.1972). 190
Fig 10.7 – Isobaths map on top of the Hercynian discordance 191
Fig 10.8 – Isobaths Map on top of the lower Triassic sandstone series (SI: Serie Inferieur) 192
Fig 10.9 – Isobaths Map on top of the lower intermediate series "2a-unit" (C-series)........................
Fig 10.10 – Isobaths Map on top of the middle intermediate series "2b-unit" (B-series)..................... 193
Fig 10.11 – Isobaths map on top of the Upper Triassic shaly sandstone"3a-unit"
(A -sandstone reservoir) 194
Fig 10.12 – Boundaries of Deposition of the 2a (C-series); 2b and lower 2c -series)
(B- series), and 3a- (A-series) sandstone units. 194
Fig 10.13 – Well location Map and three structural cross sections passing by
Hassi R'Mel field and nearby wells 197
Fig 10.14 – A NW/SE Paleogeologic and Structural cross section passing by HR-7,
HR-16, HR-1, and HR-3 198
Fig 10.15 – A NE/SW Paleogeoloic and structural cross section passing by HR-4,
HR-9, HR-1, HR -10, HR-12, and HR-8 198
Fig 10.16 – A N/S structural cross section passing by HRS-1, HR-10, HR-11,
HR-14TR-1 and NL-5 199
Fig 10-16a – A geologic Cross section of the Triasssic detritals among wells of Hassi R'Mel field...... 199
Fig 10.17 – Isopaches of Sandstones unit 2b (old B-series) in Oued Noumer field 201
Fig 10.18 – Isobaths Top of unit 3a-unit "old A-series" in Oued Noumer Area 202
Fig 10.19 – Isobaths on Top of sandstone of unit 2b-2c units "old unit B " for both
Oued Noumer and Ait Kheir Fields 202
Fig 10.20 – Isopaches of Sandstone 3a unit (Old A-series) for Oued Noumer Field 203
Fig 10.21ab – a) Isochrone map and a structural Geological Section on top of the upper
intermediate seologiceries (top of 2c-unit or "B"- series); and b) A structural
schematic section of Oued Noumer field 204

Fig 10.22 – A Structural cross section –Top Triassic Sandstone and Drill Stem test results of Oued Noumer and Ait Kheir fields ....
Fig 10.23a – A Schematic cross section showing the paleostructural evolution of petroleum migration through the Triassic reservoirs in NC East of Algeria..........................
Fig 10.23b – An explanatory NW-SE schematic cross section, passing by several oil and gas Fields. ....
Fig 11.1 – Movement of continents from Pangaea period to Nowadays 208
Fig 11.2 – A sketch diagram showing two subductiuon zones at the south Tellian and South Atlassic flexure belts.. 209
Fig 11.3 – Density Contrast with Depth between Salt and overlying Sediments. 212
Fig 11.4 – Evolution of Permian salt dome Structures in N-Germany 213
Fig 11.5 – Location Map of Salt Domes, NE Centeral Algeria 214
Fig 11.6 – Schematic Diagrams showing the evolution of salt Dome Structures in NE Centeral Algeria. 215
Fig 11.7a – Triassic Outcrops at Dj Nador Sud Area. 217
Fig 11.7b – Geographic location of Triassic Outcrops at Sersou (ex-Chellala- Reibell). 218
Fig 11.8 – A Composite Paleogeographic Map from the Tehys Ocean to the Mediterranean sea 219
Fig 11.9 – Schematic Diagrams showing evolution of Intracontinental Plates 221
Fig 11.10 – A generalized N/S Schematic Cross section showing tectonic forces and structural features In NE Central Algeria 222
Fig 11.11a – A Location of prominent structural features of Gulf of Mexico............... 223
Fig 11.11b – A Diagrammatic Map outlines of the major structural features of the southern USA and Gulf Region (Halbouty, 1970).. 223
Fig 11.12 – An idealized section showing common types of hydrocarbon traps associated with salt domes 225
Fig 12.1 – Isobaths Top Triassic Sandstone "T2-unit/A at Oued Noumer Area 230
Fig 12-2a – Isobaths Top Triassic Sandstone "T1-unit/B at Oued Noumer Area............. 230
Fig 12.2b – Isobaths of top T1-unit (B-series) of both Oued Noumer and Ait Kheir fields........ 230
Fig 12.3 – Isopaches of Net pay sandstone "T2-unit/A at Oued Noumer Area 230
Fig 12.4 – Isopaches of Net pay sandstone "T1-unit/B at Oued Noumer Area 231
Fig 12-5 – A Geographic Location of Hassi R'Mel Field 236
Fig 12-6 – A Subcrop map of the Hercynian Discordance of Hassi R.Mel Field 242
Fig 12.7 – Isobaths top Triassic Sandstone "T2-unit/A" of Hassi R'Mal Field................ 237
Fig 12.8 – Isobaths top Triassic Sandstone "T1-unit B"-series of Hassi R'Mal Field.......... 238
Fig 12.9 – Isobaths top Triassic Sandstone "T1-unit C-series" of Hassi R'Mal Field......... 238
Fig 12.10 – Isopaches Net Pay Sandstone "T2-unit/A-series" of Hassi R'Mal Field 239
Fig 12.11 – Isopaches Net Pay Sandstone "T1-unit/B-series" of Hassi R'Mal Field.......... 239
Fig 12.12 – Isopaches Net Pay Sandstone "T1-unit/C-series" of Hassi R'Mal Field 240

List of Tables

Table 4.1 – Principal properties of water-based drilling fluids................................
Table 5.1 – Geophysical well logging methods and practical applications 53
Table 5.2 – Sonic velocities and interval transit times for different matrices used in the Sonic Porosity Formula.. 62
Table 5.3 – Matrix densities of common lithologies uased in the dsensity posorsitiy formula. 64
Table 6.1 – A numeric code for theas Total Petroleum Systems - Grand Erg/Ahnet Province 89
Table 6.2 – A numeric code for the Total Petroleum Systems-Illizi Basin 89
Table 6.3 – A numeric code for the Total Petroleum Systems –Trias-Ghadames Province.................... 90
Table 6.4 – General well data of Rhourde Yacoub well (RY-1)................................ 99
Table 6.5 – Formation tops from Micropalaentological results in some wells at Sebha area 103

Table 6.6 – Stratigraphical Units in Assiut-Kharga well 110
Table 8.1 – Comparison between the old and recent classification of the Trissic formation ....
Table 9.1 – Comparison between old classifications of the Triassic formation of the
High Plateaus by (Caratini, 1970), Technoexport (1970) and Assaad, (1972).................... 167
Table 9.2 – Comparison of Trias/Lias classifications, of the High Plateaus and the
Saharan Platform between (Assaad, 1972) G. Busson (1970)... 168
Table 9.3 – Former classifications of the Triassic/Liassic formations of both the High Plateaus
and the Saharan Platform by Achab, (1971), Stoica (1972, 1978), and Becip (1978)........ 168
Table 10.1 – General Well data of the Triassic reservoir rocks in difeerent oil and gas fields............. 185
Table 10.2 – Comparison between the Old and recent Classification of the Triassic Formation.......... 192
Table 10.3 – General well data and Petrophysical characteristics of Hassi R'Mel reservoirs 196
Table 10.4 – Drill Stem Testing (DST) Results of the upper Triassic reservoirs of the
Hassi R'Mel gas Field. 197
Table 10.5 – General well and Reservoir Data of the upper Shaly Triassic Sandstones-
Oued Noumer and Ait Kheir Fields 197
Table 12.1 – Different Positions of oil/well contact due to different analyses of the Reservoir rock 232
Table 12.2 – Determination of porosity Percent of both "A" & "B" Petroleum factors for reserveses at
Oued Noumer & Ait Kheir fields 232
Table 12.3 – Permeability Measurements "average arithmatic and mediene) of
Oued Noumer area 233
Table 12.4 – Petroleum Factors for Reserves of Oil, Gas and Gazoline Fields 234
Table 12.5 – General well data of isobaths and isopaches of Oued Noumer fields 241
Table 12.6 – A Summary of Reservoir factors in Hassi R'Mel Field.. 241

PART (I)
Ch (1) - (Ch5) - Petroleum Exploration Methods

"Historical aspect of petroleum industry, occurrences; petroleum migration, accumulation, and sedimentary basins; plate tectonics; geophysical exploration methods; drilling technology, and electrical well logging".

PART I
Petroleum Exploration Methods

Chapter 1
Introduction—Petroleum Hydrocarbons

1.1 Historical Aspect

The early uses of oil date from the time of Noah, who had used asphalt to make his ark watertight; in those times, oil was used for medication, waterproofing, and in warfare. Many references are found to using pitch or asphalt, collected from the natural seepage with which the Middle East abounds.

It is interesting to note that in early petroleum drilling operations, a major American company once employed a chief geologist whose exploration philosophy was to drill in old Indian graves; whereas another oil finder in the early 1900s used to put on an old hat, gallop about the prairie until his hat dropped off, and start drilling where it landed. History records that he was very successful!

1.2 Petroleum Occurrence and Chemical Composition

Petroleum is a naturally occurring complex mixture of hydrocarbons which may be either solid, liquid, or gas, depending upon its own unique composition and the pressure and temperature at which it is confined. The liquid phase of crude petroleum exists naturally in underground reservoirs and remains liquid at atmospheric pressure after passing through surface separation facilities. Petroleum may contain numerous impurities such as carbon dioxide, hydrogen sulfide, and other complex compounds of nitrogen, sulfur, and oxygen. Because of its numerous impurities it is difficult to perform a precise chemical analysis; petroleum is therefore often classified by its base designation, e.g., paraffin base, asphalt base, or mixed base crude. Paraffin base crude is oil consisting mainly of parafffins i which leave a solid residue of wax when completely distilled.

Asphalt base crude is oil primarily composed of cyclic compounds (mostly naphthenes), which leave a solid residue of asphalt when distilled. Mixed base oils are those that fall in the middle of the other categories (Gatlin 1960).

There are three principal hydrocarbon series of petroleum:

(a) **Paraffins, or saturated hydrocarbons (alkanes)**, compounds of the general formula C_nH_{2n+2}, are chemically stable and have either straight or branched chains. All crude oils contain some paraffins, particularly as the more volatile (low boiling point) constituents.

   Fig. 1.1a is an example of the first few members of the paraffin series.

(b) **Cycloparaffins, or naphthenes**, of the general formula C_nH_{2n}, have a ring structure, e.g., cyclopropane and cyclobutane **(Fig. 1.1b)**.

(c) **Aromatics or benzene series**, of the formula C_nH_{2n-6}, are chemically active and contain the benzene ring, the simplest member being benzene **(Fig. 1.1c)**.

Many crude oils contain porphyrins, and nearly all contain nitrogen. Petroleum rotates the plane of polarized light; this property is primarily restricted to organic materials known as optical isomers; it further suggests the organic origin of petroleum, which accounts for the large quantities of carbon and hydrogen needed to form petroleum deposits. The role of anaerobic bacteria may play a part in promoting the alteration process through which organic materials are transformed into petroleum.

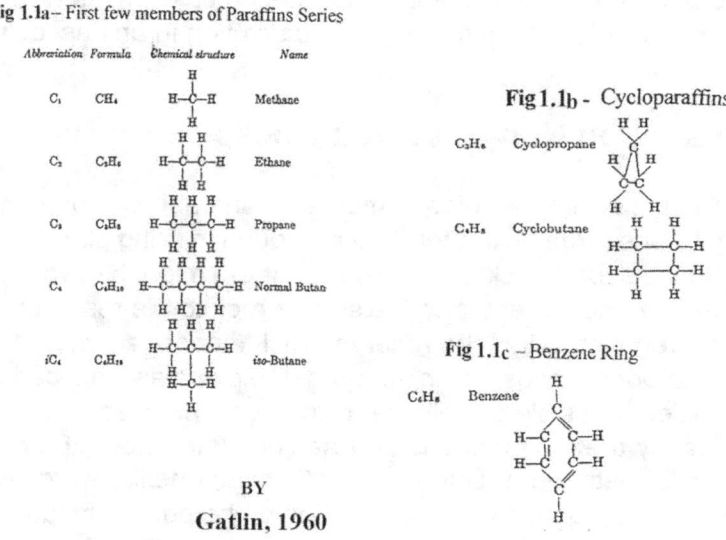

Fig. 1.1 (**a**) First few members of Paraffins Series. (**b**) Cycloparaffins. (**c**) Benzene Ring

1.3 Properties of Crude Oils

The crude oils range widely in their physical properties, e.g., color, specific gravity (0.80–0.95), and in their chemical properties. Gravity is the most effective energy causing the fluid's movement, although earth compression and capillarity also play a part. The difference in specific gravities separates the gas, oil, and water trapped below the seal or cap rock. According to the American Petroleum Institute (API), gravity is actually a measure of oil's density, and is related to specific gravity by the formula:

API gravity (in degrees) = 141.5 – (131.5)
(NB: API of 10° equivalent to sp. gr. of one).

1.4 Natural Gas—Definition

Petroleum gas, or natural gas, recently became a highly valuable product. Previously, gas produced with oil was sold primarily on a local scale, and the excess was flared. As the natural gasoline and liquefied petroleum gas (butane and propane) industry developed, the utilization of the residue gas (dry gas remaining after liquids are removed) also increased. Natural gas and its associated products are virtually as important as oil.

The content of the natural gasoline or the liquefied petroleum gas can be expressed in GPM (gallons per thousand [Latin *mille*] standard cubic feet [MCF]); if gasoline's GPM = 1–2, it is considered wet; and if it is = 0.2, it is considered somewhat dry. Gas gravity, the ratio of the density of a gas to the density of air at standard conditions, is also used as a measurement.

Natural gas is commonly measured in cubic feet; for example, if we start with a given volume of gas at a temperature of 60°F and a pressure of one atmosphere (14.7 PSI at sea level), then an increase in temperature will cause the volume of the gas to increase. Gas volumes are written in multiples of 1,000, abbreviated as M; thus, 3,555,000 cubic feet of gas would be equal to 3,555 MCF.

It is worth mentioning that the migration of gas globules would be affected by the same factors as with oil globules. Since hydrocarbon gas is soluble in water, the solubility rises with an increase in pressure. An increase in the depth of burial will drive more gas into solution in any associated oil, and vice versa.

1.5 Petroleum Hydrocarbon Non-Reservoir Rocks

The hydrocarbons found in the non-reservoir sediments presumably are produced directly, either wholly or partly, from the hydrocarbons (organic matter) that are found in living plant and animal matter. The reservoir rock is unlikely to be the source rock. In such cases, there might be two stages for the development of petroleum hydrocarbons: (a) movement from the source rock to the reservoir rock (primary migration), and (b) movement and segregation within the reservoir rock (secondary migration).

The petroleum hydrocarbons deposited in non-reservoir shales and carbonates as disseminated soluble hydrocarbon particles, possibly of colloidal or microscopic size, were thus associated with the nonsoluble organic matter. By the time diagenesis was complete, most of the petroleum hydrocarbons were probably in the form of petroleum. During and after diagenesis, water was squeezed out of the non-reservoir rocks into the reservoir rocks, and a fraction of the petroleum and petroleum hydrocarbons was entrained in the water. The movement from the non-reservoir rocks to the reservoir rocks is called the primary migration to distinguish it from the concentration and accumulation into pools of oil and gas called the secondary migration.

Carbonates and shale with an originally high content of bituminous matter become good source rocks if they do not retain their hydrocarbon potentials and are subjected to alteration into a different type of rock under high temperature and pressure. However, hydrothermal solutions of high temperatures allow the carbonate rocks to be crystallized from impervious bituminous into a porous reservoir trap rock. The consequent normal epigenetic changes of clay and shale and the alteration of carbonate rocks provide the time to develop the mother rock of hydrocarbons that migrates to, or is retained in the nearby reservoir trapped rocks (Levorsen 1954).

1.6 Petroleum Reservoir Rocks

Petroleum reservoir is the portion of the rock that contains the pool of petroleum and consists of four essential elements: (1) the reservoir rock (pay sand, oil sand, or gas sand); (2) the pore space, or porosity, or the "effective pore space"—that portion of the reservoir rock available for the migration, accumulation, and storage of petroleum; (3) the fluid content (water, oil, and gas) occupied by the effective pore space, which may be in a state of either static or dynamic equilibrium in which petroleum then occurs within a water environment; the distribution of fluids in a reservoir rock being dependent on the densities of the fluids and the capillary properties of the rock; (4) the reservoir trap that holds oil and gas in place in the pool. Rock traps are formed from a wide variety of combinations of structural and stratigraphic features of the reservoir rocks. The trap generally consists of an impervious cover, overlying and sealing reservoir rock, e.g., impervious shales, salt, or anhydrite deposits.

A petroleum reservoir rock must have two completely different properties: (1) porosity (the fluid holding capacity) of variable types of openings, e.g., the intergranular pores in the sedimentary rocks, cavities in fossil rocks such as the oolitic limestone, fractured rocks, and joints formed by solution; and (2) permeability (fluid-transmitting capacity).

1.7 Petroleum Migration and Accumulation

1.7.1 Introduction

The process of concentration and emplacement of the hydrocarbon accumulation is referred to as migration and accumulation. Petroleum and petroleum hydrocarbon migrate from the source rock into porous and permeable beds where they accumulate and continue their migration until finally trapped.

There are several forces that cause such migration: (a) **compaction** of sediments as the depth of burial increases; (b) **diastrophism** of crustal movements that cause pressure differentials and consequent subsurface fluid movements; (c) **capillary forces** that cause oil to be expelled from fine pores of source rocks by the preferential entry of water; and (d) **gravity forces** that promote fluid segregation because of the density factor.

Forces due to compaction and capillarity, influence fluid movement in horizontal strata into the porous rocks, but differences of pressure occur when the strata become tilted and cause fluid movement from localities with higher pressure to those with lower pressure.

1.7.2 Oil Migration

The **primary migration** of hydrocarbon takes place "in solution" in the pore space or adsorbed on organic or inorganic matter during its path from the source rock to the reservoir rock; in other words, petroleum hydrocarbons are entrained in the water that is squeezed out of the shales and clays during diagenesis and, to a lesser extent, in the confined water of the normal hydraulic circulation after diagenesis (Levorsen 1954). The **secondary migration** occurs when both movement and segregation of hydrocarbons occur within the reservoir rock. The effect of the hydrodynamic conditions can be extremely important to the movement of the petroleum.

Dilution by the entry of meteoric waters is improbable as an explanation for most oil accumulations; the migrating solution, on entering the reservoir rock, intermingles with water already in that rock, and a soap solution is effectively diluted and leads to the release of oil.

Because nearly all petroleum pools in oil producing regions exist within an environment of the confined (formation) water, petroleum migration is directly related to hydrology, fluid pressure, and water movement. The interstitial water content of the reservoir rocks generally shows a hydrodynamic fluid pressure gradient and moves in the direction of the lower fluid potential; the rate varies with the magnitude of the difference in fluid potential and the transmissibility of the aquifer.

1.7.3 The Rôle of Connate Water

At the time of complete diagenesis, reservoir and non-reservoir rocks are filled with confined (connate) water in which regional circulation patterns develop. This might be static if no fluid potential exists, but might continue to change if fluid pressure gradients change due to diastrophism, mountain building, erosion, deposition, etc. The flow of the connate or confined water may be either up or down the dip or slope, because the rate and direction of flow of confined water is proportional to the rate of change of the head of the hydrodynamic fluid potential gradient (measured in relation to sea level). This is not proportional to the rate of change in hydrostatic pressure along the flow path, because it can flow from an area of low fluid pressure to one of high fluid pressure, provided the head is lower in the direction of the flow. **Fig. 1.2** is a diagrammatic profile along the direction of flow and pressure gradient of confined water (Levorsen 1954).

1.7.4 Differential Entrapment of Petroleum Hydrocarbons—Gussow Theory

Under ideal lithological and structural circumstances with widespread permeable rocks, Gussow's theory of differential entrapment explains how oil and gas were separated during updip regional migration through a series of structural traps. Gussow's theory discusses the occurrence of oil and gas in a nearby fold trap as oil is eventually forced out at the bottom of the fold, either because additional oil and gas enter the trap or because a loss of pressure brings about an increase in gas volume. When the trap is full to the spill point, an additional volume of oil or gas will force the excess oil out at the bottom of the fold, and will move up the dip and be caught in the next trap. This procedure is repeated, and oil continues to migrate and be trapped, until the final up-dip trap contains connate water. The theory of differential entrapment is a useful stimulus to explanation thinking, particularly when information from bore holes becomes available **(Fig. 1.3,** adapted from Levorsen 1954).

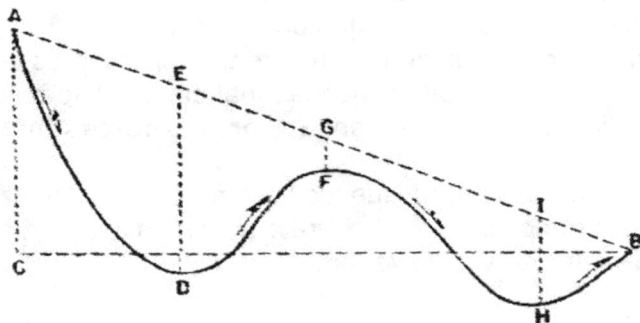

Fig. 1.2 Diagrammatic sketch showing the path water takes in flowing from intake area A to outlet area B as it passes across synclines and anticlines. The reservoir pressure at F would raise the water to G; this fluid pressure is less than at D, where the fluid pressure would raise the water to E, or at H, where the fluid pressure would raise the water to I. The surface AB is the potentiometric surface; its slope governs the overall flow of water from A to B. (Levorsen, 1954)

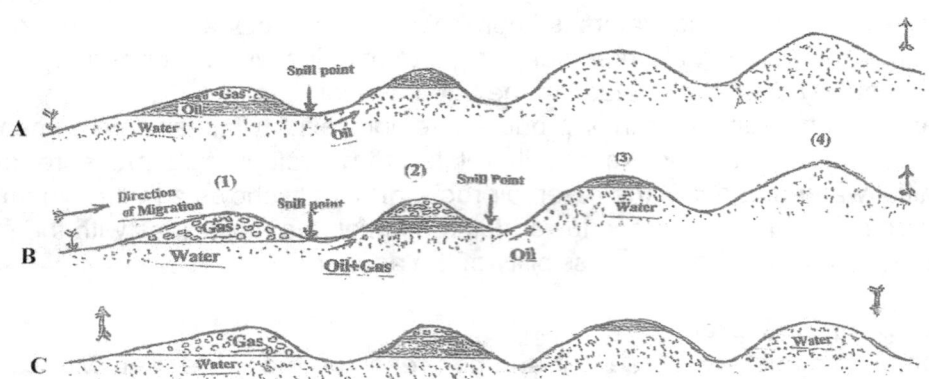

Fig. 1.3 Gussow Theory depends on differential entrapment of petroleum. In general, oil and gas moves from left to right up the regional dip, where oil is forced out at the spill point and is caught at the next trap; Stage **(A)**; shows two steps in a row: when oil-gas contact reaches the spill point, oil passes up dip to trap #2; whereas, trap #3 & #4 are full of connate water; **Stage (B)** shows Trap #1. full of gas, and spilled all its oil up the dip to trap#2 which is partly full of oil and gas; the excess bypasses up the dip to trap #3; in **Stage (C)** trap #1 remains the same, Trap#2 is full of oil and gas; trap #3 is partly full of oil and water; trap #4 remains the same and the process continues, where the tectonics finally shaped the current structures. (adapted from Levorsen, 1954)

A **trap** is an area of low potential energy toward which the buoyant oil and gas move from areas of high energy, either aided or hindered by the movement of water in the aquifer. The oil and gas separate according to their differential density and remain in the trap, because the geologic environment (such as faulting, folding, or the rate of water flow through the reservoir rocks) does not change.

1.7.5 Petroleum Accumulation

There are many factors that lead to secondary migration of petroleum hydrocarbons through the reservoir rocks and their accumulation into pools, such as the entrained particles, their buoyancy, tilted oil/water contacts, stratigraphic barriers, vertical migration, etc.

There is a wide variation in the character of the trap that may occur because of structural or stratigraphic factors, or a combination of both.

Structurally, entrapped petroleum can be released in either of two different ways or a combination of both: (a) the faulting up of reservoirs against impervious sections; and/or (b) the accumulation of hydrocarbons into local or regional apexes of the newly created reservoir's up dip against unaltered impervious rocks. Secondary migration of oil and gas can be provided by reducing the hydrostatic head through erosion.

Should the reservoir rock become tilted, there will be a greater chance of another migration (a tertiary migration), because the buoyancy factor due to the overall height of the oil or gas masses will be increased. Tertiary migration, and ultimately oil loss, might occur during the active structural history of an area.

A **stratigraphic trap** in a reservoir rock may occur in a structural oil field or a gas field, sealed by an impermeable barrier, which has been tilted by an earth movement, but not folded or faulted in such a way that could affect the accumulation process. An isolated offshore sand bar completely surrounded by muddy sediments may contain a stratigraphic accumulation if hydrocarbons have entered it from these muddy sediments and are unable to migrate further. A coral reef can form a stratigraphic trap if it is surrounded by muddy sediments and sealed by an impermeable muddy cap rock.

Stratigraphic traps may be divided into two types: primary stratigraphic traps, which form during the deposition or diagenesis of the rock, including those formed by lenses, facies changes, and reefs, while secondary stratigraphic traps result from later causes such as solution and cementation, but chiefly from unconformities.

The lower boundary of the reservoir, either wholly or partly, is the plane of contact (oil–water or gas–water contact), simply given as O/W or G/W.

Leakages of hydrocarbons from accumulations are referred to as seepages or shows and occur either by exposure of the reservoir rock as a result of erosion, or from a buried reservoir rock that faces "permeable" faults, or joints which allow fluid movement through the cap rock. In such cases, although the reservoir rock is exposed, asphaltic matter from the crude oil in the reservoir rock might clog the rock pores at the surface, thereby greatly reducing the rate of loss of oil (Levorsen 1954). On the other hand, gas might escape under water; and in some cases both gas and water move through clayey rocks, forming a gas-charged mud, that issues as a mud volcano on the surface.

1.7.5.1 Other Factors Affecting Oil Accumulation

Time Factor: The time needed to form an accumulation depends on many factors, including the rate at which oil is entering the reservoir rock, and the time at which the release of oil occurs. Lateral and upward movements result in a better local accumulation., It took approximately one million years for oil and gas to originate and migrate, but it might take a much shorter time to accumulate into pools, possibly only thousands or even hundreds of years.

Temperature and Pressure Factors: The temperature of a reservoir may fluctuate and can be approximately 75°C (167°F), with a maximum of 163°C (325°F) in the sediments of an oil region. The **fluid pressures** within the reservoir rock may also fluctuate during the life of the reservoir, depending on the geologic history of the region, and might range from one atmosphere up to 1,000 atmospheres. The geologic history of the trap may vary widely—from a single geologic episode to a combination of many phenomena extending over a long period of geologic time.

An **increase in the depth of burial** will drive more gas into solution in any associated oil, and vice versa. Natural gas is much more soluble in oil than in water. However, a reduction in the depth of burial may, by releasing gas from solution in the oil, cause the volume of the hydrocarbons to exceed the capacity of a trap. Excess hydrocarbons would move from the trap to escape or to accumulate elsewhere; the possibility of hydrocarbon-release by increase in salinity may also occur in many oilfields; the brines are more saline than the normal seawater; if such a feature develops early in the history of the rocks, then as regards time, it would be compatible with the requirement that the oil should appear at an early date in the reservoir rock.

Fluids during compaction move upwards through the sediments, where their exact path is determined by the geometry and permeability of the encountered beds. When an oil mass or gas mass of suitable height has been built up, it will be capable of rising because of its buoyancy within the reservoir rock; and the movement could be aided by favorably directed water flow.

The natural energy that is present in the reservoir and available to move the oil into the producing wells is the potential energy of reservoir pressure, which is stored mainly in the compressed fluids, and its amount depends largely on the fluid potential, or head, of the reservoir fluids; and to a lesser extent on the compressed rocks that form the reservoir rock (Levorsen 1954).

1.8 The Capacity of Oil and Gas Traps

The law of gases discusses how the volume of a gas varies in direct proportion to the absolute temperature, but inversely with pressure ($V \propto T/P$); natural gases can be trapped in a pool in a separate reservoir and under a single pressure system. The capacity of a trap to hold gas is a direct function of the reservoir pressure; i.e., when a trap is completely full of gas, the present accumulation of gas could not have occurred until the present pressure was reached. Because the reservoir pressures in most areas are related to the depth of burial, the trap could not have been filled until the present depth of overburden was deposited. In a trap full of gas at a depth of 5,000 ft (1,592.4m), the gas could not all have entered the trap until its pressure had reached 143 atmospheres (the average pressure at that depth); in a lower pressure environment, less gas would have been required to fill the trap.

Oil has its greatest mobility when its viscosity is the lowest and its buoyancy is the highest—in other words, at its saturation pressure. Upward movement might take place through well-developed structures of reservoir rocks because of the buoyancy of the fluid or because of the density factor. More gas was formed in place from oil or organic matter, being evolved at the expense of the heavier hydrocarbons by biochemical, thermal, or catalytic cracking, as the environmental conditions of the reservoir changed during geologic time (Levorsen 1954):

$(Gas_{Sp.Gr} = 0.0007; Oil_{Sp.Gr} = 0.7 - 1.0;$
$Water_{Sp.Gr} = 1.0 - 1.2).$

1.8.1 General Review
a) Oil has its greatest mobility when its viscosity is the lowest and its buoyancy is the highest, i.e. at its saturation pressure; b) Gas could not have entered a trap until its pressure had reached 143 atmospheres and a burial of 5000ft (or <2000ms). For a lower pressure, less gas would have been required to fill the trap; c) The trap is full to the spill point (measuring point at which an additional volume of oil or gas would force the excess oil out through the highest bounding syncline); d) A saturated oil pool of a free gas cap was attained, when the accumulation is complete at the time of the present reservoir pressure and the present depth of burial were reached either by deposition or by erosion.

References

Levorsen AI (1954) Geology of petroleum. W. H. Freeman, San Francisco, pp 540, 541, 574; Fig. 12.2, p 547

Gatlin C (1960) Petroleum engineering—drilling and well completion. Englewood, New Jersey, Prentice-Hall, p 19

Cited References

Vassilliou, M.S.(2009)-A to Z of the petroleum industry (electronic Resource). TN865/2009eb;E.Book

Rihard Mistrach,, Kate Orff (2012) -Petrochemical America-Marvin H. Sterne Library -Circulating Collection:F370M65, UAB Libraries.

Advances in Petrochemicals and Polymers- Selected from Contributions from "2nd., 2007", International Conference on Advances in Petrochemicals and Polymers (ICAPP, 2007), in Bangkok, (Thailand); June 25-28, 200. Symposium ed.; Pitt Supaphol, Bangkok (2008) ;UAB Mervin H. Sterne LibraryTitles

Richard A. Esposito ; Peter M. Walsh.. "et al, advisors", (2010)- Business for commercial-Scale carbon dioxide. Sequetration (electronic Resource); with focus on storage capacity and enhanced oil recovery in Citronelle Dome -UAB Libraries Mervin H. Sterne Libraries TA7. T42 E7762b-Sterne Thesis Collection (Electronic). Larry W. Lake, Univ. Texas, Austin (1989) - Enhanced Oil Recovery, Prrentice Hall, Englewood Cliffs, N.J.0763

Elsevier (1972) -Oil and Carbonate rocks - editors: (George, V. Chilinar Robert W. M. annon & Herman H.Ricke III.

John M. Hunt (1996) -Geochemistry & Geology (2nd Ed.)-Most important : Applications (part four): Glossary (pp6333-706) ; rreferences (pp642-707) ; Name index (pp 710- 714); subject index (pp715-743).

L. P. Dake (1978) -Fundamentals of Reservoir Engineering ; Development in Petroleum Science; Elssevier publishing CO.INC

Chapter 2
Plate Tectonics and Sedimentary Basins

2.1 Scope

The outer part of the earth's crust, the lithosphere, is constructed predominantly of sedimentary rocks of Mesozoic, Tertiary, and Quaternary deposits extending to a depth of about 12 km on land and only 1 km or less on the ocean floors.

The crust itself is the bedrock of the outer layer of the earth and is relatively thin, 20–40 km thick under the continents, composed mainly of acidic igneous rocks (granites, granodiorites, etc.); but is much thinner, about 8 km thick or less, under the oceans, where it is composed mainly of basaltic rock. The crust differs in composition and thickness depending on whether it underlies the continents or the oceans. The crust is floating generally on a **solid mantle** about 3,000 km in thickness, and is composed of very dense rocks surrounded by outer shell layers of the ultra dense and highly viscous liquid rock of the asthenosphere. The **core**, which is the innermost solid iron mass, with a radius of about 3,500 km, is surrounded by the mantle.

During the past three decades (since 1967), the theory of plate tectonics has dominated geological thinking. According to this theory, the surface of the earth's crust (under both continents and oceans) is made up of rigid, aseismic lithospheric blocks which are slowly and intermittently in motion. There are major and minor plates, together with more complex smaller units. The plates are separated by linear "active bands," in which vulcanicity and seismic activity seem to be concentrated (**Fig. 2.1**; Komatina 2004).

The plate tectonic theory of continental displacement replaced the geosynclinal theory of the development of the folded mountain ranges of the earth's crust; the term "geosyncline" was originally used to describe the long, narrow, subsiding depressions in which thick sediments accumulated and were affected by orogenesis (violent folding and uplift accompanied by volcanic activity). In general, the average rate of recent plate movements is usually from 1 to 2 mm/yr (Komatina 2004). There are two forms of tectonic movements:

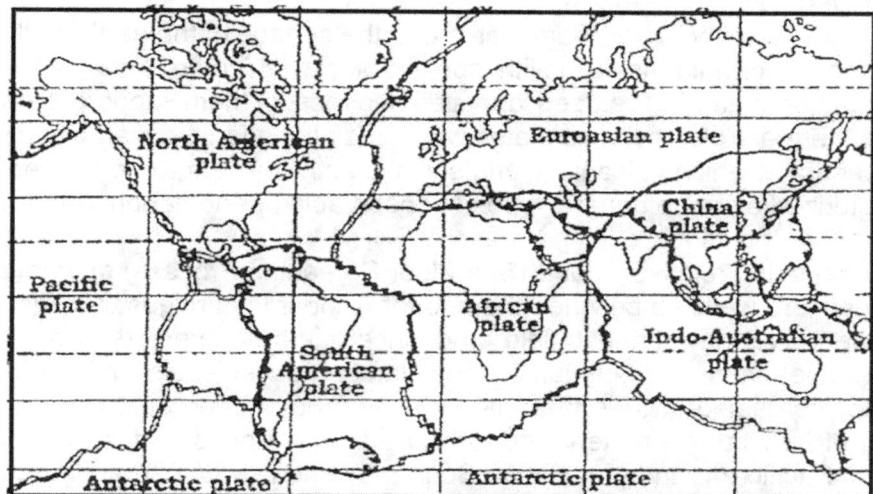

Fig. 2.1-Position of main tectonic plates.
(Komatina, 2004)

Fig. 2.1 Position of main tectonic plates (Komatina, 2004) downwellings) and orogenic movements (mountain formation on "geosynclinal" areas), which have led to Alpine and Mesozoic mountain ranges where tectonic activity was manifested in powerful seismicity and the development of volcanism, e.g., in the Alpine-Mediterranean zone in Eurasia.

- A **passive tectonic system** is characteristic of a large part of the Precambrian platforms, and applies to the entire territory of North American and Russia, and the large central zone of Africa, e.g., the Algerian Saharan platform, where the relief is level, low-lying in some areas and elevated in others.

2.2 Plate Tectonics and plate Interactions

2.2.1 Introduction

The unifying concept that has drawn continental drift, sea-floor spreading, seismic activity, crustal structures, and volcanic activity into a coherent model of how the outer part of the earth evolves. The theory proposes a model of the earth's upper layers in which the colder, brittle, surface rocks form a shell (the lithosphere) overlying a much less rigid asthenosphere. The shell comprises several discrete, rigid units (tectonic plates), each of which has a separate motion relative to the other plates. The plate margins are most readily defined by present-day seismicity, which is a consequence of the differential motions of the individual plates. The model is a combination of continental drift and sea-floor spreading. New lithosphere plates are constantly forming and separating, and so being enlarged, at constructive margins (ridges); while the global circumference is conserved by the subduction and recycling of material into the mantle at destructive margins (trenches). This recycling results in andesitic volcanism and the creation of new continental crust, which has a lower density than oceanic crust and is more difficult to subduct. Many features of the Earth's history are explicable within this model, which has served as a unifying hypothesis for most of the earth sciences. Previous mountain systems are now recognized as the sites of earlier subduction, often ending with continental crustal collisions; the movement of plates has been used with varying success in interpreting orogenic belts as far back as the early Proterozoic. Plate motions are driven by mantle convection and are likely to have occurred throughout earth history, although the resultant surface features are likely to have changed with time.

2.2.2 Plate Tectonics Features (By Hobson, G.D. & Tiratsoo, E.N. (1975)

Developed since 1967, the Plate tectonic theory states that the surface of the earth (both continents and oceans) is made of a mosaic or rigid, aseismic lithospheric blocks which are slowly and intermittently in motion, about a dozen major "plates" have been identified and most of them support at least one massive continental craton; but there are also oceanic plates which mainly underlie ocean basins; the plates are separated by "linear bands" in which vulcanicity and seismic activity seem to be concentrated and their boundaries are characterized by three general types of observable phenomenon:

- When the plates are moving relatively apart (Sea-Floor Spreading Process) magnetic material, from the underlying lithosphere wells up between them to form linear lava ridges; it takes up the direction of magnetism of the prevailing magnetic field, and since this has reversed its polarity many times during the past, parallel linear "Mid-Ocean-Ridges", are produced, being magnetized in opposite directions. Sea-floor spreading implies that the continents have been split and spread apart as a result of upwelling of new crustal material from the Earth's mid-ocean ridges, which are known to exist in the Atlantic, Pacific and Indian oceans. Such movement would be sufficient to produce the phenomena of continental drift in the course of the geological time, e.g. the separation of Africa and south America, and the associated growth of the Atlantic Ocean.
- Plates can also apparently slide past one another along lines of lateral or "transform" faults without and resultant generation or destruction of plate material. Thus, the mid-ocean ridges in the Atlantic and Pacific Oceans, which run roughly north/south, are broken into series of segments, each about 200 miles long and offset relative to each other; at the offset points, other ridges branch off roughly at right angles to the main ridge, known as "transform fault zones", which project above the deep ocean floors, as well as being earthquake and volcano zones.
- When plates converge, there are several possible results, depending on whether oceanic or continental type plates are involved:
  - When two oceanic plates collide, one plate is thrust under the other, and the lower plate descends towards the mantle, where it is partially destroyed by heat.
  - Where two continental-type plates slowly converge, major mountain ranges (e.g. Himalayas) would be formed as a result of the squeezing of the sediments carried on the under thrust plate.
    There is certainly persuasive evidence to support the theory that a major process of continental break-up and movement started in the Triassic Times, about 200 million years ago. At that time, there were only two continents in existence: the "super-continent" of Gondwanaland", and "Laurasia"; both were separated by the Tethys Ocean and surrounded by long sedimentary troughs, which were sites of considerable seismic and volcanic activity. For some 40 million years, a general northward drift of the continental masses continued intermittently; then about 160 million years later, Gondwanaland began to break up into a number of separate components. That period of breakup was accompanied by a general out pouring of basaltic material; the formation of down-warped basins which were invaded by shallow seas, lead to the production of great thickness of evaporites and the spread of the Tethys- Ocean, southwards along the edges of the new continents.

The striking similarity of shape of the West coast of Africa and the east coast of south America has for long, encouraged the belief that they were once united and have since moved apart. The evidence for the relative movement or "drift" of continents, which was first developed as a geological theory in 1912, has been widely studied and lead to the final conclusion that the continents must have moved in relation to both the rotational and the magnetic poles, as well as to reach each other., over a period of at least 200 million years (most probably in Late Triassic Time).

2.2.3 Plate Interactions

In plate tectonic theory, composite continents are assembled by crustal collisions that occur when the consumption of oceanic lithosphere beneath arc-trench systems results in the closure of an oceanic basin.

The arrival of a continental block at a subduction zone where the intervening oceanic lithosphere was consumed, will thus throttle subduction, and the position of the previous subduction zone will be taken by a crustal suture belt marking the line of tectonic juxtaposition of the two continental blocks involved in the crustal collision **(Fig. 2.2)**. Examples of continental collisions in which one of the continental margins has a marginal arc-trench system and the other has a rifted-margin sediment prism (Dickinson 1974).

All plate interactions that involve construction of new lithosphere or consumption of old lithosphere as a result of large horizontal motions of plates, result into significant vertical motions of the lithosphere. There are three basic causes of subsidence or uplift as a result of plate interactions: (1) changes in crustal thickness; (2) thermal expansion or contraction of the lithosphere; and (3) broad flexure of plates of the lithosphere in response to local tectonic or sedimentary loading.

From a kinetic point of view, there are three kinds of plate junctures, analogous to the three classes of faults as defined by relative displacements:

(a) **Divergent plate junctures** (analogous to normal faults)—A separation of two plates (sea-floor spreading) occurs and causes rupture of the intact old lithosphere, which in turn results in intercontinental rifting when an incipient rift crosses a continental block. The rate of spreading may average several centimeters per year and might produce over geologic time the continental drift that represents the separation of Africa and South America and the associated growth of the Atlantic Ocean.

The plates diverge or move relatively apart and result in sea-floor spreading, in which magmatic material from the underlying lithosphere wells up in between to form linear lava ridges—parallel, linear "mid-ocean ridges"—which are found in the Atlantic, Pacific, and Indian Oceans.

(b) **Transform plate junctures** (analogous to strike-dip faults)—One plate slides laterally past the other along a transform or deep fault, without accretion or consumption. However, hybrid plate boundaries also occur in some areas, as in oblique slip faults, where some component of extensional or contractional motion occurs along a transform, and hence the two terms *transtension* and *transpression* are used to describe the interaction.

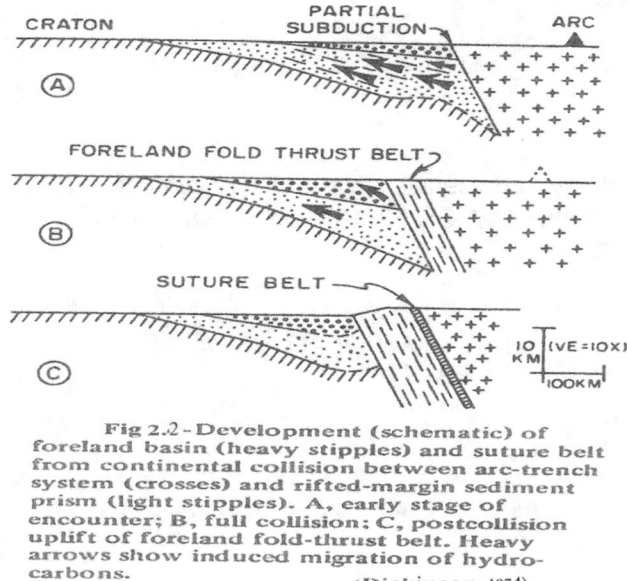

Fig. 2.2 Development (schematic) of foreland basin (heavy stipples) and suture belt from continental collision between arc-trench system (crosses) and rifted-margin sediment prism (light stipples). An early stage of encounter; B, full collision; C, postcollision uplift of foreland fold-thrust belt. Heavy arrows show induced migration of hydrocarbons. (Dickinson, 1974)

In accordance with the transform movements of the plates, the mid-ocean ridges in the Atlantic and Pacific Oceans, which run roughly north/south, are broken into a series of segments, each about 200 miles long, together with their related offsets, at the points of which are "transform fault zones," formed roughly at right angles to the mid-ocean ridges, and projecting above the deep ocean floors; these become earthquake and volcano zones.

- **Convergent plate junctures** (analogous to thrust faults)—When two oceanic plates collide, one plate is thrust at an angle beneath the other, and dives down into the mantle where it is partially destroyed by heat. Convergent junctures are sites of plate consumption where oceanic lithosphere formed previously at a divergent plate juncture, descends into the mantle. Part of the slab continues to sink to a depth of about 700 km, where it comes to rest; the part which rises to merge with the upper plate is converted to low-density magma.

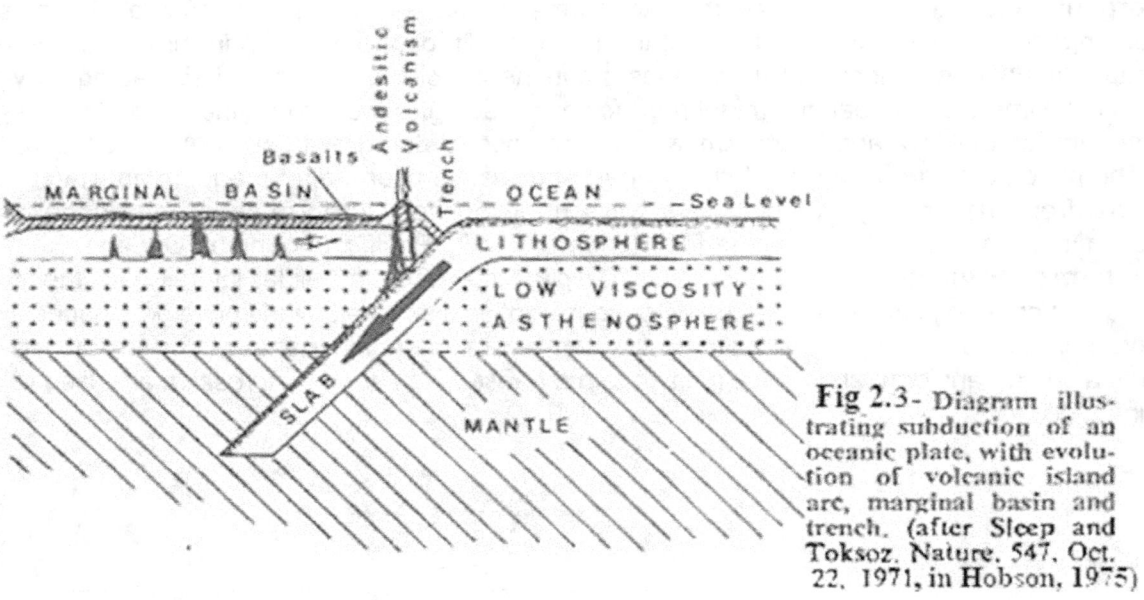

Fig 2.3- Diagram illustrating subduction of an oceanic plate, with evolution of volcanic island arc, marginal basin and trench. (after Sleep and Toksoz, Nature, 547, Oct. 22, 1971, in Hobson, 1975)

Fig. 2.3–A diagram of a subduction of an oceanic plate, with evolution of volcanic island arc, marginal basin and trench (after Sleep and Toksoz, Nature, 547, Oct.22, 1971, in Hobson 1975)

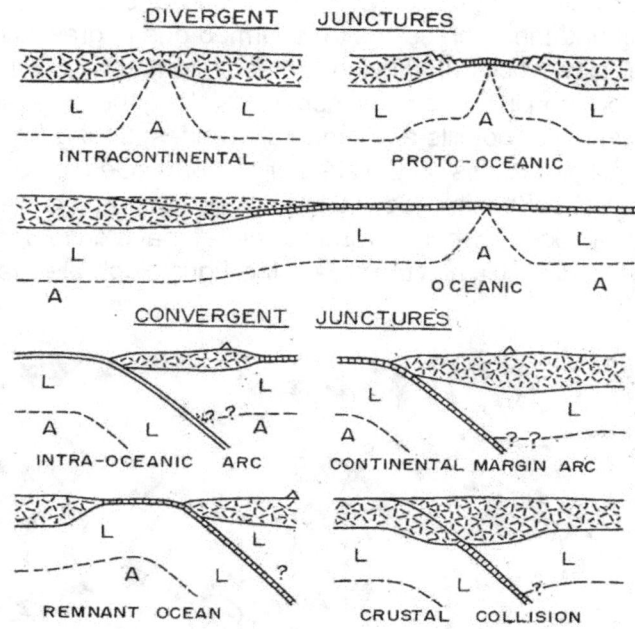

Fig 2.4-Principal kinds of plate interactions at divergent(above) and convergent (below) plate junctures showing relations of lithosphere (L) and asthenosphere (A) as well as crust (ornamented)
(Dickinson, 1981)

Fig. 2.4 Principal kinds of plate interactions at divergent (above) and convergent (below) place junctures showing relations of lithosphere (L) and asthenosphere (A) as well as crust (ornamented) (Dickinson, 1981)

2.3 Continental Shelves

The earth's crust consists of two types, continental and oceanic, below which lie the vastly more bulky and heavier mantle and core. The continental crust, which is three to four times as thick as the oceanic crust, includes all the major land masses with submerged borders, known as continental shelves. The boundary zones between the continents and the oceans form the continental margins, which are made up of three components: (1) the continental rises, which largely comprise the "fans" of continental sediments; (2) the gently dipping shelves 5–250 miles in width, which constitute the submerged edges of the continents; and (3) the steeper slopes, about 10–30 miles wide, the bases of which mark the transition zone between the continental-type crust and the oceanic-type crust.

Under the 1958 Geneva Convention, the continental shelf is legally defined as "the sea bed and subsoil of the submarine areas adjacent to but outside the area of the territorial sea to a depth of 200 m or, beyond that limit, to where the depth of the superjacent waters admits the exploitation of the natural resources of such areas."

From the edge of the continental shelf, the surface of the continental crust slopes gradually downwards to the deep oceanic basins, which are underlain by the thinner oceanic-type crust. The thickness of the continental crust decreases gradually below the continental slope until the thinner oceanic crust is reached **(Fig. 2.5)**.

Fig. 2.5–The world's continental shelves include the major seas, lakes, and some marine platforms within the 200-m depth range (Petroleum Press Service 1951).

In some oil field provinces, a broad "geosynclinal belt" of much sedimentation, which typically evolves to a large asymmetrical structural basin by being subjected to orogenesis, is the consequence of developments that have taken place over a long period of time. Sedimentary basins of such type are termed active when the floors have continued to sink and the depressions thus formed due to gravitational forces have filled with sediments. On the gentler side, older rocks underlying the sediments may outcrop on a part of one of the major shields where faulting is common in some belts, especially along the hinge line. On the more disturbed side (of the mobile belt side), there are foothills and mountain ranges; strong folds and overthrusting occur. The thickness of the sediments formed is less on the shelf or platform side than elsewhere.

Fig 2.6–An asymmetrical structural basin "geosyncline trough" showing a maximum thickness that lies between the shelf side and the mobile rim where strong structures have developed; examples of this broad pattern are the Arabian shield of Saudi Arabia and the area of the Tigris-Euphrates valley of Iraq (Hobson 1975).

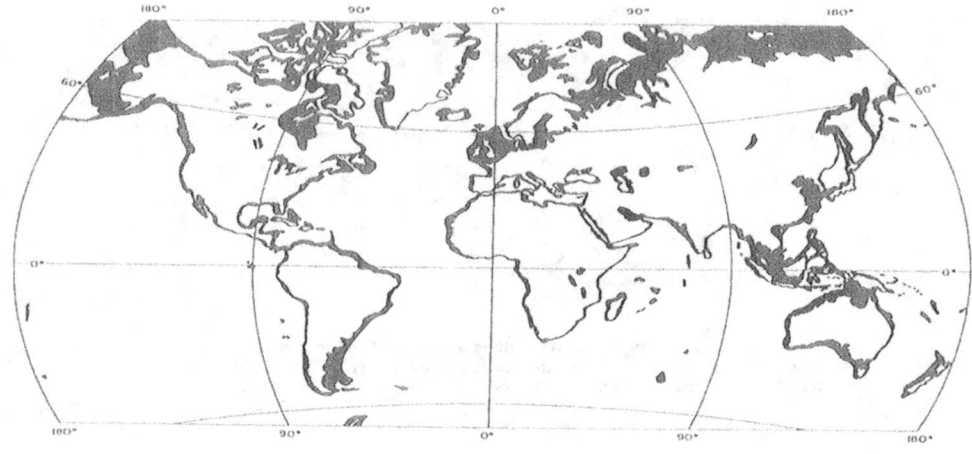

Fig. 2.5 World continental shelves (Petrol. Pres Service, 1951)

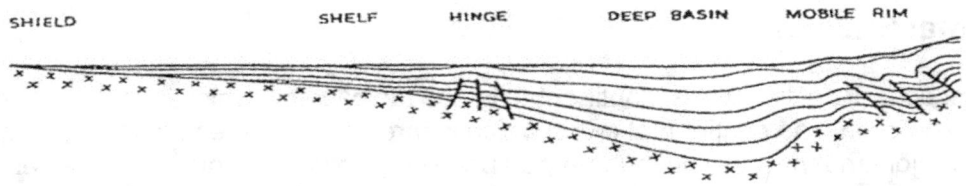

Fig. 2.6 A cross section of sedimentary basin its length of several hundred miles, and the thickness of the sediments overlying the basement could be a maximum of 10 miles. It lies between a stable shield (foreland) and a mobile belt. (Modified from Hobson, 1975).

2.3.1 Development of Sedimentary Basins

Plate tectonics theory is generally related to lateral and vertical motions of plates that form sedimentary basins. A sedimentary basin develops as an accumulated prism of strata, resulting either from subsidence of the basin floor, or from uplift of confining basin margins. Each basin has its own unique history connected to a particular sequence and combination of plate interactions and depositional conditions. During basin evolution, the stratigraphic fill of the basin took place because of the activity of the depositional systems. Folded and faulted structures within the basin were formed because of either tectonic or sedimentary evolution. Extensional deformation usually produces normal faults and tilted blocks, whereas contractional deformation produces folds and thrust faults.

Plate movements have lately been considered to result from the presence of "hot spots," which represent a number of thermal centers, and are fixed in the upper mantle, from which "plumes" of hot material rise intermittently to burn holes in the overlying crust. Consequently, the continental plates are pushed away from these "hot spots" by the creation of new ocean floor; and when the movement of the lithosphere above the plume occurs by the process of sea-floor spreading, a "plume scar" is left on the crust in the form of a line of volcanic cones.

2.3.1.1 Geodynamic Settings of Sedimentary Basins
Sedimentary basins can occur in two general kinds of geodynamic settings.

(1) **Rifted settings occur** where divergent plate motions and extensional structures dominate and are later enhanced by thermal decay as time passes and may be augmented mostly by flexures in response to sedimentary loading.
(2) **Orogenic settings** occur initially where convergent plate motions and contractional structures dominate; subsidence occurs initially by plate flexures, either because of plate consumption or because of local tectonic thickening of crustal profiles, possibly augmented by sedimentary loading and other thermotectonic effects. The lithosphere may exchange locations between rifted and orogenic settings, and hence the name composite basins, in terms of plate tectonic settings. Basins of rifted settings, in which tectonic evolution is dominated by extensional plate motions and crustal rifting, and those of orogenic settings, in which tectonic evolution is dominated by contractional plate motions and orogenic deformation, include several idealized subgroups.

Oceanic basins evolve regularly from nascent phases dominated by extensional tectonics into remnant phases dominated by contractional tectonics. Similarly, the sedimentary assemblages along rifted continental margins include phases deposited within protoceanic rifts underlying the younger miogeoclinal prism, which may in turn be covered and flanked by the deposits of a progradational continental embankment. These rifted-margin sedimentary associations may later be covered in part, with the onset of orogeny, by the foreland deposits of retroarc, or peripheral basins as arc (Fig. 2.9) or as collision (Fig. 2.10, Dickinson 1981).

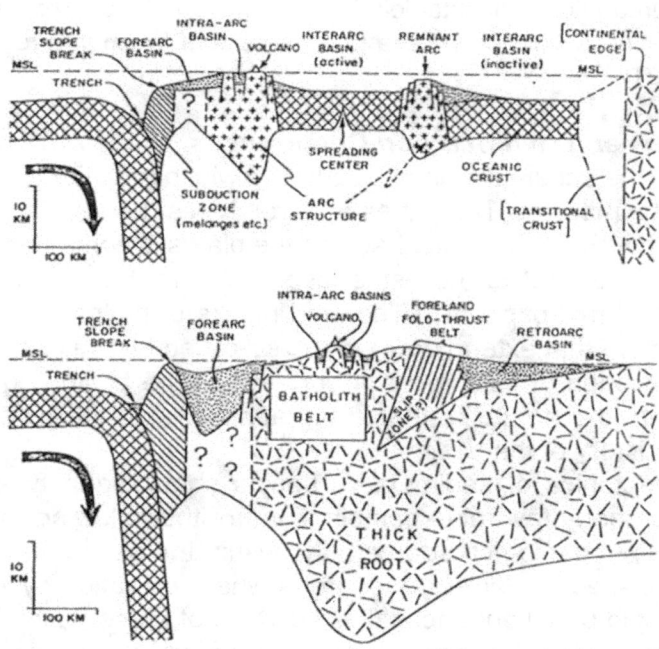

Fig. 2.7 Schematic diagrams (vertical exaggeration 10X) to illustrate sedimentary basins associated with intraoceanic (above) and continental margin (below) magmatic arcs. (Dickinson, 1981)

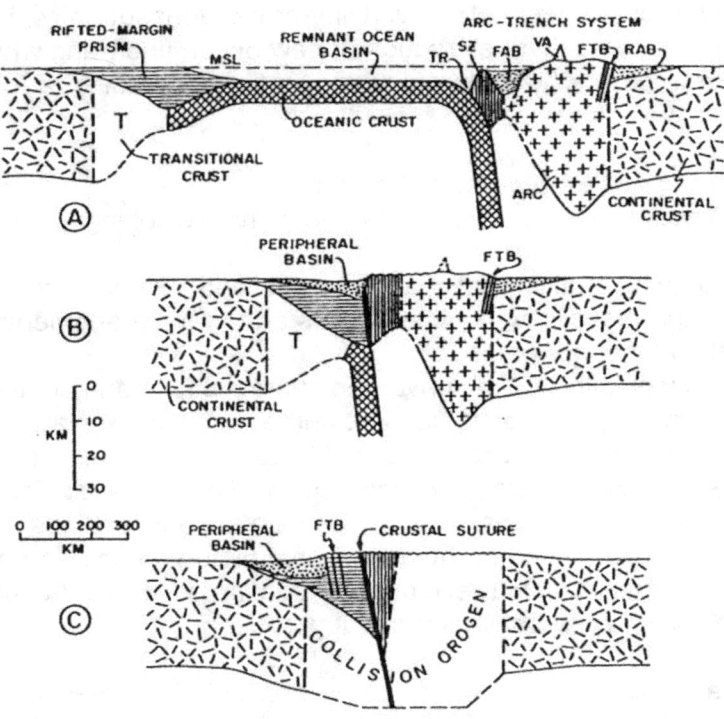

Fig. 2.8 Schematic diagrams (vertical exaggeration 10X) to illustrate sedimentary basins associated with crustal collision to form intercontinental suture belt with collision orogen. Symbols: TR, trench; SZ, subduction zone; FAB, forearc basin; RAB, retroarc basin; FTB, foreland fold-thrust belt. Diagrams A-B-C represent a sequence of events in time at one place along a collision orogen marked by diachronous closure; hence, erosion in one segment (C) of the orogen where the sutured intercontinental join is complete could disperse sediment longitudinally past a migrating tectonic transition point (B) to feed subsea turbidite fans of flysch in a remnant ocean basin (A) along tectonic strike. (Dickinson, 1981)

2.3.1.2 Sedimentary Basins and Hydrocarbon Occurrences

Dickinson (1981) discussed the accumulation of oil and gas within the thick prisms of sediment deposited along many rifted continental margins. The plates apparently rest upon a mobile layer called the asthenosphere, within which the geothermal gradient across the plates is controlled mainly by the conductivity between the basal temperature (that of the asthenosphere) and the surficial temperature (that of the troposphere [atmosphere and hydrosphere]). The rigid blocks of plates are spherical caps, or arcuate slabs, of the earth's lithosphere that extend down to the so-called low velocity zone of the upper mantle. The amount, nature, and location of fluid hydrocarbons within the basin is governed by the geometric shape and size of the basin, the nature of the stratigraphic fill and the types of structures, and most importantly by the thermal history of the basin.

Analysis of the plate tectonics of sedimentary basins should identify styles and times of basin evolution and trends that embody favorable combinations of the following four important attributes for hydrocarbon occurrence. (1) Organic-rich source beds within the sedimentary sequence are generally fine-grained sediments deposited in black-bottom areas where oxidation by aerobic decay during early diagenesis is limited, and rapid burial enhances the likelihood of preservation. Basins in which sedimentary fill includes the favorable sites that harbor the largest sources of potential hydrocarbons exist in oxygen-minimum zones on open slopes and in low-oxygen zones within saline non-marine or marginal marine environments. (2) Fluid hydrocarbons are normally generated by appropriated heat for thermal maturation of liquid hydrocarbons or thermal gas. (3) Permeable migration paths for the concentration of fluid hydrocarbons should have access to gather any hydrocarbons produced. The most effective carriers are tilted conduit beds of well-sorted sandy strata contained beneath impermeable sealing beds of shales or eruptive rocks. (4) Petroleum and petroleum hydrocarbons can be contained in porous reservoir beds (whether clastic or carbonate), and are confined by impermeable capping beds or traps that may be formed either by stratigraphic enclosures or by structural features of tectonic or sedimentary origin.

Conceivably, plate tectonics may lead to a comprehensive theory of hydrocarbon genesis. Plate interactions fully explain the causes of subsidence, the sequencing of depositional events, the development of structural features, and the timing of thermal flux—all of which in turn can eplain conditions in the geologic record of sedimentary basins (Dickenson 1981).

2.3.2 Rifted-Margin and Sediment Prisms

In plate tectonic theory, ocean basins are initiated by continental separations that begin with **intracontinental rifting**. When complete separation of continental fragments is achieved, the raw edges of continental blocks along rifted continental margins are in preferred positions to receive thick prisms of sediment, composed both of debris washed off the adjacent continental blocks and of biogenic calcareous materials that were deposited in shoal areas at the flanks of the adjacent oceans.

The growing rifted-margin sediment prism imposes sufficient load on the lithosphere near the interface between old continental and new oceanic crust to induce downbowing of the lithosphere by flexure **(Fig. 2.9,** Walcott 1972). The flexure tilts the surface of the continental block seaward to allow landward parts of the rifted-margin sediment prism to encroach as much as 100–250 km beyond the initial continental edge, which may be depressed as much as 2.5–5 km. Continued growth of the rifted-margin sediment prism may in time allow its oceanward edge to advance well into the adjacent oceanic basin until the continental shelf, and perhaps even part of an accretionary coastal plain, come to stand fully above the oceanic basement. The type of rifted-margin sediment prism developed by wholesale sedimentary progradation of the continental edge has been aptly termed a continental embankment (Dietz 1963).

Accumulations of petroleum in rifted-margin sediment prisms are well known. The natural association of offshore source beds and near shore reservoir beds of various kinds is inherent.

The following two factors combine to exert a pumping action that drives fluid hydrocarbons updip from offshore source beds into favorable reservoir beds. (1) The **progressive loading** of offshore source beds because of depositional growth of the rifted-margin prism; and (2) The **progressive seaward tilting** of the continental edge and the successive layers of the rifted-margin prism because of flexural bending of the lithosphere under the growing sedimentary load offshore. These two factors are probably most influential within fully developed continental embankments, where potential conduit beds, intercalated with sealing beds, connect offshore source beds with inshore facies suitable as reservoir beds without interruption by thin slope facies (Dickenson 1974).

In conclusion, the largest reserves of petroleum are inferred to accumulate where conditions are most favorable for long-distance updip migration of oil from offshore source beds into attractive reservoirs in nearshore deposits along or near platforms margins.

The requisite regional dips and overburden loads are attained within the sediment prisms, deposited along rifted continental margins, and also beneath foreland basins associated with foreland fold-thrust belts adjacent to major orogens.

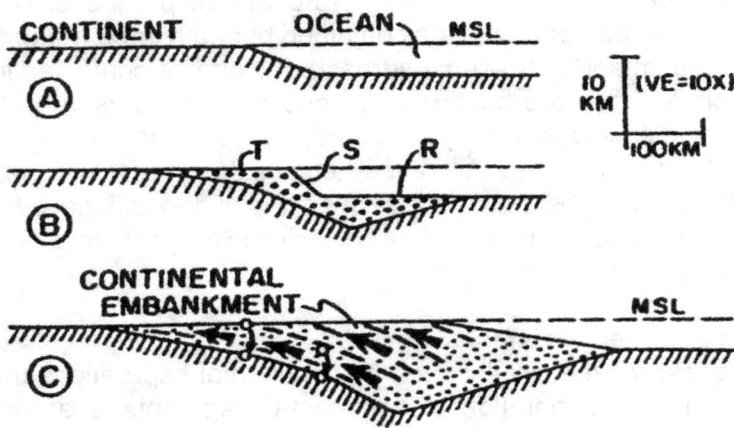

Fig. 2.9 Growth (schematic) of rifted-margin sediment prism (stippled). A) rifted continental margin without sedimentation; B) continental terrace (T) slope (S) rise (R) configuration; C) progradational continental embankment; small arrows show rotational tilt of basement along continental margin downward toward ocean basin; heavy arrows show updip migration of hydrocarbons parallel to bedding (dashes). (Walcott, 1972).

Apparently, optimum relations for oil concentration thus occur in the Middle East, where a rifted margin sediment prism was drawn partly into a subduction zone, now a suture belt marking the site of continental collision, and was partly buried beneath a foreland basin sequence (Dickenson, 1979).

Hobson (1975) discussed the convergence of plates according to a different approach, depending on whether oceanic- or continental-type plates are involved:

- A **coastal mountain range** is formed along the leading edge of the continental plate, when it is advancing towards and overriding a relatively stationary oceanic plate.
- **Island arcs and subduction zone trenches** are formed when an oceanic plate is advancing and passing beneath a relatively stationary continental plate.
- **Major mountain ranges** are formed when two continental-type plates slowly converge, as a result of the squeezing of the sediments carried on the underthrust plate (e.g., the Himalayas in India).

2.3.3 Foreland Basins

Coney (1973), noted that the back-arc thrusting is associated with the development of foreland fold-thrust belts **(Fig 2.9)**, and foreland basins that occur in continental interiors behind continental margin arc-trench systems. Partial subduction, in this case related to strictly limited under thrusting of the cratonic foreland beneath the rear flank of the arc, may thus influence oil migration along the orogenic flanks of all pericratonic foreland basins, regardless of whether the adjacent orogeny is a collision orogeny or arc orogen.

The largest reserves of petroleum are inferred to accumulate where conditions are most favorable for long-distance updip migration of oil from offshore source beds into attractive reservoirs in near shore deposits along or near platform margins. The requisite regional dips and overburden loads are attained within the sediment prisms deposited along rifted continental margins, and also beneath foreland basins associated with foreland fold-thrust belts adjacent to major orogens. Apparently, optimum relations for oil concentration thus occur in the Middle East where a rifted-margin sediment prism was drawn partly into a subduction zone, now a suture belt marking the site of continental collision, and was partly buried beneath a foreland basin sequence (Dickenson 1979).

2.4- Historical Aspect of the Sub-Sea Floor fluid

Davis and Elderfield (2004) submitted a concise review of the relatively brief history of hydrogeology, which began shortly after the revival of plate tectonic theory little more than thirty years ago. It describes the nature and important consequences of fluid flow in the sub-seafloor, ending with a summary of how the oceans are affected by the surprisingly rapid exchange of water between the crust and the water column overhead.

The importance of fluid circulation below the seafloor and the exchange of water between the crust and the oceans can be easily appreciated by considering that the oceanic crust constitutes the most extensive geological formation on earth, and that hydrologic activity within it extends from mid-ocean ridges to subduction-zone accretionary prisms. The upper part of the crust is characterized by very high permeability, and it is host to huge fluxes of water. At the ridge axes occur the most spectacular and directly observable manifestations of flow; and at the seafloor, heat from magmatic intrusions, drives high-temperature springs **(Fig. 2.12)**

2.5 Subduction and Oil Migration

Petroleum and natural gas pools are abundant within the thick prisms of sediment deposited along many rifted continental margins. The rifted-margin sediment prisms are drawn down against and beneath the suture belts formed by crustal collisions that assemble composite continents. During such partial subduction, fluid hydrocarbons may be driven updip away from the subduction zones to accumulate in reservoirs along adjacent platform margins and within growing foreland fold-thrust belts, e.g., the immense petroleum accumulations of the Persian Gulf southwest of the Zagros suture belt. Similar oil migration in response to partial subduction may have influenced the distribution of petroleum in foreland basins in general; if so, strategies for exploration can perhaps be improved by taking this factor into account. Dickinson (1974) assumed that there were possible direct relations between plate movements accompanying subduction and the opening of migration paths for petroleum.

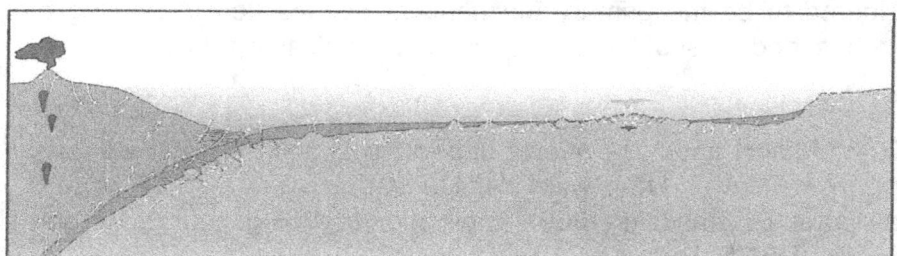

Fig. 2.10 Schematic cross-section depicting various types of sub-seafloor fluid flow, ranging from topographically driven flow through continental margins and consolidation-driven flow at subduction zones to thermal buoyancy driven flow at mid-ocean ridge axes and in the oceanic crust beneath broad regions of the oceans. Oceanic igneous crust and sediments are shown light gray and darker gray, respectively, and magma is shown in black: flow is depicted by white arrows. The figure was originally prepared for the integrated Ocean Drilling Program Initial Science Plan. 2003–2013. and is reproduced courtesy of JOI, Inc. (Devis and Elderfield, 2004)

2.6 - Convection Currents and Subduction Zones- Update Review

Eugene Huter and Warren E. Teasdale (USGS, 1989) discussed the convection currents that move its extreme heat in the deep mantle of the earth, outward to the earth's surface where it is conducted into space.

Convection currents of the mantle material of the earth (hot magma) which form of hot lighter fluids (liquids and gases), than cold fluids, are in constant motion, and their molecules increase their velocity as well as the number of collisions with higher temperature, forcing the molecules further apart and lightening the fluid; Convection currents then form when the hotter (lighter) fluids float upward through colder (heavier) fluids like a stiff liquid, and conversely, colder fluids drop down through hotter fluids.

Convection currents of hot magma rise to the surface and move laterally beneath the crust until the magma cools and sinks back toward the center of the earth.

Spreading Centers - or rifts- are present in most of the world's ocean basins, where the currents move laterally under the crust, and impart frictional forces that pull the crust apart. Fresh magma then rises to fill the void, cools, and forms replacement crust; e.g. an active rift zone called the mid-Atlantic ridge is located in the Atlantic ocean between South America and Northwest Africa.

Subduction zones- take place as the spreading centers generate new crust, while, the existing crust simultaneously consumed elsewhere in the world, knowing that the thicker and stiffer continental crust over-rides the weaker oceanic crust, forcing it back into the earth where it melts into magma.. Compressive forces distort the continentalcrust and push up mountains, often releasing volcanism.

The west coast of South America is an example of an active subduction zone, which created the off-shore Chilean Trench.

It is worthy mentioning that the movements and distortions of the earth's crust described above are part of the theory of plate tectonics. The earth's crust is visualized as a number of distinct plates **(see Fig 2.1)**, whose margins are constantly extended by spreading centers and trimmed by subduction. **Plate tectonics** is considered to be the primary mechanism responsible for creating the earth's mountain ranges, ocean trenches, and most of the earthquakes .and volcanic activity.

References

Komatina MM (2004) Medical geology—effects of geological environments on human health. Fig 2.12; ISBN 0-444-51615-8, Elsevier; USA, www.elsevier.com

Eugene Huter and Warren E. Teasdale (1989)- Application of drilling, coring, and samplingTechniques to test holes and wells **(USGS, Denver, CO)**.

Dickinson WR (1981, 1974) Subduction and oil migration. Geology 2/1974, pp 1519–1540, 1950

Dietz RS (1963) Wave base, marine profile of equilibrium, and wave-built terraces: a critical appraisal. Geol Soc Am Bull 74:971–990, *in* Dickinson (1974)

Davis E, Elderfield H (2004) Hydrogeology of the oceanic lithosphere. Cambridge University. Press, New York, NY, Fig 1; publicity@Cambridge.org.

Hobson GD (1975) Introduction to petroleum geology—modern tectoinc theories. three para: pp.240, 241; Figs 69 & 70/p242; Scientific Press Ltd., Beaconsfield, England

Walcott RI (1972) Gravity, flexure, and the growth of sedimentary basin at a continental edge.Geol Soc Am Bull 83:1845–1848, *in* Dickenson (1974)

Falcon NL (1973) Exploring for oil and gas. *in* Modern petroleum technology, 4th ed. Hobson GD and Pohl Weds (1973)

Coney, P.J., (1973) - Plate Tectonics of marginal foreland thrust-fold belts Geology, v.I, p.131-134.

Dunnington, H.V. (1985) – Generation, migration, accumulation, and dissipation of oil in northern Iraq, in Weeks, L.G., ed., Habitat of oil: Tulsa, Okla., Am. Assoc.Petroleum Geologists, p.1194-1251.

Cited References

Jia, ChengZao (2012)- Characteristics of Chineese Petroleum Geology (electronics resource)-geological Features and exploration Cases of Stratigraphic, foreland, and deep formation traps. UAB Labraries-Mervin H. Sterne Library/TN876.C5.

Chapter 3

Surface Geophysical Petroleum Methods, Remote Sensing, and Satellite Images in Salt Dome Exploration

3.1 Introduction

Surface geophysical techniques determine density, magnetic, and acoustical properties of a geologic medium. Three geophysical methods used in petroleum exploration comprise magnetic, gravimetric, and seismic (including refraction/reflection) techniques. The magnetic and gravity methods are used only in primary surveys where little is known of the subsurface geology and/or the thickness of sediments of potential prospective interest. The seismic reflection method is universally used for determining the underground geological structure of a reservoir rock in a certain area. The method(s) selected will depend on the type of information needed, the nature of the subsurface materials, and the cultural interference.

3.2 Magnetic Survey

A magnetic survey is primarily used to explore for oil and minerals. Magnetic exploration is based on the fact that the earth acts as a magnet. Any magnetic material placed in an external field will have magnetic poles induced upon its surface. The induced magnetization (sometimes called polarization) is in the direction of the applied field, and its strength is proportional to the strength of that field. The location of an area in relation to the magnetic poles is measured by the inclination of the earth's field or the "magnetic inclination" (USGS, U.S. Army, 1998).

Aeromagnetic Surveys were developed in wartime to overcome the problem of detecting submerged submarines from aircraft; they have since gained considerable success in petroleum and mineral exploration. Both the aeromagnetic surveys and the airborne radioactivity surveys can be classified as "remote sensing techniques." The advantages of the use of airborne surveys over ground instruments are the speed of the surveys and the possibility of reaching otherwise inaccessible areas.

3.3 Gravimetric Survey

Variations in gravity depend upon lateral changes in the density of earth materials in the vicinity of the measuring point. Many types of rocks have characteristic ranges of densities, which may differ from other types that are laterally adjacent. Driscoll (1986) stated that the earth's gravitational attraction at a particular site is a function of the density of the surface sediments and the underlying rock units. The density variations may be attributed to changes in rock type (porosity or grain density), degree of saturation, fault zones, and the varying thickness of unconsolidated sediments overlying the bedrock. Thus an anomaly in the earth's gravitational attraction can be related to a buried geological feature, e.g., a salt dome or other deposit which has limited horizontal extent. Actually, all geophysical surveys concentrate on the discovery of "anomalies" in the rocks which overlie or surround possible petroleum accumulation.

3.4 Seismic Exploration Survey

3.4.1 General

The word "seismic" refers to vibrations of the earth, including both earthquakes and artificially created sound waves that penetrate into the earth. Sounds measured are in the frequency range of about 10–100 cycles s^{-1}. The depths investigated for a sound to travel into the earth and return are as much as 16 km.

Seismic investigations depend on the fact that elastic waves (or seismic waves) travel with different velocities in different rocks. It is possible to determine the velocity distribution and locate subsurface interfaces where the waves are reflected or refracted by generating seismic waves at a given point and observing the times of arrival of those waves at a number of other points (or stations) on the surface of the earth.

Seismic surveys which are based on the velocity distribution of artificially generated seismic waves in the ground are produced by hammering on a metal plate, by dropping a heavy ball, or by using explosives. Energy from these sources is transmitted through the ground by elastic waves, which are so called because, when the waves pass a given point in the rock, the particles are momentarily displaced or disturbed, but immediately return to their original position or shape after the wave passes.

3.4.2 Seismic Refraction Methods

Seismic refraction methods were originally developed to locate concealed masses of salt plugs (e.g., in Algeria, Mexico, and Germany), and to trace major anticlinal axes in massive limestones (e.g., in Iran) which could not be located with surface geology. In both cases, the resulting shock waves caused by the explosives travel faster through salt and limestone than through the associated sedimentary rocks. Seismic refraction is also a useful reconnaissance tool for determining the depth of a high velocity metamorphic or igneous basement below a small sedimentary basin, etc.; each geologic formation has a characteristic seismic velocity that affects the arrival time.

3.4.3 Seismic Reflection Method

The seismic reflection method depends on the echo sounding principle and is a special tool for oil and gas exploration; it records reflected shock waves from a number of successive beds and their angle of inclination along the line of observation. The method uses a seismic wave produced by a weight dropping, a hammer blow, or another seismic source that is reflected off the bedrock and returns directly to the geophone, where the elapsed time is recorded. Hammer stations are usually at 9.1 m or less from the geophone to maximize the reliability of the reflected wave energy. The operator strikes a hammer plate at five to ten sites that are within 9.1 m of the geophone. The seismic signals received from these sites are summed automatically by the seismograph, canceling out the surface waves and other extraneous impulses, and the primary reflected wave is prominently displayed on the cathode ray tube (Driscoll 1986).

3.4.3.1 Multiple Reflections and Marine Exploration

Special ships allow rapid surveying in marine exploration where the presence of water as the medium for inducing the shock waves give remarkably good results. A system known as vibroseis employs a non-impulsive sound source with transmitters like huge loudspeakers with their diaphragms pressed against the ground. Because the seismic signal is quite unrecognizable until it is "pulse-compressed" by the correlation process, the vibroseis input can be regarded as a degenerate form of seismic impulse. Degeneration of a seismic impulse often occurs naturally in its travel through the earth, the commonest form being reverberation in a "ringing layer." Ringing is often a big problem in marine exploration. The shot has two very good reflecting interfaces immediately above and below it, namely, the sea surface and the sea bed. The downward traveling impulse from the shot is followed closely by a reflection from the sea surface, and when reaching the sea bed, it is partially reflected back to the surface, only to be once again sent on a downward course. Therefore the initial downward wave has three "ghosts" following it, leading to an endless process; this process continues as part of the more general problem of multiple reflections (Tarrant 1973).

3.4.3.2 Seismic Profiling of Diapiric Structures

The reflecting seismic process records reflections from acoustic discontinuities in the subsurface by generating a sound wave near the surface and detecting the reflected energy return from subsurface discontinuities.

Features associated with salt domes identified by seismic sections are radial faulting, the doming of overlying strata, the dip of sediments on the flank, nonconformity and wedging effects, and the development of rim synclines. A modeling process permits accurate interpretation of these features, e.g., piercement diapirs, salt dome structures at different stages of development, and associated geologic phenomena, thus increasing success in drilling for hydrocarbon reserves.

3.5 Remote Sensing and Land-Satellite Images in Salt Dome Exploration

Evaporites, including salt domes, are formed by the evaporation of brines because of dryness in arid conditions; they are normally interbedded with carbonate rocks together with red and green shales in cyclical sequences. In some parts of the world, buried evaporite beds lie several hundreds of meters beneath the ground surface, and have generated salt plugs or salt domes which have moved upwards through the overlying beds and probably appear on the surface in "diapiric" or piercing plastic flow, e.g., the Zechstein of North Germany, the Triassic of the Gulf region and Algieria, as well as the Miocene of the Suez Canal of Egypt.

Normal aerial photographs can be used to detect certain geological and ecological features peripheral to many ore deposits or oil and gas fields. Landsat images can be used by adapting remote sensing methods to the images; early space imagery displaced extended structural elements such as closed anticlines, domes, intrusive bodies, folded mountain belts, fault zones, and regional joint patterns.

Remote sensing data obtained from satellite images are most beneficial to proper and more accurate interpretations of the earth's surface. The use of aircraft to obtain data in locating a resource target considerably reduces the cost of exploration, but the use of spacecraft to obtain remote sensing information reduces the overall cost of the ground survey even further; and therefore "the higher we go, the deeper we can see" (Trollinger 1968).

References

Driscoll FG (1986) Groundwater and wells. Johnson Division, UOP Inc., St. Paul, Minnesota 55165, 3-para/ pp. 168–177.

Tarrant LH (1973) Geophysical methods used in prospecting for oil, in Hobson GD and Pohl Weds. (1973). Modern petroleum technology, p.81, Applied Science Publishers Ltd; England: The Institute of Petroleum Geology.

Trollinger WV (1968) Surface evidence of deep structure in the Delware Basin: "Delware Basin Exploration", Guidebook. West Texas Geol. Soc. Pub. #68-55, pp. 87–104; in Halbouty MT (1967) Salt domes.

USGS, U.S. Army (1998) Earth Science Applications. National Training Center, Fort Irwin, California (http://wrgis.wr.usgs.gov./docs/geologie)

Chapter 4
Drilling Technology in Petroleum Geology

4.1 Introduction

There are two drilling methods used in the petroleum industry: **(1)** The **cable tool method**; the first oil well, drilled in 1859 to a depth of 65 ft.(~21ms), was first employed by the early Chinese in the drilling of brine wells; some cable tool rigs are still working in parts of Europe as well as in the USA (**Pic, 4.1**). **(2)** The **rotary drilling method**, started by a French civil engineer in 1863, is the most common method that performs a rotary grinding action. ; Rotary drilling methods are much more effective in drilling shallow, unconsolidated sands than the cable tool operations.

A cable tool rig is a percussion drilling apparatus consisting of the drill string, which is mainly composed of the drill bit, made of a heavy steel bar; the drill stem, a cylindrical steel bar screwed directly above the bit; jars of heavy steel links, to produce a sharp upward blow on the tools; and tool joints that connect these parts.

In a standard cable tool rig, there are three rig lines or cables: the drilling line, the sand line, and the calf or casing line (used to run the casing into the well). The derrick floor supports the drilling line, the sand line on which the bailer is run, and the casing line (Brantly 1952; Fig. 4.1).

The cable tool rig can be used to drill wells up to about 170 m, but it is generally confined to much shallower operations. In cable tool drilling, the hole is kept partly filled with water to soften the formation and prevent caving. At depth, when the bottom of the hole becomes full of cuttings, the rock bit should be withdrawn and a bailer run to remove the cuttings.

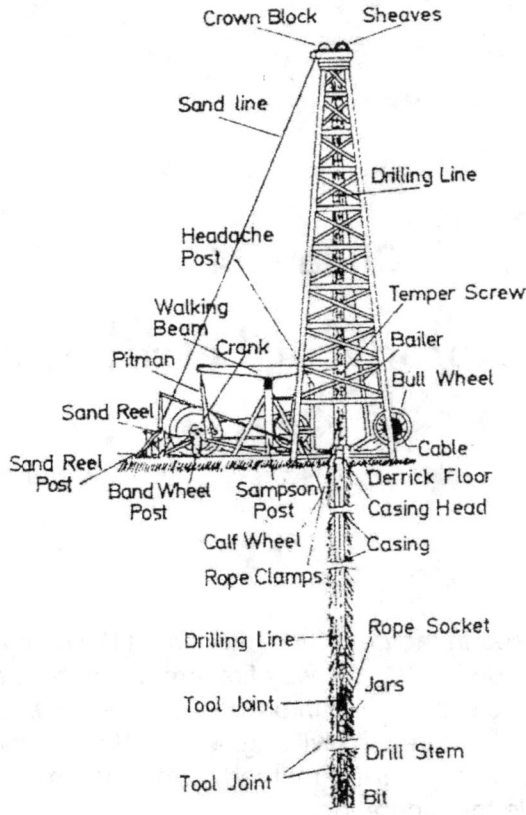

Fig. 4.1 Standard cable tool drilling system (Brantly, courtesy AIME)

A **rotary drilling rig** (API 1954; Fig. 4.2) comprises six components: (1) The **derrick and substructure**. The derrick provides the vertical clearance necessary for raising and lowering the drill string into and out of the borehole during drilling operations, whereas the substructure is the support on which the derrick rests. (2) The **Mud Pump** circulates the drilling fluid at the desired pressure and volume. **(3)** The **drawwork** or **hoist** is the control center from which the driller operates the rig. It houses the drum and is the key piece of equipment on a rotary rig; it includes the clutches, chains, sprockets, engine throttles, etc. **(4)** The **Rotary Drill String** consists of the Kelly joint, drill pipe, tool joints, and drill collars. The drill pipe furnishes the necessary length for the drill string and serves as a conduit for the drilling fluid. Between the drill pipe and the rock bit are the drill collars, which are heavy-walled steel tubes to furnish the compressive load on the bit, thus allowing the lighter drill pipe to remain in tension. The drill string is an extremely expensive rig component and must be replaced periodically to avoid failure due to material fatigue that may result from corrosion and//or improper care and handling; The Kelly joint is the topmost joint in the drill pipe and is commonly square, hexagonal, or even octagonal. It passes through snugly fitting, properly shaped bushings in the rotary table, allowing the table's rotation to be transmitted to the entire drill string. **(5) Rock bits** are designed to give optimum performance in various formation types **(pic 5.A-1)**. (6) The **Drilling Line** affords a means of handling the loads suspended from the hook during all drilling operations.

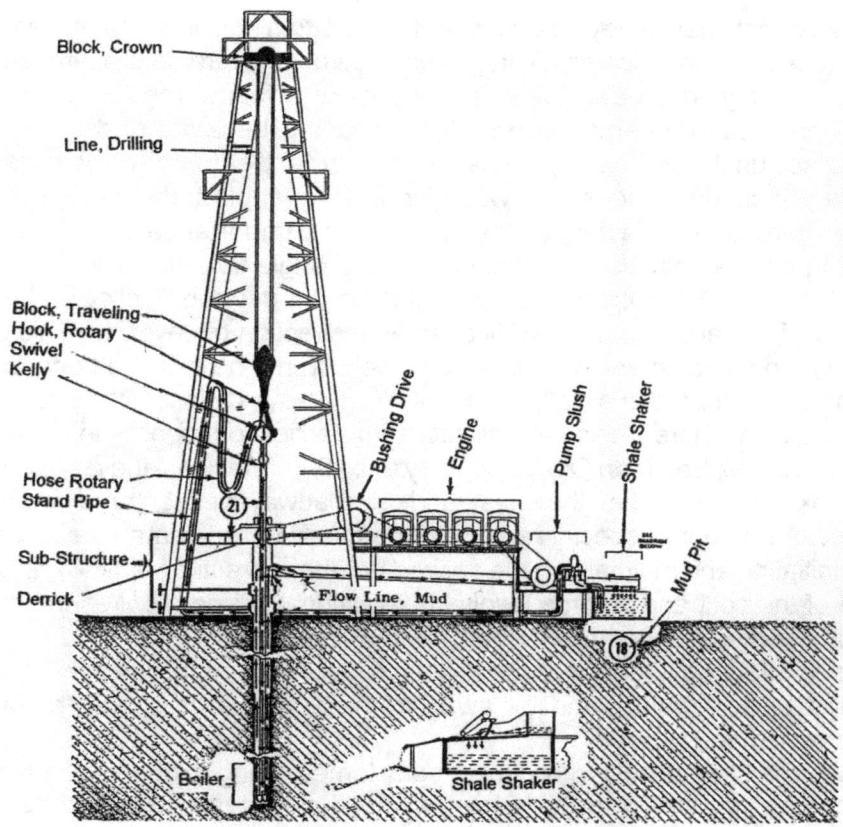

Fig. 4.2 Basic components of a rotary drilling rig. Courtesy API (1954)

In rotary drilling, the well is drilled by a rotating bit to which a downward force is applied; the bit is fastened to and rotated by a drill string, which is composed of high quality drill pipe and drill collars, with joints added as drilling progresses. The cuttings are lifted from the well by the drilling mud fluid, which is continuously circulated down, inside the drill string through the nozzles in the bit, and upward in the annular space between the drill pipe and the borehole.

The returning mud fluid discharged at the surface enters into a segregated sedimentation tank or pit that affords a sufficient period to allow cuttings— separation and any necessary treating—then the mud is picked up by the pump suction to repeat the cycle.

Turbo drilling of high speed rotation is used in Europe to avoid the rotation of a long, slim drilling column and its removal from the hole when the cutting tool is worn out. By using the turbo drill, the mud should be kept in good condition and free of sand or any other abrasive particles.

4.2 Petroleum Drilling Operations

4.2.1 Discussion

Petroleum wells are considered vertical wells in rotary drilling operations, although in practice borehole deviation exists and 3°–5° is commonly specified as the maximum acceptable deviation in vertical boreholes; in the early days, some wells happened to run into each other during drilling. In drilling operations after the development of reliable directional surveying equipment, the bottom hole location of many of the original wells deviated from the surface location; accordingly large reserves of recoverable oil were bypassed.

Subsurface mapping relying on depth measurements was found impossible to depend on; depths obtained from drill pipe measurements were quite misleading because of the deviation of wells from the verticality. It was not until 1950, when a later fundamental treatment of hole deviation was carried out successfully, that principal consideration was given to the angle between the hole and the vertical; whereas the compass direction of deviation was of secondary importance.

Directional drilling has recently been practiced to control angle deviation from the vertical in a borehole. This can be accomplished by Totco instruments that induce the deviation, followed by survey instruments that record the amount of deviation from the vertical and the direction of deviation. This process is applied in oil or gas reservoirs, when it is desirable to drill several wells from one location because of surface climatic conditions, or to extinguish an oil fire in a well; it is mainly used to avoid placing the bottom of a borehole at an inaccessible surface location where a vertical well site is impractical or impossible because of topographical and/or legal problems (Fig. 4.3; Eastman Oil Well Survey Co. 1960). Besides, the required bottom of the drilled hole may be under a lake or river, just off the coast, under a roadway, or in an area where legal access to the ideal location is barred. A number of specialized tools, commonly called primary deflection tools, are used in directional drilling for initiating and maintaining the desired borehole direction; but they are used as infrequently as possible because of the cost and rig time involved. Directional drilling requires the measurement of both vertical and horizontal directions, accomplished with various devices.

One device used to effect deviation is known as a **whipstock**—a tapered steel wedge with a concave groove on its inclined face; it guides the bit away from the previous course of the well towards the direction that faces the inclined groove.

Figure 4.4 shows directional drilling tools on salt domes using modern whipstocks and surveying instruments which can be grouped into three basic deflection patterns or types, depending on the geologic factors, the economics, and the desired final results (Halbouty 1967; modified from Cook, 1957).

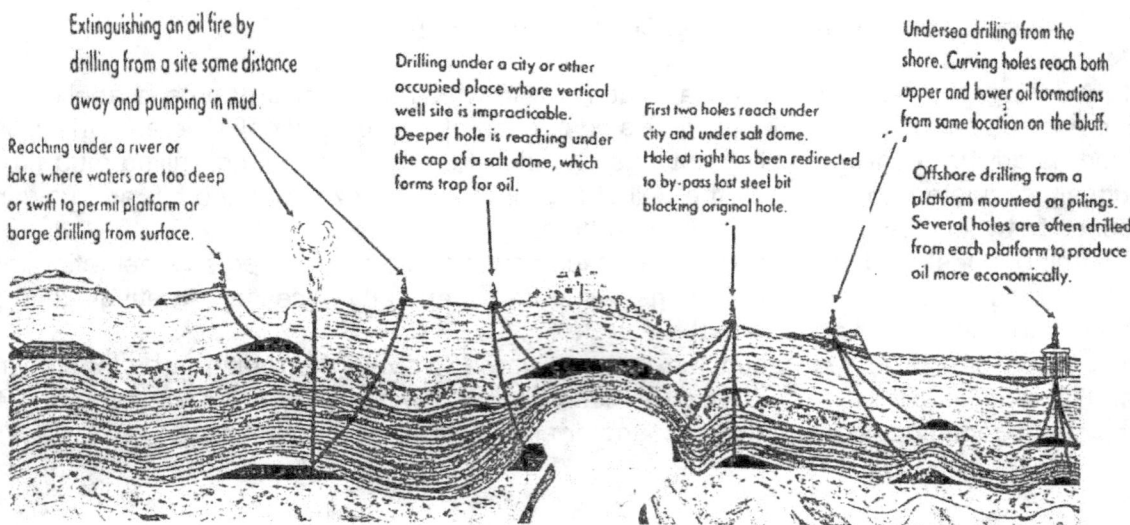

Fig. 4.3 Specific applications of directional drilling. Courtesy Eastman Oil Well Survey Company

Directional Drilling on Salt Domes

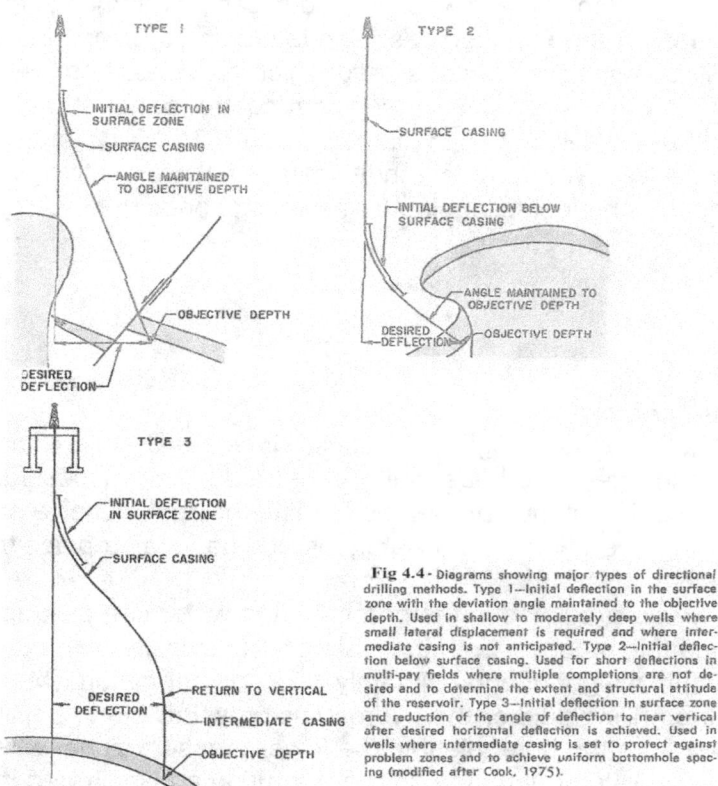

Fig 4.4 - Diagrams showing major types of directional drilling methods. Type 1—Initial deflection in the surface zone with the deviation angle maintained to the objective depth. Used in shallow to moderately deep wells where small lateral displacement is required and where intermediate casing is not anticipated. Type 2—Initial deflection below surface casing. Used for short deflections in multi-pay fields where multiple completions are not desired and to determine the extent and structural attitude of the reservoir. Type 3—Initial deflection in surface zone and reduction of the angle of deflection to near vertical after desired horizontal deflection is achieved. Used in wells where intermediate casing is set to protect against problem zones and to achieve uniform bottomhole spacing (modified after Cook, 1975).

Fig. 4.4 Diagrams showing major types of directional drilling methods. Type 1 – Initial deflection in the surface zone with the deviation angle maintained to the objective depth. Used in shallow to moderately deep wells where small lateral displacement is required and where intermediate casing is not anticipated. **Type 2** – Initial deflection below surface casing. Used to show deflections in multi-pay fields where multiple completions are not desired and to determine the extent and structural attitude of the reservoir. **Type 3** – Initial deflection in surface zone and reduction of the angle of deflection to near vertical after desired horizontal deflection is achieved. Used in wells where intermediate casing is set to protect against problem zones and to achieve uniform bottom hole spacing (Halbouty, 1967, modified from Cook, 1957).

4.2.2 Types of Drilling Operations

1. **Offshore Drilling**—The basic drilling operations in both onland and offshore wells are exactly the same. The only difference is in the procedure of connecting the well to the rig, which is either floating or standing above the location. There are four basic forms of offshore rigs: fixed multi-well platforms, mobile jack-up platforms, drill-ships, and semi-submersible platforms. In offshore drilling operations, seawater is economically used as the base fluid for the salt-mud system. Seawater mud is effective in drilling shallow offshore domes; however, additional treating agents are necessary to overcome contamination by calcium and magnesium ions.
2. **Coring Techniques**—Core drilling is necessary to obtain and examine larger, unbroken pieces of reservoir rock samples which are obtained either from the bottom during drilling or from the side of the borehole wall after drilling:

    (a) **Bottom coring**, at the time of drilling, utilizes an open center bit that cuts a doughnut-shaped hole, leaving a cylindrical plug or core in the center which rises inside the core barrel and is then captured and raised to the surface for analysis. Diamond core bits are preferable; although they cost much more, they drill more total footage and when worn out, can be returned to their supplier for salvage.

(b) **Sidewall coring after drilling** is often desirable to obtain core samples from a particular zone or zones already drilled. A special device is used which provides a hollow bullet that imbeds in the formation wall and is then fired from an electrical control panel at the surface. Samples of this type, are normally 3/4 or 113/16 inch in diameter, and ¾ to 1 inch long. Sidewall coring is widely applied in soft rock areas, where hole conditions are not conductive to drill stem testing (DST); geologists can depend on electric logging for selecting zones to be sampled.

4.3 Drilling Fluids

4.3.1 Properties of Drilling Muds

Drilling mud properties vary with geologic conditions, since the degree of change in mud properties depends on the nature of the geologic beds being penetrated. For successful well completion in soft rock drilling areas, precise control of mud properties requires using expensive and complicated chemical mixtures. In hard rock drilling operations, plain water, air, and gas may replace the drilling fluids (see API Recommended Practices No. 29 1949).

When only "clean" water is circulated in the borehole, the water can pick up clay and silt and form a natural drilling mud. During this process, both the weight and viscosity of the drilling fluid increases. It is possible to attain a maximum weight of approximately 1.32 kg/l per gallon when drilling in natural clays.

When the weight of the drilling fluid or the hydrostatic pressure in the borehole exceeds that of the reservoir rock, fluid moves from the borehole into the lower pressure zone, where the incorporated fine particulate matter can be deposited during the drilling operation and infiltrated into the pore space of the zone, together with any solids added to the drilling fluid; an impermeable and hard "mud cake" or "filter cake" is thus formed on the wall of the borehole, when bentonitic drilling-mud is used.

Viscosity (or the resistance to flow) is an important property of the drilling fluid. In combination with the velocity of the circulated fluid, viscosity controls the ability of the fluid to remove cuttings from the borehole; but it has no relationship to density. In the field, it is measured by the time required for a known quantity of fluid to flow through an orifice of special dimensions. Most liquid drilling muds, either colloidal or emulsions that behave as plastic fluids (non-Newtonian), differ from those of (true) Newtonian fluids such as water and light oils, in that their viscosity is not constant, but varies with the rate of shear—on which in turn depends on the ratio of shearing stress to the rate of shearing strain.

Yield point and **gel strength** are two additional properties that are considered in evaluating the characteristics of drilling mud. **Yield point** is a measure of the amount of pressure that must be exerted by the pump, upon restarting after a shutdown, to cause the drilling fluid to start to flow. **Gel strength** is a measure of the ability of the drilling fluid to maintain suspension of particulate matter in the mud column when the pump is shut down. There is a close relationship between viscosity, yield point, and gel strength (Assaad et al. 2003).

4.3.2 Composition and Nature of the Drilling Fluids

A typical drilling mud fluid consists of a continuous liquid base, a dispersed gel-forming phase (e.g., colloidal solids and/or emulsified liquids for the required viscosity), and wall cake; other inert dispersed solids (e.g. weighing materials, sand, and cuttings); and various chemicals necessary to control properties within desired limits. The general functions of drilling fluids are to cool and lubricate the bit and drill string; to remove and transport cuttings from the bottom of the hole to the surface; to suspend cuttings during times when circulation is stopped; to control encountered subsurface pressures; and to wall the borehole with an impermeable mud cake.

4.3.3 Different kinds of Drilling Mud Fluids:

(a) **Saltwater muds**—Sodium bentonite does not form a satisfactory colloid in salt water. The clay mineral attapulgite (known as salt clay), together with hydrates, forms a stable suspension in salt water and is used in saline water in about the same manner as bentonite in freshwater.

Salt concentrations neutralize the electric charge on dispersed bentonite particles, allowing the formation of particle aggregates which are larger than colloidal size (flocculation). Saltwater muds are used in drilling through salt beds and in localities of abundant saltwater (e.g., in offshore drilling, and in swamp and seaside locations). In general, the difference between freshwater and saltwater muds is the type of clay used in the gel-forming phase.

In offshore drilling operations, seawater, which is more economically used as the base fluid for the salt-mud system, is also effective in drilling shallow offshore domes.

(b) **Freshwater muds**—The basic ingredients are freshwater and suspended clays. Certain clay minerals, if ground to colloidal size and mixed with water, readily hydrate (adsorb water) to form stable colloids. Sodium bentonite, mainly composed of the clay mineral montmorillo-nite, yields higher viscosity at lower clay content than do other clay minerals.

It is a common practice to pretreat the mud system with calcium; such muds are known as lime base muds, in which calcium is allowed to remain in the mud because it tolerates other flocculating salts (up to 50,000 ppm of sodium chloride).

(c) **Oil base muds**—These are generally composed of varying materials: high flash diesel oil; oxidized asphalt (a colloidal fraction that provides the wall building property); a combination of an organic acid and an alkali (forming an unstable soap that governs the viscosity and gel strength of the mixture); and various stabilizing agents, plus 2–5% water.

Oil base muds are expensive, and their main uses are: (a) drilling and coring of possible productive zones; (b) drilling of bentonitic (heaving) shales that continually hydrate, swell, and slough into the hole when mixed with water; (c) deep drilling operations in high temperature environments where solidification may occur; (d) as a perforating fluid to be added (a few barrels) to prevent contamination of the section after being perforated; and (e) remedial operations on producing wells to avoid other drilling problems such as a stuck pipe and corrosion.

4.3.4 Mud Systems in Salt Structures

A complete technological cycle in mud engineering has been performed since Stroud (1925) introduced weighted drilling fluids to control blowouts. Mud systems became increasingly complex thereafter. Many organic and inorganic additives were added to gel muds to maintain the desired mud qualities under varying conditions. However, there has been recently a tendency to return to simpler mud systems that accomplish all the desired effects of former complex muds with fewer additives.

There are two basic types of mud systems—one designed for drilling shallow domes, and the other for deep salt structures; if salt is drilled at very shallow depths, a saturated saltwater mud is mixed in the surface system to be utilized as a spud mud. However, if salt domes are several thousand feet deep (but still in the shallow or intermediate depth range), a freshwater mud system can be used until the cap salt rock is penetrated; then the mud is converted to a saturated salt mud. Both native or gel mud freshwater may require certain inorganic additives to develop the proper alkaline values which promote hydration and dispersion of the penetrated shales. The commonly used freshwater clays or gels are adversely affected by high salt concentration; therefore, a special saltwater clay, attapulgite, should then be used.

Saltwater muds have salt concentrations above 10,000 ppm, or one percent salt. Normal freshwater muds with either high or low pH values maintain good physical properties up to that concentration of salt. Saturated saltwater muds, which are characterized by high gel strength, are usually added to the system if abnormal viscosity and high gel strength create a problem (Halbouty 1967).

4.3.5 Salt Dome Drilling

Drilling on salt domes can be classified into three main types: (1) **salt dome drilling** on crests of shallow domes, where lost circulation is a common problem that occurs in drilling highly cavernous cap rock over most shallow domes; (2) **extended drilling capabilities to medium depths** on the flanks of shallow, piercement domes; and (3) **drilling of deep domal and anticlinal salt structures**, involving greater risks than shallow drilling and with many potential complications and problems—therefore, specialized mud systems must be programmed to ensure the best chance for a successful operation. Problems of deep drilling wells in or above salt structures include lost circulation, heaving shales, salt water flows, abnormally high pressure zones, and high temperature.

4.4 Drilling Hazards

Improper control of mud properties can create various drilling hazards:

(a) **Enlargement of boreholes in salt beds** of considerable thicknesses can cause very difficult fishing operations in case of drill string failure.
(b) **Heaving shale problems** in shale sections rich in bentonite or other hydrate clays continually adsorb water, swell, and slough into the hole; in other words, upon hydration, the shale forming the wall of the bore hole may disintegrate and fall into the hole, sticking the drill pipe and causing a severe drilling hazard. In hydratable shales, where heavy muds with a high content of suspended solids directly affect the viscosity, the control of viscosity by dilution requires the addition of large quantities of barite, which greatly raises the cost of a mud program.
(c) **Blowouts** occur when the encountered formation pressure exceeds the mud column pressure, allowing the formation fluids to blow out of the hole. Blowout preventers are connected to the top of the wellhead and are used to shut in and control the well in the event of a blowout of gas or oil, while drilling in the reservoir rock which encounters higher pressures than that exerted by the column of mud in the hole. Blowout preventers consist of two or three ram preventers and one big-type preventer; their working pressure is 3,000, 5,000, or 10,000 lb/sq inch, depending on the conditions of the well and the expected pressures of the reservoir.
(d) **Partial or complete lost circulation**, defined as the loss of substantial quantities of a part or of the whole mud, might occur in an encountered formation; though it is always required that the mud column pressure must exceed the formation pressure. Undesirable effects of lost circulation are expected, such as lack of lithological information, since no cuttings are obtained; the possibility of a stuck drill pipe, resulting in a fishing job; a consequent loss of drilling time; and cost increases.

4.5 Drill Stem Testing (DST)

4.5.1 Scope

A drill stem test is carried out when oil shows are traced in the cuttings; this can occur many times during the drilling of an exploratory well. It allows the productive capacity of an encountered formation to be tested in a borehole full of drilling mud. The testing tool assembly attached to the lower part of the string of the drill pipe (or drill stem) is lowered and placed opposite to the reservoir rock, isolated by an expandable packer(s), relieved of the mud column pressure, and allowed to produce through the drill stem. The hydrostatic pressure of the mud column inside the hole is always greater than the formation pressure of the production zone to be tested (USEPS 1977, Edwards and Winn 1974, Kilpatrick 1955). **Fig 4.5 (Black, Courtesy AIME).**

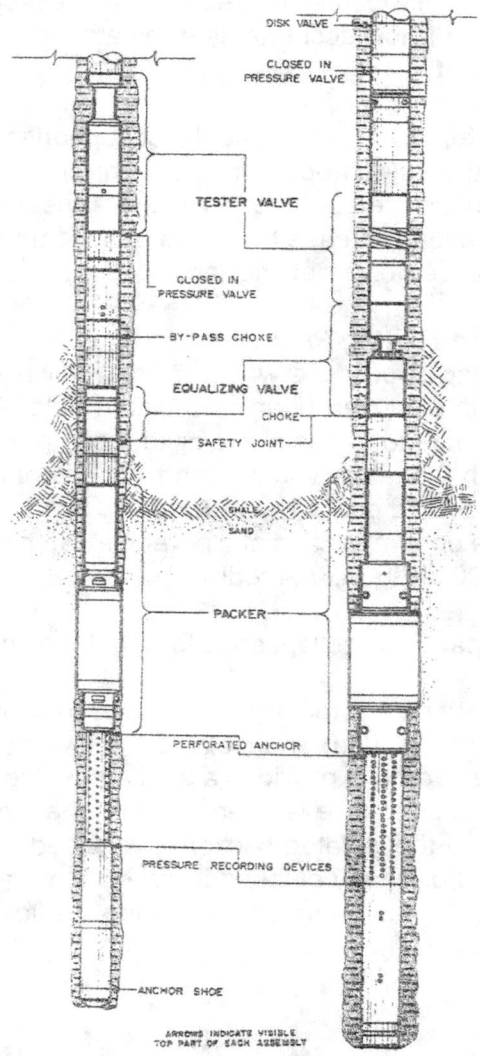

Fig. 4.5 Typical conventional drill-stem test tools (After Black, Courtesy AIME).

The basic drill stem test tool assembly consists of the following::

1. A rubber packer(s) expanded against the hole to separate the annular sections above and below the encountered zone.
2. A tester valve (upper pressure recorder) to control mudflow into the drill pipe (or exclude mud) during entry into the hole and to allow formation fluids to enter during the test. A hydraulic jar is fixed underneath the tester valve to increase the pressure to 8 psi, for releasing the packer after the test.
3. An equalizing bypass valve to allow mud pressure equalization across the packer(s) in and out of the DST assembly after completion of the flow test.
4. A pressure recorder for obtaining quantitative results during the period of flow in pressure (FIP) or pressure building (close-in pressure, CIP) following the period in which the formation fluids are allowed (FIP). Pressure recorders furnish a complete record of all events that may occur during a particular test and are in the form of a graph of pressure versus time.

Two pressure recorders are usually located to measure the pressure inside and outside the perforated anchor. The two measurements allow accurate determination of whether or not the perforations have become plugged during the test.

A perforated anchor is the extension below the tool that supports the weight applied to set the packer. It rests either on the bottom of the hole in open-hole tests or on cement plugs that have been spotted at the desired location. Safety joints are equipped to afford a means of unscrewing the drill string at a point convenient for fishing operations, in case the packers become stuck. Another safety joint is added underneath to prevent the DST assembly from sticking.

4.5.2 The Halliburton Formation Testing Procedure

The Halliburton Formation Testing Procedure was the first known method to determine the potential productivity of a reservoir rock in either an open or a cased hole. The testing procedure requires the opening of a section of the borehole to atmospheric or reduced pressure. The testing string is lowered into the hole or the drill pipe with the tester valve closed to prevent entry of well fluid into the drill pipe, leading to an undesired fishing job and possibly a stuck drill pipe.

The procedure for testing the bottom section of a borehole can be summarized as follows (Haliburton Oil well Cementing Company, 2001; **Fig. 4.6a,** modified):

(a) While going in the hole, the packer is collapsed, allowing the displaced mud to rise, as shown by the arrows **(Fig. 4.6a, step #1).**
(b) After the drill stem reaches bottom, and the necessary surface preparations are completed, the packer is then set (compressed and expanded) and isolates the lower zone (or the desired zone) from the rest of the open hole; in other words, it provides a seal above the zone to be tested (step #2).
(c) The bypass is closed as the tester valve is opened; here, the packer supports the hydrostatic pressure load of the well fluid, and the isolated section is exposed through the open tester valve to the low pressure inside the empty or nearly empty drill pipe, allowing the formation fluid to enter and the flowing formation pressure (FIP) can be measured during the flow period (step #3).

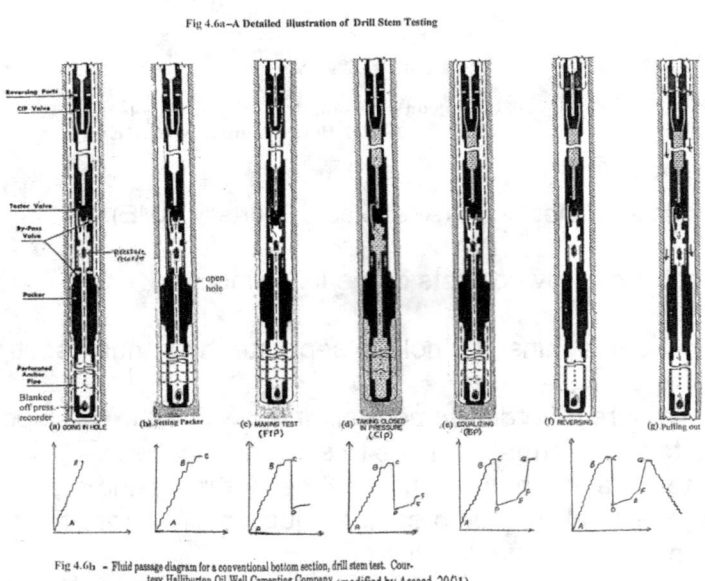

Fig. 4.6 (a) A detailed illustration of drill stem testing. **(b)** Fluid passage diagram for a conventional bottom section, drill stem test. Courtesy Halliburton Oil Well Cementing Company (modified by Assaad)

At the end of the test, the tester valve is closed (CIPV), trapping any fluid above it; this makes possible the measurement of the static formation's "built-up in-pressure (CIP)" (step #4).

(d) After the final closed-in period, the bypass valve is opened to equalize the pressure (EP) across the packer (step #5).
(e) Formation fluid received during the test can be removed from the drill pipe by reverse circulation before the pipe is removed from the borehole. This reversal is performed by closing the blowout preventers and pumping mud down the annulus; the mud then enters the drill pipe through the reversing ports, thereby displacing any formation fluids in the pipe. The recovered fluids may be sampled as they are discharged at the surface (step #6).

Finally, the setting weight is taken off and the packer is pulled free (step #7). The fluid content of each successive pipe section is examined when it is removed. The graphic charts are absolutely essential to get the accurate interpretation of test results. **Fig. 4.6b** shows a record of fluid passage, a graph of pressure versus time for a conventional bottom section of a drill stem test (Haliburton 2001). Tests through perforations are sometimes required to retest a reservoir rock after casing has been emplaced. **Fig. 4.7, 4.8, 4.9, 4.10, 4.11, 4.12, 4.13 and 4.14**, show common types of drill stem testing.

4.6 General Remarks

4.6.1 The DST—Procedure

The following is a further explanation of the steps given in **Fig 4.6a of** the DST procedure:

(1) Initial hydrostatic pressure (IHP) is exerted by mud column. (2) Initial closed-in pressure (ICIP). (3) Initial flow pressure (IFP) is the lowest pressure recorded just after the tool is opened. (4) Final flowing pressure (FFP) is the pressure just before the tool is closed. (5) Final closed-in pressure (FCIP). (6) Final hydrostatic in-pressure (FHP).

– The choke size of the selected bottom hole surface orifices, depends on the test's anticipated conditions; the bottom choke is used primarily as a safety measure and should be considerably large enough to minimize surface pressure in a flowing test conditions.

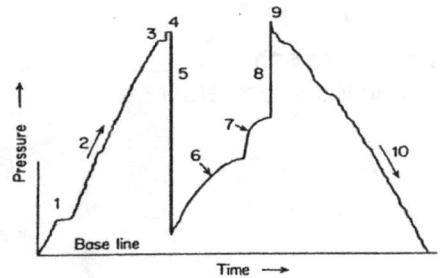

1. Putting water cushion in drill pipe
2. Running in hole
3. Hydrostatic pressure (weight of mud column)
4. Squeeze created by setting packer
5. Opened tester, releasing pressure below packer
6. Flow period, test zone producing into drill pipe
7. Shut in pressure, tester closed immediately above packer
8. Equalizing hydrostatic pressure below packer
9. Released packer
10. Pulling out of hole

Fig 4.7- Normal sequence of events in successful drill stem test.(After Kirkpatrick, courtesy *Petroleum Engineer*)

Fig. 4.7 Normal sequence of events in successful drill stem test. After Kirkpatrick, courtesy *Petroleum Engineer*

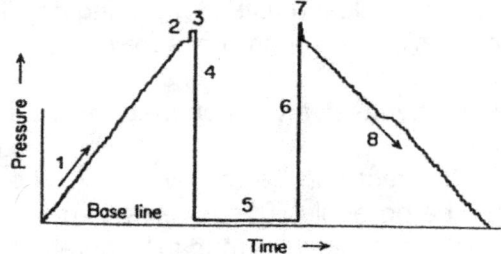

1. Running in hole
2. Hydrostatic pressure (weight of mud column)
3. Squeeze created by setting packer
4. Opened tester, releasing pressure below packer
5. Flow period, test zone open to atmosphere
6. Closed tester and equalizing hyd. pressure below packer
7. Pulled packer loose
8. Pulling out of hole

Fig 4.8 – *Dry* test. After Kirkpatrick,[1] courtesy *Petroleum Engineer*.

Fig. 4.8 *Dry* test. After Kirkpatrick,[1] courtesy *Petroleum Engineer*

- The water cushion of a certain length or head of liquid is placed inside the drill pipe rather than running it dry, to reduce the external pressure on the drill pipe in deep holes and/or to reduce the pressure drop on the formation and across the packer(s) when the DST assembly is first opened.
- The length of the DST depends greatly on some observations during the test; e.g., in hard formations, the drill pipe is not liable to be stuck and the flow test is often several hours long; the shut-in period after the flow test should be long enough to permit establishment of a stabilized static pressure.

4.6.2 Required Conditions and Reasons for Carrying Out a DST in a Petroleum Reservoir Formation

- The formation pressure of oil or gas should be maintained during drilling and the mud fluid pressure should exceed that of the formation.
- Saturated cuttings and core samples of gas or oil.
- Porous and permeable core samples.
- Presence of oil or gas in the drilling mud.
- Shows of fluorescence in cuttings and core samples.
- Identification of petroleum from electric log interpretations.

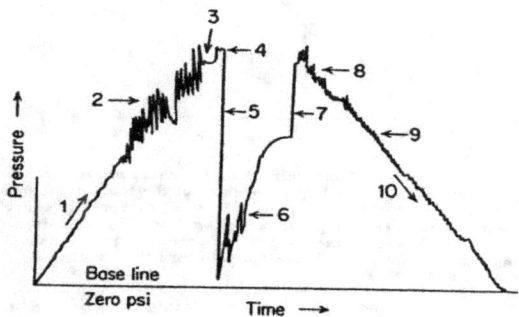

1. Running in hole
2. Indicates bad hole condition, scraping wall cake
3. Hydrostatic pressure (weight of mud column)
4. Squeeze created by setting packer
5. Opened tester, releasing pressure below packer
6. Indicates plugging of perf. anchor or tester
7. Closed tester and equalizing hyd. pressure below packer
8. Indicates swabbing due to bad hole condition
9. Indicates less swabbing, better hole condition
10. Pulling out of hole

Fig 4.9 - Effect of poor hole condition. After Kirkpatrick, courtesy *Petroleum Engineer*.

Fig. 4.9 Effect of poor hole condition. After Kirkpatrick, courtesy *Petroleum Engineer*

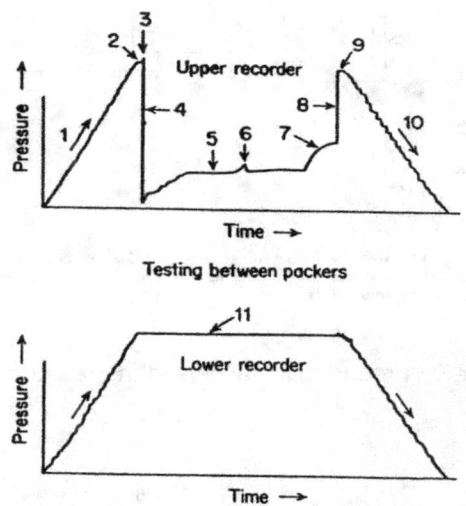

1. Running in hole
2. Hydrostatic pressure (weight of mud column)
3. Set packer
4. Opened tester, releasing pressure between packers
5. Test zone flowing
6. Rise in pressure due to closing in to change chokes
7. Shut in bottom hole pressure
8. Equalizing hydrostatic pressure between packers
9. Hyd. pressure at conclusion of test
10. Pulling out of hole
11. Lower recorder, below bottom packer shows no drop in pressure, proves bottom packer is holding

Fig 4.10 - Straddle packer test. After Kirkpatrick, courtesy *Petroleum Engineer*.

Fig. 4.10 Straddle packer test. After Kirkpatrick, courtesy *Petroleum Engineer*

Applications of Petroleum Tools for Field Geologists

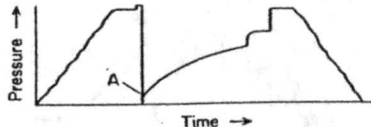

Fig. 4.11 Effect of pressure recorder inertia. After Kirkpatrick, courtesy *Petroleum Engineer*

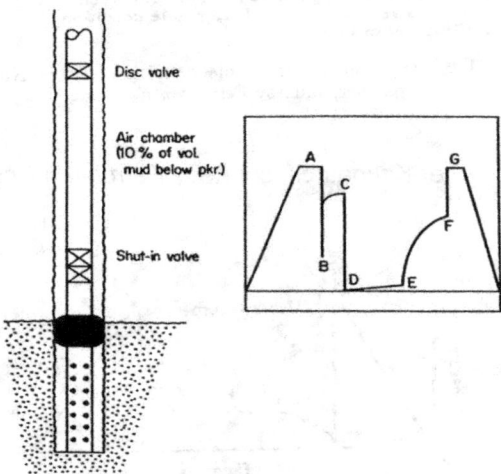

Fig. 4.12 Schematic tool arrangement for procuring initial closed-in pressure. Point C on chart is taken as reservoir pressure. Courtesy Johnson Testers

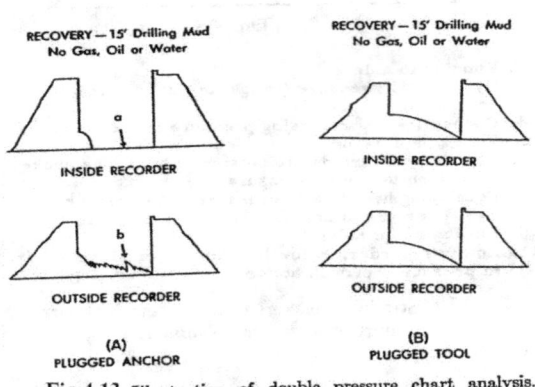

Fig. 4.13 Double Pressure charts – Anchor plugged

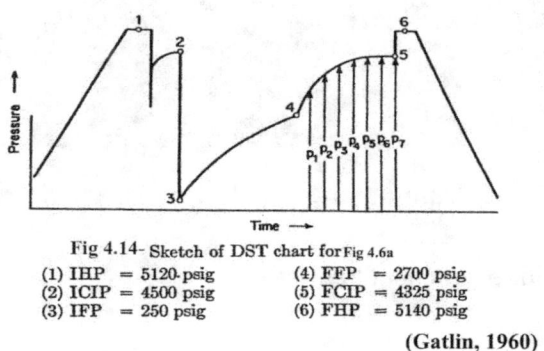

Fig 4.14- Sketch of DST chart for Fig 4.6a
(1) IHP = 5120 psig (4) FFP = 2700 psig
(2) ICIP = 4500 psig (5) FCIP = 4325 psig
(3) IFP = 250 psig   (6) FHP = 5140 psig

(Gatlin, 1960)

Fig. 4.14 Double Pressure charts – Choke plugged

Appendix 4.A- see images:
Selected Graphs of Drill Stem Testing in Different Reservoir Rocks and Interpretations:
(Figs. 4.15, 4.16, 4.17, 4.18, 4.19, 4.20, 4.21, 4.22, 4.23, 4.24, 4.25, 4.26, 4.27, 4.28, 4.29, 4.30, 4.31, 4.32, 4.33, 4.34, 4.35, 4.36, 4.37, 4.38, 4.39.

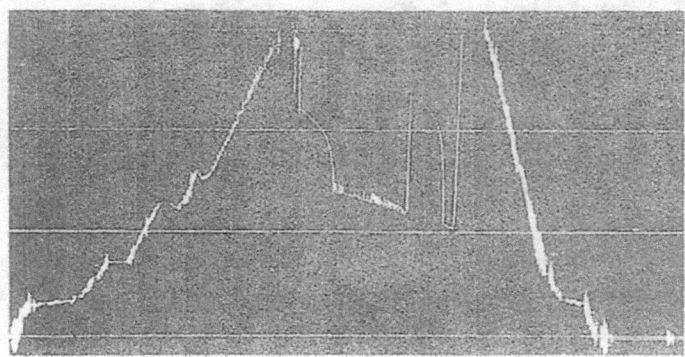

Fig. 4.15 Incorrect base line and shut off during flow (Ligne de base incorrecte et bouchage pendant débit)

Fig. 4.15 Incorrect base line and shut off during flow (Ligne de base incorrecte et bouchage pendant débit)

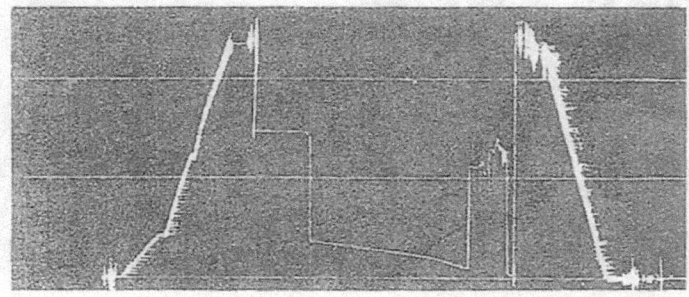

Fig. 4.16 Leaking dual closed—in-pressure valve "DCIPV" (Fuite au Vanne de fermeture et ouverture)

Fig. 4.16 Leaking dual closed—in-pressure valve "DCIPV" (Fuite au Vanne de fermeture et ouverture)

Fig. 4.17 Clock running away (Montre marchant très vite)

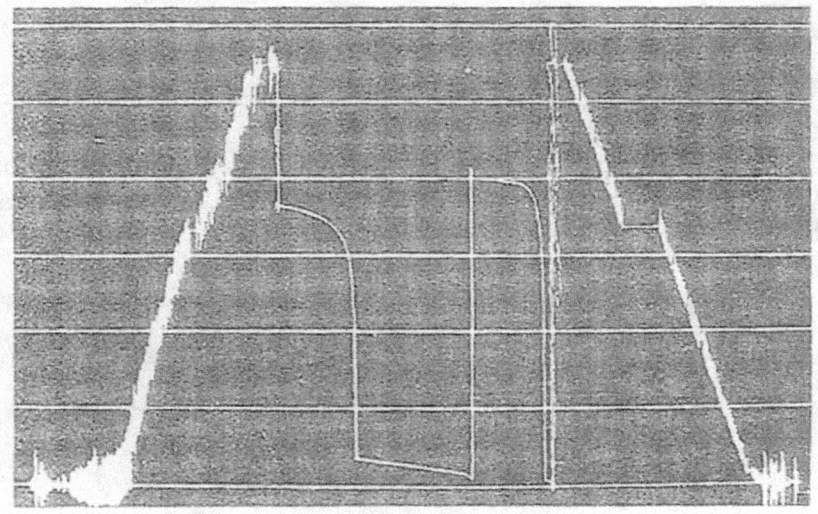

Fig. 4.18 Initial rapid built up pressure (Premiere remontèe de pression surcharge)

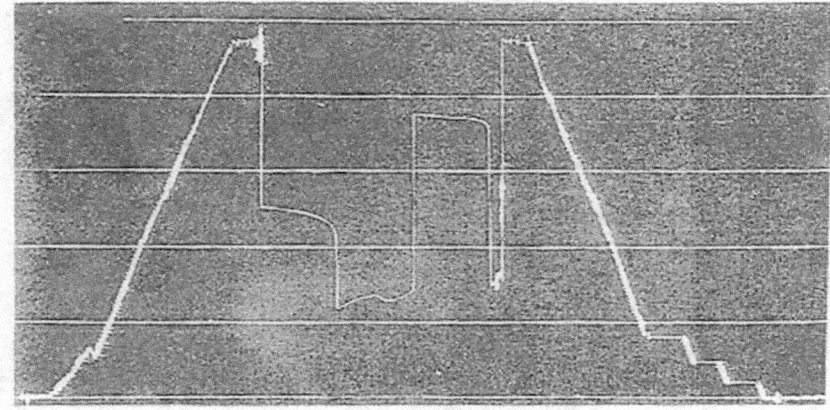

Fig. 4.19 Depletion of gas reservoir (Reservoir de gaz en depletion)

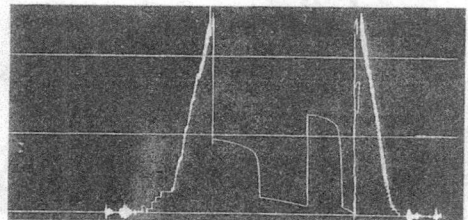

Fig. 4.20 Depletion of oil reservoir (Reservoir de liquide en depletion)

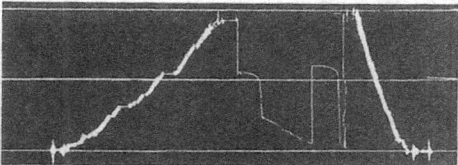

Fig. 4.21 Plugging flow period with a uniform segment (Bouchage pendant dèbit)

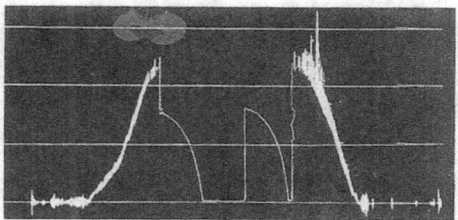

Fig. 4.22 Low permeability formation with high reservoir pressure (Formation á faible permèabilitè avec grande pression de reservoir)

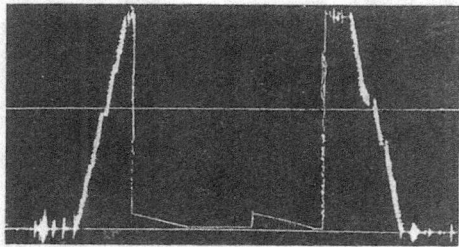

Fig. 4.23 Low permeability formation with low reservoir pressure (Formation á faible permèabilitè avec grande faible pression de reservoir)

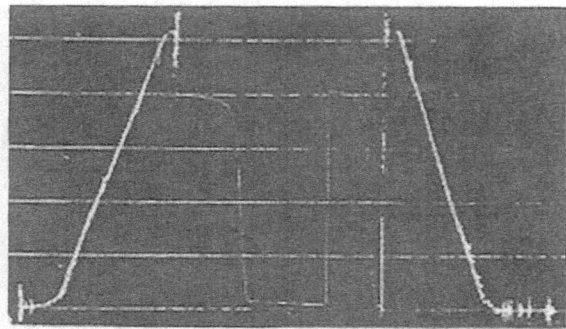

Fig. 4.24 The "S" curve indicates a possible after-production effect of bypass gas (Courbe en "S" indiquant l'effect d'une venue de gaz possible apres-production)

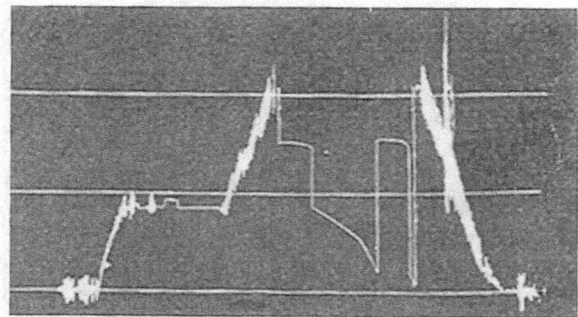

Fig. 4.25 Transition of filling up of the drilling pipes of the well (Transition de remplissage de massestiges aux tiges de forage)

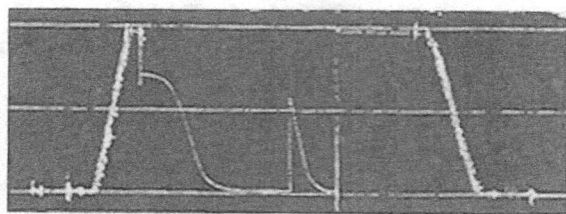

Fig. 4.26 The "S" curve indicates a possible vertical permeability (Courbe en "S" indiquant une permèabilitè verticale possible)

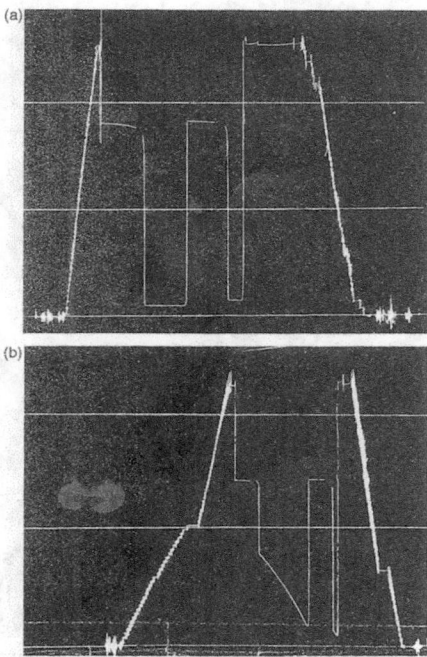

Fig. 4.27 (a) Low productivity and high damage (Faible productivitè + grande endommagement) **(b)** A high productivity and high damage (Grande productivitè + Grande endommagement)

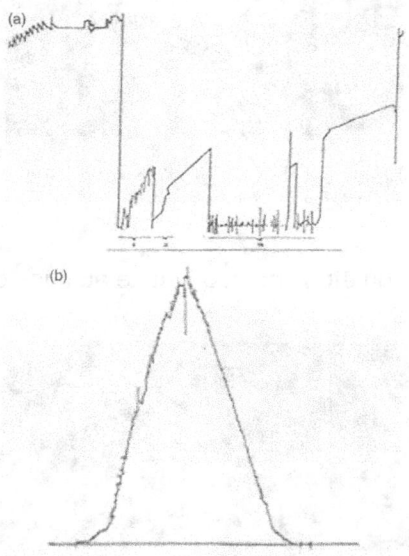

Fig. 4.28 (a) 1—Sliding of packer; 2—reanchoring of packer/test opened/and sliding of packer; 3—partial plugging of "fringes" (in French:1— Glissement du packer; 2—Reancrage du packer/Test ouvert/et glissement depacker 3—Bouchage partial des crepines). **(b)** Leaking of packer (Fuite au packer)

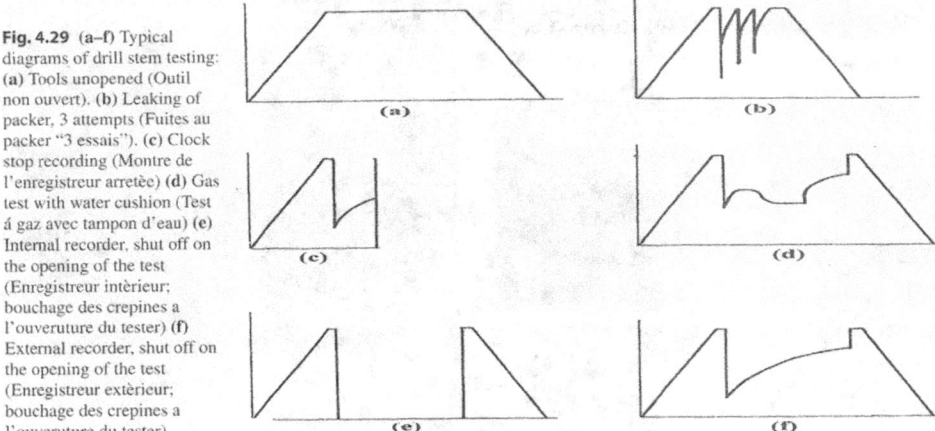

Fig. 4.29 (a–f) Typical diagrams of drill stem testing/translation to English: **(a)** Tools unopened (Outil non ouvert). **(b)** Leaking of packer, 3 attempts (Fuites au packer "3 essais"). **(c)** Clock stop recording (Montre de l'enregistreur arretèe) **(d)** Gas test with water cushion (Test á gaz avec tampon d'eau) **(e)** Internal recorder, shut off on the opening of the test (Enregistreur intèrieur; bouchage des crepines a l'ouveruture du tester) **(f)** External recorder, shut off on the opening of the test (Enregistreur extèrieur; bouchage des crepines a l'ouveruture du tester)

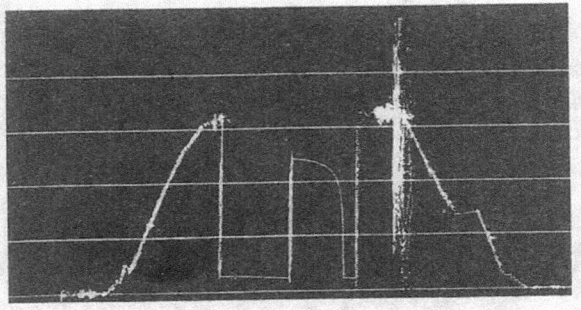

Fig. 4.30 Leaking of the drilling pipes. (Trou ètroit indiquè et fuite au tiges de forage)

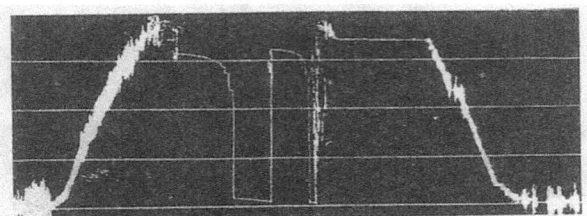

Fig. 4.31 A recorder of stair stepping (Enregistreur: stair stepping)

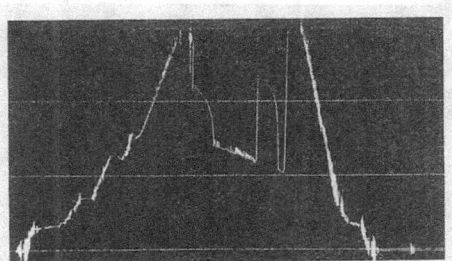

Fig. 4.32 Incorrect base line and plugging during flow (Ligne de base incorrecte et bouchage pendant dèbit)

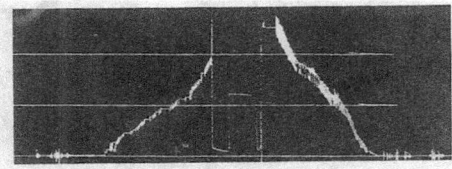

Fig. 4.33 Clock stopped (Arrêt de la montre)

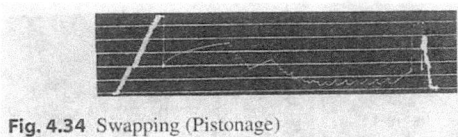

Fig. 4.34 Swapping (Pistonage)

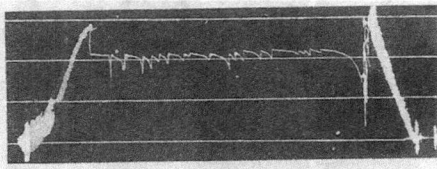

Fig. 4.35 Plugged formation through flow testing (Débit bouché)

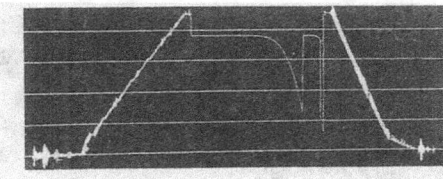

Fig. 4.36 Equalized flow (FSIP = FFP) (Débit égalisé)

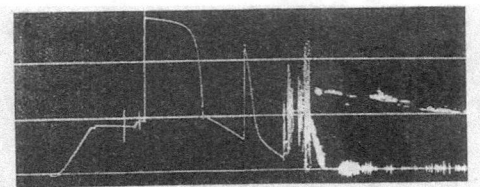

Fig. 4.37 The after flow pressure conceals ISIP (Courbe en "S" dévellope par compression de fluide due a l'ancrage du packer assez plus haut)

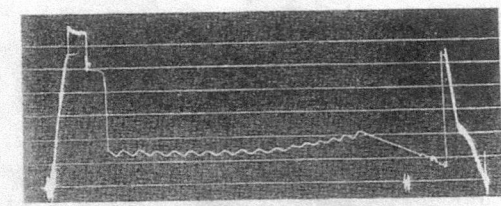

Fig. 4.38 Flowing in leads (Dèbit en coup)

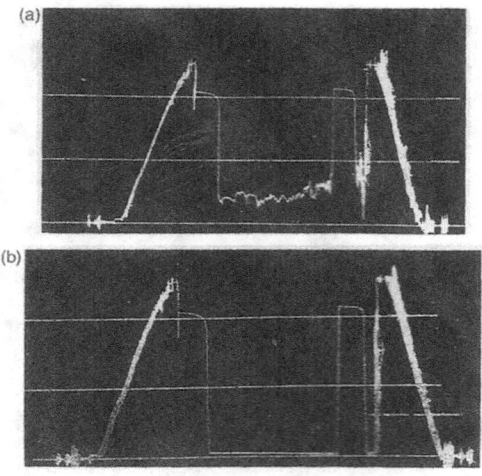

Fig. 4.39 Bottom pressure measurement show plugging of perforations (Diagramme enregistreur du bottom). **(b)** Top pressure measurement show plugging of perforations (Diagramme enregistreur du top)

References

Assaad F. et al (2003) Field Methods for Geologists and Hydrogeologists, Drilling and Testing–Soil samplers, Ch (6)/p122 Springer, Heidelberg, Germany.

Edwards AG and Winn RH (1974) A summary of modern tools and techniques used in drill stem testing; presented at the dedication of the US East-West trade centerTulsa, Oklahoma, EPA

Kilpatrick CV (1955) An integrated summary of formation evaluation criteria, AIME T.P. 595-G, presented at formation evaluation symposium, Houston Univ.

Brantly JE (1952) Rotary drilling handbook, 5th ed. New York, Palmer Publications, pp10 –235 *in* Petroleum Engineering, Gatlin (1960); Fig 4.1/p41

Eastman Oil Well Survey Co., *in* Petroleum Engineering, Gatlin (1960), Fig 9.21/p157 API, R.P. 9B, 2nd (1954) American Petroleum Institute, New York, *in* Petroleum Engineering, Gatlin (1960); Fig 5.1/p53

Main WC (1949) Detection of incipient drill-pipe failures. API drilling and production recommended practices, No. 29

Black WM (1956) A review of drill-stem testing techniques and analysis. Jr. of Petroleum Technology, p21, *in* Petroleum Engineering, p256/Fig 13.3, Gatlin, (1960)

Kilpatrick CV (1954) Formation testing—the petroleum engineer, pB-139; Figs 13.5 –13.9/p258–259 & Fig 13.11/p260: *in* Petroleum Engineering, Gatlin (1960)

Edwards SH and Miller CP (1939) Discussion on the effect of combined longitudinal loading and external pressure on the strength of oil well casing. API Drilling and Production Practices, p483

Halbouty MT (1967) Salt domes—Gulf Region, United States and Mexico. Figs. 9.1/p145; Gulf Publishing Co., P.O. Box 2608, Houston, Texas 77001 Stroud BK (1925) Use of barytes as a mud-laden fluid. *In* Oil World, p29

Appendix 4.A-1- Steel Tricone Bits, Foremost Mobile Catalogue- 1988-Mobile Drilling CO.

Chapter 5
Geophysical Well Logging Methods of Oil and Gas Reservoirs & Well Log Interpretation

5.1 Introduction

Several texts, manuals, field books, and guidebooks discussed different geophysical log methods to define productive and nonproductive reservoirs in the petroleum business.

The properties of penetrated formations and their fluid contents are recorded by geophysical logs to measure their electrical resistivity and conductivity, their ability to transmit and reflect sonic energy, their natural radioactivity, their hydrogen ion content, their temperature and density, etc. The logs are then interpreted in terms of lithology, porosity, and fluid content **(Table 5.1, USEPA 1977)**.

The basic concepts of well log interpretations and the factors affecting logging measurements are presented by Asquith and Gibson (1982) in several examples of oil and gas reservoirs, together with log interpretations given by Schlumberger (1972) and Dresser Industries (1975).

5.2 Borehole Parameters and Rock Properties

The parameters of log interpretations are determined by electrical, nuclear, or sonic (acoustic) logs. Rock properties that affect logging measurements are porosity, permeability, water saturation, and resistivity, that were used in terms and symbols in well log interpretations (Schlumberger 1974).

Table 5.1 Geophysical well logging methods and practical applications (USEPA 1977)

	Method	Property	Application
1	Spontaneous Potential(SP) log	Electrochemical and electrokinetic potentials	Formation water resistivity; (Rw) shaliness (sand or shale).
	Non-focused electric log	Resistivity	Water and gas/oil saturation; porosity of water zones; Rw in zones of known porosity; formation resistivity (Rt); resistivity of invaded zone (Ri).
3	Focused and non-focused micro- resistivity logs	Resistivity	Resistivity of the flushed zone (Rxo); porosity; bed thickness.
4	Sonic log	Travel time of sound	Rock permeability.
5	Caliper log	Diameter of borehole	Without casing.
6	Gamma Ray	Natural radioactivity	Lithology(shales & sand)
7	Gamma-Gamma	Bulk density	Porosity, lithology.
8	Neutron-Gamma	Hydrogen content	Porosity with the aid of hydrogen content

Symbols and Interpretations:

1. **Borehole diameter (d_h)** is the outside diameter of the drill bit and describes the borehole size of the well.
2. **Drilling Mud (R_m)** is a circulating fluid having a special viscosity and density to help remove cuttings from the well bore, lubricate and cool the drill bit, and keep the hydrostatic pressure in the mud column greater than the formation pressure.
3. **Mud filtrate (R_{mf})** is the fluid of the drilling mud that filters into the formation during invasion, whereas its solid particles of clay minerals are trapped on the side of the borehole and form the mud cake (R_{mc}).
4. **Water saturation** is an important log interpretation concept as it determines the hydrocarbon saturation of a reservoir by substracting the water saturation from the value of 1 (where 1.0 = 100% water saturation):

$$\text{Water Saturation (Sw)} = \frac{\text{formation of water occupying pores}}{\text{Total pore space in the rock}}$$

5. **Invaded Zone**—consists of a flushed zone (R_{xo}) "close to the borehole, a few inches from the well bore" where the mud filtrate has completely flushed out the formation's hydrocarbons; **a transition zone or annulus zone** (R_j), where a formation's fluids and mud filtrate are mixed, occurs between the **flushed zone** (R_{xo}) and the **un-invaded zone** (R_t). The depth of invasion of the mud filtrate into the invaded zone is referred to as the diameter of invasion "di and dj" and depends on the permeability of the mud cake and not upon the porosity of the rock **(Fig. 5.1**; Schlumberger 1977, AAPG 1982).

NB: An equal volume of mud filtrate can invade low porosity and high porosity rocks, if the drilling muds have equal amounts of solid particles which coalesce and form an impermeable mudcake, acting as a barrier to further invasion.

When oil is present in the flushed zone, the degree of flushing by mud filtrate can be determined from the difference between water saturations in the flushed (S_{xo}) zone and the uninvaded (S_w) zone. Usually, about 90% of the oil is flushed out (S_{xo} = 90%); the remaining oil is called residual oil saturation (ROS):

☐ Resistivity of the zone

○ Resistivity of the water in the zone

△ Water saturation in the zone

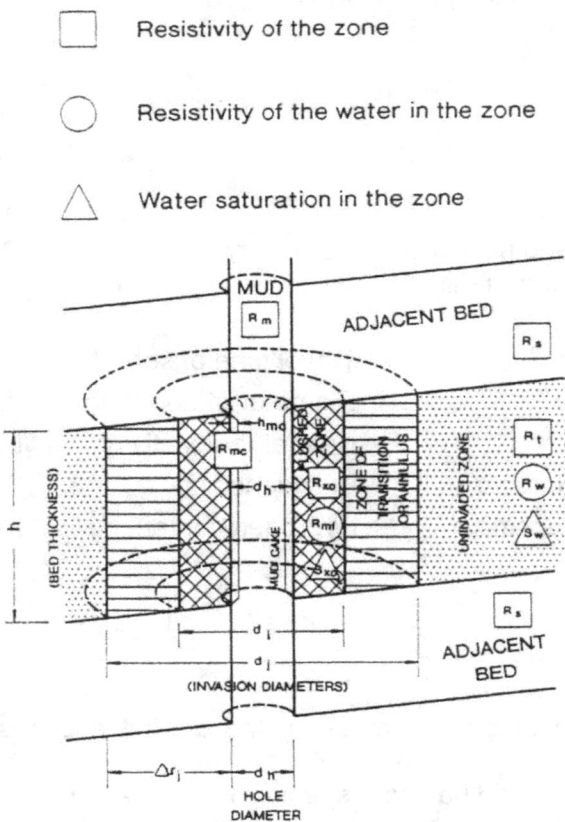

Fig 5.1 - A DIAGRAM SHOWS THE INVASION OF FLUIDS THROUGH THE SURROUNDING ROCK; THE CYLINDRICAL NATURE OF THE INVASION IS ALSO SHOWN BY DASHED LINES.
After Asquith, G. and Gibson, C., 1982. "Courtesy Schluberger Well Services, Copyright 1971."
(AAPG, 1982)

Fig. 5.1 A Diagram Shows the Invasion of Fluids through the Surrounding rock: The Cylindrical nature of the Invasion is also Shown by Dashed lines (After Asquith, G. and Gibson, C., 1982). "Courtesy Schlumberger Wall Services, Copyright 1971." (AAPG, 1982).

Degree of flushing by mud filtrate (ROS);
$S_{10} = \{S_w - S_{xo}\}$; or $S_{10} = (1.0 - S_{xo})$.

6. **Uninvaded zone (R_t)**—is located beyond the invaded zone. In hydrocarbon-bearing reservoirs, there is always a layer of formation water on grain surfaces. Water saturation of the uninvaded zone (S_w) is an important factor in hydrocarbon reservoir evaluation:

$S_h = \{1.0 - S_w\}$.

Where:
S_h = hydrocarbon saturation, or the fraction of pore volume filled with hydrocarbons.
S_w = water saturation in the uninvaded zone, or the fraction of pore volume filled with formation water.

The ratio between the water saturation of the uninvaded zone (S_w) and that of the flushed zone (S_{xo}) is an index of hydrocarbon movability.

5.3 Resistivity Measurements by Well Electric Logs

5.3.1 Definition

Resistivity, measured by electric logs, is a basic measurement of a reservoir's fluid saturation and is a function of porosity, type of fluid, (e.g., hydrocarbons, salt, or fresh water), and rock type; resistivity measurements are used to detect hydrocarbons and estimate the porosity of the reservoir. Both rock and hydrocarbons act as insulators, but salt water is conductive. Resistivity is measured in ohm-meters:

$$\text{Resistivity } R = \frac{r.A}{L} \quad (1)$$

where:

R = resistivity (ohm-meter$_2$/meter- or ohm- meters)
r = resistance (ohm)
A = cross sectional area of measured substance (m²)
L = length of substance being measured (ms).

Resistivity can be used by logging tools to detect hydrocarbons and estimate the porosity of the reservoir. Cornard Schlumberger in 1912 began the first experiments which led eventually to the development of the modern day petrophysical logs; the first electric log was run by a French engineer in 1927, and in 1942 Archie with Shell Oil Co. presented a paper to the AIME in Dallas, Texas, which set forth the concepts used as the basis for modern quantitative log interpretation.

Archie submitted an experiment that the resistivity of a water-filled formation (R_0), filled with water of a resistivity R_w, can be related by means of a formation resistivity factor:

$$\text{Archie's formula:} \quad R0 = F_x R_w \quad (2)$$
$$\text{or, } F = R_0/R_w$$

where:

R_o = resistivity of water-filled formation (100% water saturated);
F = Formation resistivity factor;
R_w = Formation water resistivity.

Also, Archie's formula stated that formation factors can be related to porosity by:

$$F = 1.0/\phi^m \text{ (m = cementation exponent).} \quad (3)$$

The cementation exponent varies with grain size and its distribution, and with the complexity of the paths between pores, known as turtuosity, for which the value is proportional with the m value.

5.3.2 Annulus and Resistivity Profiles—Hydrocarbon Zone

The annulus profile, which reflects a temporary fluid distribution, occurs between the invaded zone and the uninvaded zone, denotes the presence of hydrocarbons, and can be detected only by an induction log run soon after a well is drilled; however, it should disappear with time. In the annulus zone, beyond the outer boundary of the invaded zone, the pores are filled with residual hydrocarbons (R_h) and formation water (R_w), and results in an annulus profile that causes an abrupt drop in measured resistivity. The annulus profile is important for petroleum geologists because it only appears in hydrocarbon bearing zones. In such cases, the mud filtrate invades hydrocarbons, which move out first; then formation water is pushed out in front of the mud filtrate, forming an annular (circular) ring at the edge of the invaded zone. Log resistivity profiles illustrate the values of the invaded and uninvaded zones.

Figure 5.2 shows a horizontal section through a permeable hydrocarbon-bearing formation and the related resistivity profiles which occur when there is invasion by either freshwater- or saltwater-base muds (AAPG 1982):

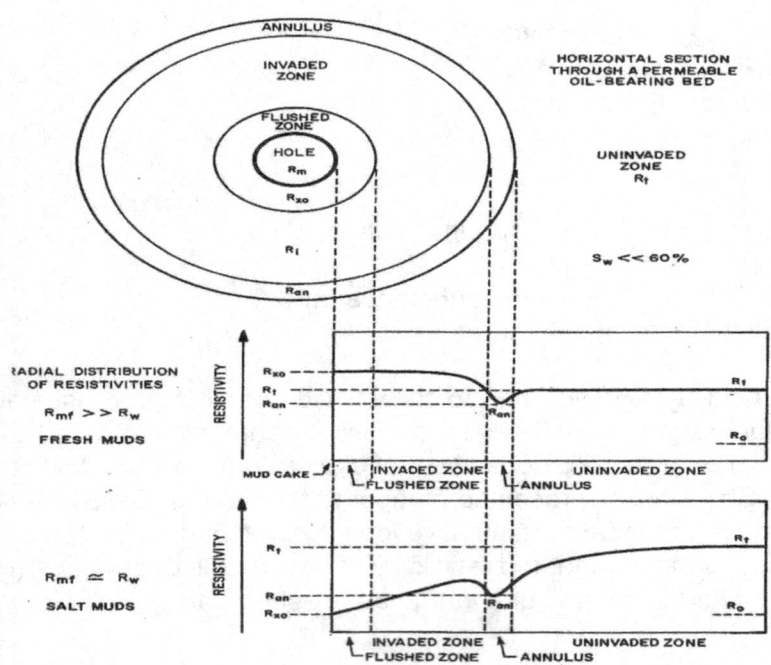

Fig 5.2 - Resistivity Profile- Hydrocarbon Zone
(AAPG, 1982)

Fig. 5.2 Resistivity Profile-Hydrocarbon Zone (AAPG, 1982)

(1) In **freshwater muds**—the resistivity of both the mud filtrate (R_{mf}) and the residual hydrocarbons (R_h) is much greater than that of the formation water (R_w); therefore, the resistivity profile is built up where the shallow flushed zone (R_{xo}) is comparatively high (the flushed zone has both mud filtrate (R_{xo}) and some residual hydrocarbons (R_h).

The **medium invaded zone** (R_i) has a mixture of mud filtrate (R_{mf}), formation water (R_w), and some residual hydrocarbons (R_h). Such a mixture causes high resistivity readings; a hydrocarbon zone invaded with freshwater mud results in a resistivity profile where the shallow (R_{xo}), medium (R_i), and deep (R_t) resistivity tools, all record high resistivities. In some instances, the deep resistivity is higher than the medium resistivity due to the annulus effect. The **uninvaded zone** causes higher resistivity than if the zone had only formation water (R_w), because hydrocarbons are more resistant than formation water (R_w).

So: "$R_t > R_o$"; "R_t"—is normally somewhat less than that of the flushed (R_{xo}) and invaded "R_i" zones. Generally, $R_{xo} > R_i >$ which is more or less than R_t.

(2) In **saltwater muds**- Because the resistivity of the mud filtrate (R_{mf}) is approximately equal to the resistivity of formation water ($R_{mf} \sim R_w$), and the amount of residual hydrocarbons (R_h) is low, then the resistivity of the flushed zone (R_{xo}) is low. As more hydrocarbons mix with mud filtrate in the invaded zone, the resistivity of the invaded zone (R_i) begins to increase.

Generally: $R_t > R_i > R_{xo}$.

5.4 Formation Temperature (T_f)

Formation temperature is important in log analysis because the resistivity of the drilling mud (R_m), the mud filtrate (R_{mf}), and the formation water (R_w) vary with temperature. The formation temperature can be calculated by defining the slope or temperature gradient from the following equation (Schlumberger 1974, Wylie and Rose 1950):

$$m = \frac{y - c}{x} \quad (4)$$

where:

m = temperature gradient
y = bottom hole temperature (BHT)
c = surface temperature
x = total depth (TD)

5.5 Specific Log Types

Specific log types such as SP, resistivity, porosity, and gamma ray logs are discussed in detail by Asquith and Gibson (1982).

5.5.1 Spontaneous Potential Logs (SP)

A spontaneous potential log is a record of direct current (DC) voltage differences between the naturally occurring potential of a moveable electrode running down the borehole and that of a fixed electrode located at the surface, measured in millivolts. The SP log, influenced by parameters affecting the borehole environment, is used only with conductive (saltwater-based) drilling muds, to detect permeable beds and their boundaries, and to determine formation water resistivity (R_w) and the volume of shale in permeable beds. The magnitude of SP deflection is due to the difference in resistivity between mud filtrate (R_{mf}) and formation water (R_w), and not to the amount of permeability.

The response of shales is relatively constant and follows a straight line called a shale baseline. SP curve deflections are measured from the shale baseline.

The detection of hydrocarbons by the suppression of the SP response is another use of the SP curve. The presence of shale in a permeable formation reduces the SP deflection. In hydrocarbon-bearing zones, the amount of SP reduction is greater than the volume of shale and is known as "hydrocarbon suppression" (Hilchie 1978). The SP curve can be suppressed by thin beds, shaliness, and the presence of gas. The spontaneous potential curve **(Fig. 5.3)** shows SP deflection from the shale baseline, with resistivity of the mud filtrate (R_{mf}) much greater than that of formation water (R_w); at the top of the diagram, the static spontaneous potential (SSP), is the maximum deflection possible in a thick, shale-free, and water-bearing (wet) sandstone for a given ratio of R_{mf}/R_w.

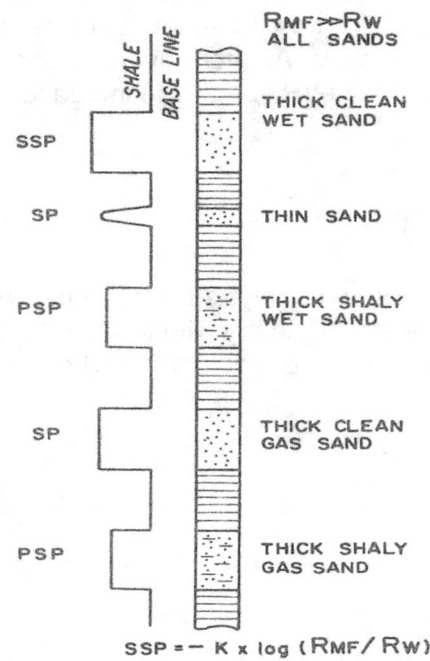

Fig 5.3- SP deflection with resistivity of the mud filtrate much greater than formation water (Rmf >> Rw) (AAPG, 1982)

Fig. 5.3 SP deflection with resistivity of the mud filtrate much greater than formation water (Rmf >> Rw) (AAPG, 1982)

SSP is a necessary element for determining accurate values of R_w and volume of shale. In the diagram, SP shows the response of the spontaneous potential curve to the presence of thin beds and/or the presence of gas. Pseudo-static-self spontaneous potential (PSP) is the spontaneous potential curve response if shale is present.

In salt dome drilling, well logging is carried out in salt mud (conductive) or oil mud (nonconductive), and can be accomplished with satisfactory results. Salt muds affect the SP and the resistivity curves on electric logs. With a highly conductive fluid in the borehole, the SP tends to flatten out or may even reverse. In nonconductive or oil base muds, the SP and conventional resistivity (excluding the induction-conductivity curves), are totally ineffective.

5.5.2 Resistivity Logs (R)

A resistivity log is an electric log that can be used to determine resistivity porosity and hydrocarbon versus water-bearing zones, and can define permeable zones. Because the reservoir rock's matrix is non-conducive, the ability of rock to transmit a current is mainly a function of water in the pores. Hydrocarbons, the same as the rock's matrix, are nonconductive; so when the hydrocarbon saturation of the pores increases, the rock's resistivity also increases.

The resistivity log can be used to determine a formation's water saturation (S_w) by applying the Archie equation (Archie 1942):

$$S_w = \frac{(F \cdot R_w)^{1/n}}{R_t} \quad (5)$$

where:

S_w = water saturation,
F = formation factor ($a/\varnothing^m$),
a = turtuosity factor,
m = cementation factor,
R_w = resistivity of formation water,
R_t = true formation resistivity as measured by a deep reading resistivity log,
n = saturation exponent (~2.0).

There are two basic types of resistivity logs that measure formation resistivity: namely, induction and electrode logs.

5.5.2.1 Induction Electric Devices

There are two types of induction devices: the induction electric log and the dual induction focused log.

The **Induction Electric Log (IEL)** is an electric log that measures conductivity; "or actually true resistivity from conductivity"; it is composed of three curves: short normal, induction, and spontaneous potential (SP). The short normal tool has an electrode spacing of 16 inches and can record a reliable value for resistivity from a bed thickness of four feet. The short normal curve is usually recorded in track #2; the induction device measures electrical conductivity using current generated by coils that produce an electromagnetic signal which induces current in the formation.

The "IEL" is normally used with fresh-water drilling mud ($R_{mf} > 3R_w$); but because it does not require the transmission of electricity through drilling fluid, it can therefore run in air-, oil-, or foam-filled boreholes.

The **Dual Induction Focused Log (DIFL)** is used in freshwater drilling muds and consists of a deep reading induction device (R_{ILD} that measures R_t) similar to the IEL, a medium-reading induction device (R_{ILM} measures R_i), and a shallow reading (R_{xo}) focused laterolog (similar to short normal); the shallow reading laterolog may be either a Laterolog-8 (LL-8) or a spherically focused log (SFL). The dual induction focused log is used in formations deeply invaded by mud filtrate; when freshwater drilling muds invade through a hydrocarbon-bearing formation (Sw << 60%), there is high resistivity in the flushed zone (R_{xo}), the invaded zone (R_i), and the uninvaded zone (R_t).

Figure 5.4 is an example of "DIFL" in a hydrocarbon-bearing formation zone, shown on the right side of the log (Tracks #2 and #3); it is a logarithmic scale from 0.2–2000 ohm-meters from left to right. It comprises three resistivity values:

1. **deep induction log resistivity curves (R_{ILD})**, measure the true resistivity (R_t) of the uninvaded zone (at depths of 8748–8774 ft), where the curves read a high resistivity (~50), because hydrocarbons are more resistant than the formation saltwater ($R_t > R_o$).
2. **Medium induction log resistivity curves (R_{ILM})**, measure the resistivity of the invaded zone (R_i = ~60); the curve records a high resistivity reading due to a mixture of mud filtrate (R_{mf}), formation water (R_w), and residual hydrocarbon (R_h) in the pores. Such resistivity is equal to or slightly more than the ILD curve; whereas, in the annulus zone, the ILM may record slightly less than the ILD curve.

3. The **spherically focused log resistivity curves (R_{SFL})**, measure the resistivity of the flushed zone (R_{xo} = ~125); the curve reads a higher resistivity than the deep (ILD) or medium (ILM) induction curves, because the flushed zone (R_{xo}) contains both the mud filtrate and residual hydrocarbons (R_h). The following ratios are also needed for the tornado chart calculations: R_{SFL}/R_{ILD} = 3 and R_{ILM}/R_{ILD} = 1.2.

5.5.2.2 Electrode Resistivity Logs

Electrode logs, which are the second type of resistivity measuring devices, are used with salt-saturated drilling muds ($R_{mf} \sim R_w$); examples of electrode resistivity tools are: normal, lateral, Laterolog, Microlog, Microlaterolog, Proximity Log, and Spherically Focused Log. In salt water drilling muds, when it invades a hydrocarbon-bearing zone (S_w << 60%), there is low resistivity in the flushed zone (R_{xo}), an intermediate resistivity in the invaded zone (Ri), and high resistivity in the uninvaded zone (R_t); the reason for the increase in resistivity deeper into the formation, is the increase of the hydrocarbon saturation (S_h).

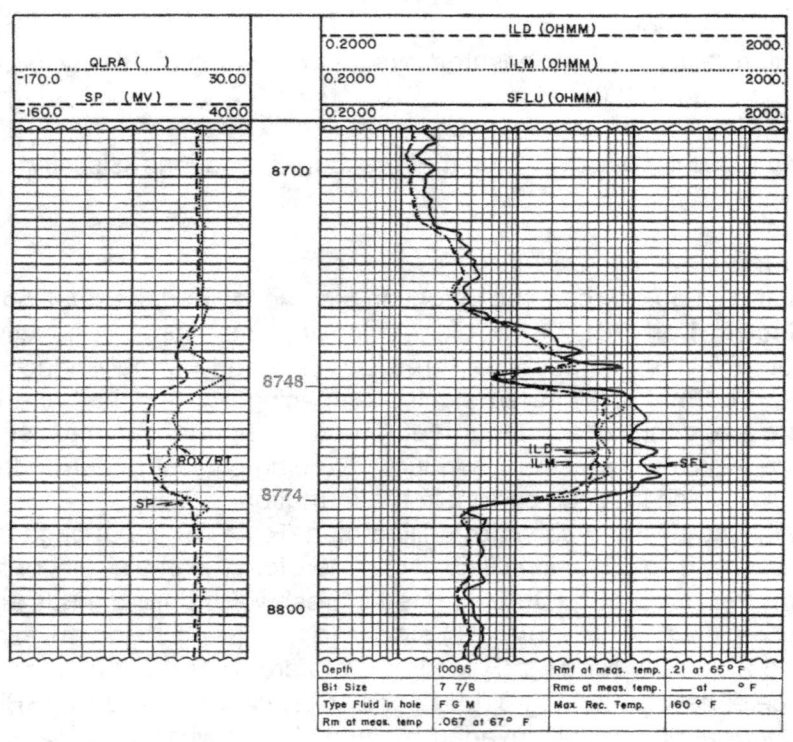

Fig 5.4 - Dual Induction Focused Log Curves through a hydrocarbon-bearing zone
(using freshwater-based drilling mud)
(Schlumberger, AAPG, 1982)

Fig. 5.4 Dual induction focused log curves through a hydrocarbon-bearing zone (using freshwater-based drilling mud) (Schlumberger, AAPG, 1982).

The **Laterolog** is designed to measure true formation resistivity (R_t) in boreholes filled with saltwater muds (where $R_{mf} \sim R_w$).

The **Dual Laterolog (LLD)/ Microspherically Focused Log (MSFL)**—consists of three curves: a deep reading (or R_t)- resistivity device (R_{LLd}), and a shallow reading (or R_i) resistivity device (R_{LLs}); both are displayed in tracks #2 & #3 of the log on a 4-cycle logarithmic scale from 0.2 to 2000, values increasing to the right. The Microspherically Focused Log (R_{SFL}) is a pad type, focused electrode log that has a very shallow depth of investigation, and measures resistivity of the flushed zone (R_{xo}). If the MSFL is run together with the Dual Laterolog (R_{LLD} & R_{LLS}; **Fig 5.5)**, the three curves (i.e. deep, shallow, and MSFL) are used to correct (for invasion), the deep resistivity (R_{LLD}) to true formation resistivity (R_t). A tornado chart is necessary to determine the diameter of invasion (d_i), and the ratio of R_t/R_{xo}. (Asquith and Gibson, AAPG 1982):

(a) **Deep Laterolog resistivity (R_{LLD})** dashed line measures deep resistivity of the formation or the true resistivity (R_t); at the sample depth of 9,324ft, the true resistivity reads a value of 16.0;

(b) **Shallow Laterolog resistivity,** the dashed and dotted line LLD, represents **R_{LLS}** and measures the shallow resistivity of the formation or the resistivity in the invaded zone (R_i). The shallow Laterolog (LLS) records a lower resistivity than the deep Laterolog (LLD) because the invaded zone has a lower hydrocarbon saturation (S_h) than the uninvaded zone (R_t); at the sample depth of 9,324ft, the resistivity (R_i) reads a value of 10.0.

(c) **Microspherically focused log (MSFL) resistivity curves** measure the resistivity of the flushed zone (R_{xo}):solid line); the curve records low resistivity because of both the saltwater mud filtrate (R_{mf}) and the residual hydrocarbon saturation (R_h) in the flushed zone (R_{xo}). The uninvaded zone (R_t) has a high resistivity, the invaded zone (R_i) has a lower resistivity, and the flushed zone (R_{xo}) has the lowest resistivity; at the sample depth of 9324 ft., it reads a value of ~3.5).

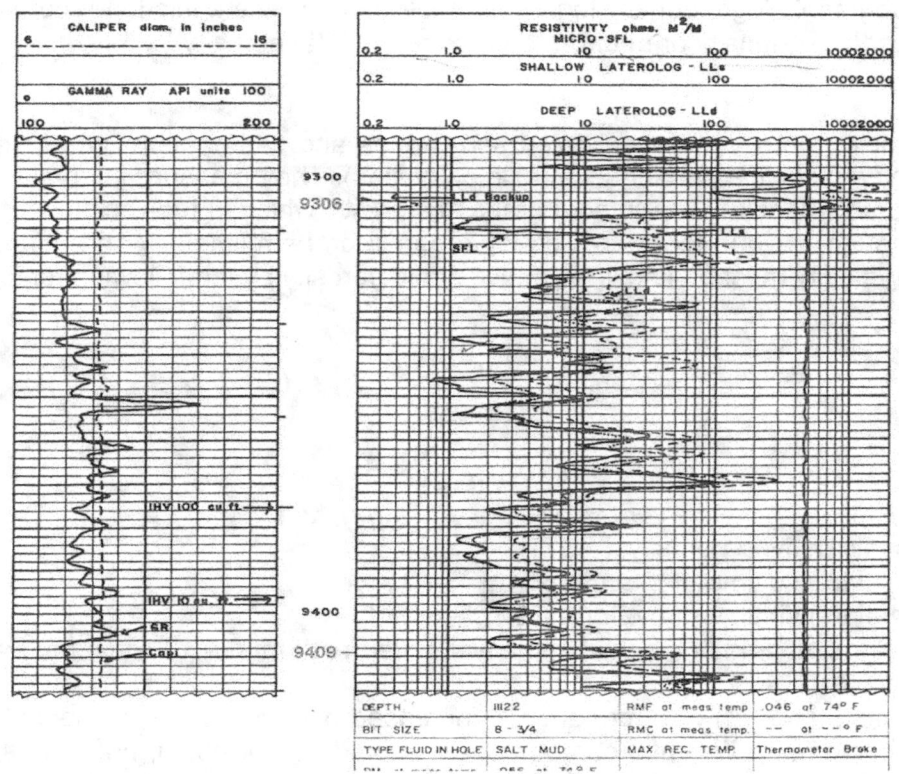

Fig 5.5 - **Dual Laterolog-Microspherically Focused Log (MSFL) curves through hydrocarbon-bearing zone**
(Saltwater-based drilling mud)

(Schlumberger, AAPG, 1982)

Fig. 5.5 - Dual Laterolog (DLL)/Microspherically-Focused Log (MSFL) curves through hydrocarbon-bearing zone, using Saltwater-based drilling mud/ **($R_{mf} \sim R_w$)**.
(By Schlumberger, AAPG, 1982)

5.5.2.3 Micrologs (ML)

The Microlog is a pad type resistivity device that primarily detects the mudcake by two resistivity measurements: the micro normal device which investigates 3–4inches (or 7–9 cm) into the formation (R_{xo}), and the micro inverse which investigates approximately 1–2 inches (or 2.35–4.5 cm) and measures the resistivity of the mud cake (R_{mc}). The detection of the mudcake by the ML indicates that invasion has occurred and the formation is permeable. Permeable zones show up as positive separation when the micro normal curves read higher resistivity than the micro inverse curves. Shale zones show no separation or negative separation (micro normal< micro inverse).

The ML does not work well in saltwater-based drilling muds ($R_{mf} \sim R_w$), or gypsum-based muds because the mudcake may not be strong enough to keep the pad away from the formation. Where the pad is in contact with the formation, positive separation cannot occur. **The Microlaterolog (MLL) and Proximity log (PL)**, like the Microspherically Focused log (MSFL), are pad type focused electrode logs, designed to measure the resistivity in the flushed zone (R_{xo}). The MLL should run only with saltwater-based drilling mud because it is strongly influenced by mudcake thicknesses greater than ¼ inch (~2/3 cm); (Hilchie 1978). The Proximity Log which is more strongly focused than the Microlaterolog, is designed to investigate deeper so it can be used with freshwater-based drilling muds where mudcake is thicker.

5.5.3 Porosity Logs

Porosity logs include sonic logs, density logs, and neutron logs. The sonic log records matrix porosity, whereas the nuclear logs (density or neutron) determine the total porosity.

5.5.3.1 Sonic Logs

Sonic Logs determine porosity in consolidated sandstones and carbonates with intergranular porosity (grainstones) or intercrystalline porosity (sucrosic dolomites). They measure interval transit time (Δt) in microseconds per foot (μsec/ft). The interval transit time (Δt), which is the reciprocal of the velocity of a compressional sound wave in feet per second, depends on both lithology and porosity. Therefore, a formation's matrix velocity must be known to derive sonic porosity by the following formula **(Table 5.2)**:

$$\varnothing\text{sonic} = \frac{\Delta t_{log} - \Delta t_{ma}}{\Delta t_f - \Delta t_{ma}} \quad (6)$$

where:

$\varnothing$sonic = sonic derived porosity,
Δt_{ma} = interval transit time of the matrix,
Δt_{log} = interval transit time of the formation,
Δt_f = interval transit time of the fluid in the well bore (fresh mud = 189; salt mud = 185).

The interval transit time (Δt) of a formation is increased because of the presence of hydrocarbons (i.e., the hydrocarbon effect). The hydrocarbon effect should be corrected, because the sonic derived porosity will be too high otherwise. The following empirical corrections for the hydrocarbon effect were proposed by Hilchie (1978):

$$\varnothing = \varnothing\text{sonic} \times 0.7 \text{(gas)}$$
$$\varnothing = \varnothing\text{sonic} \times 0.9 \text{(oil)}$$

Table 5.2 Sonic Velocities and Interval Transit Times for different Matricies used in the Sonic Porosity Formula (after Schlumberger 1972)

	V_{ma}(ft/sec.)	Δt_{ma}(µsec./ft)
Sandstone	18,000–19,500	55.5–51.0
Limestone	21,000–23,000	47.6
Dolomite	23,000–26,000	43.5
Anhydrite	20,000	50.0
Salt	15,000	67.0
Casing (Iron)	17,500	57.0

5.5.3.2 Density Logs

The formation density log is a porosity log that measures the electron density of a formation, and it helps identify evaporite minerals, detect gas-bearing zones, determine hydrocarbon density, and evaluate shaly sand reservoirs and complex lithologies (Schlumberger 1972). The density logging device is a contact tool that consists of a medium-energy gamma ray source that emits gamma rays into a formation. Gamma rays collide with electrons in the formation; the collisions result in a loss of energy from the gamma ray particle. Electron density can be related to bulk density (ρb) of a formation in gm/cc.

Formation bulk density "ρ_b" is a function of matrix density, porosity, and density of the fluid in the pores (salt mud, fresh mud, or hydrocarbons). To determine density porosity by calculation, the matrix density and type of fluid in the borehole must be known. The formula for calculating density porosity **(Table 5.3)** is:

$$\varnothing_{den} = \frac{\rho_{ma} - \rho_b}{\rho_{ma} - \rho_f}$$

Where:

$\varnothing_{den}$ = density derived porosity
ρ_{ma} = Matrix density
ρ_b = formation bulk density
ρ_f = fluid density (1.1 salt mud, 1.1 fresh mud, and 0.7 gas)

5.5.3.3 Neutron Logs

Neutron logs are porosity logs that measure the hydrogen ion concentration in a formation. In clean formations (i.e., shale-free) that are saturated with oil or water, the neutron log measures liquid-filled porosity. Neutrons are created from a chemical source in the Neutron logging tool. The chemical source may be a mixture of americium and beryllium which will continuously emit neutrons. These neutrons collide with the nuclei of the formation material, resulting in the loss of some of its energy. Because the hydrogen atom is almost equal in mass to the neutron, maximum energy loss occurs when the neutron collides with a hydrogen atom.. Therefore, the maximum amount of energy loss is a function of a formation's hydrogen concentration. Because hydrogen in a porous formation is concentrated in the fluid-filled pores, energy loss can be related to the formation's porosity. Neutron porosity is lower in gas-filled pores than in oil-or water filled-pores because there is less concentration of hydrogen in gas compared to oil or water (gas effect).

Table 5.3 Matrix densities of common lithologies used in the Density Porosity Formula ` (after Schlumberger 1972)

	ρ_{ma} (gm/cc)
Sandstone	2.648
Limestone	2.710
Dolomite	2.876
Anhydrite	2.977
Salt	2.032

Neutron log responses vary, depending on: a) differences in detector types; b) spacing between source and detector, and c) lithology (sandstone, limestone, and dolomite). These variations in response, can be compensated for by using the appropriate charts; noticing that the neutron logs must be interpreted from the specific chart designed for a specific log (i.e. Schlumberger charts for Schlumberger logs and Dresser Atlas charts for Dresser Atlas logs), because the neutron logs are not calibrated in basic physical units as other logs.

The first modern neutron log was the Sidewall Neutron Log, which has both the source and detector in a pad that is pushed against the side of the borehole. The most modern of the Neutron logs is a compensated Neutron logs, which has a neutron source and two detectors, being less affected by borehole irregularities. Both the Sidewall and Compensated Neutron logs can be recorded in apparent limestone, sandstone or dolomite porosity units.

5.5.3.4 Combination Neutron-Density Logs

The combination Neutron-Density log is a combination porosity log, and consists of Neutron and Density curves recorded in tracks #2 and #3 and a Caliper and Gamma Ray log in track #1 (the Gamma Ray scale reads from 0 to 100 API Gamma Ray units and the caliper measures a borehole size from 6 to 16 inches); The combination Neutron-Density log is used as a porosity device, determines lithology, and detects gas-bearing zones. (Fig5.6).

Both the neutron and density curves are normally recorded in limestone porosity units with each division equal to either 2- percent or 3-percent porosity; however, sandstone and dolomite porosity units can be also recorded (Schlumberger, 1972)

The density porosity (ϕ_D) is represented by a solid line and the Neutron Porosity (ϕ_N) is represented by a dashed line.

Fig 5.6 - An example of a combination Neutron-Density Log with gamma ray log and caliper; the log illustrates the log curves and scales of a combination log., but it also used for picking values from charts 5.6a & 5.6b.

The true porosity can be determined first by reading the apparent limestone porosities from the neutron and density curves.The scale for both is the same, ranging from -10% to +30% in increments of 2%, and is measured in limestone porosity units.

The values of the apparent limestone porosities are crossplotted on a neutron-density porosity charts (Chart 5-6a or 5-6b) to find true porosity ; the first chart (chart 5-6a) is used to correct neutron--density log porosity for lithology where there is freshwater-based drilling mud (Rmf >3Rw) , and the second chart (chart 5-6b) is used where there is salt-water based drilling mud (Rmf~Rw),:

5.5.3.4.1 Examples and Interpretation

Example (1): True porosity, obtained from the apparent limestone porosities of the neutron and density curves.. (**Fig 5.6**), shows at depth 9,310 ft., the apparent neutron porosity: ϕ_N = 24% and density porosity

ϕ_D = 9%; these values are **crossplotted** on a Neutron-Density porosity chart **(charts 5.6a or 5.6b)** to find true porosity. In the example from Fig 5.6 the true porosity using the ccrossplotted neutron-density porosities at depth 9,310ft. can be corrected according to the type of the drilling mud, either with freshwater-based drilling mud (Rmf >3Rw),/ (using **Chart 5.6a**), where the intersection shows a true porosity value 16.5%. or using **(chart 5.6b; for** saltwater based drilling mud (Rmf~Rw), where .the values meet at a point just off the lithology curve for dolomite, and the intersection shows a true porosity value of 17%.

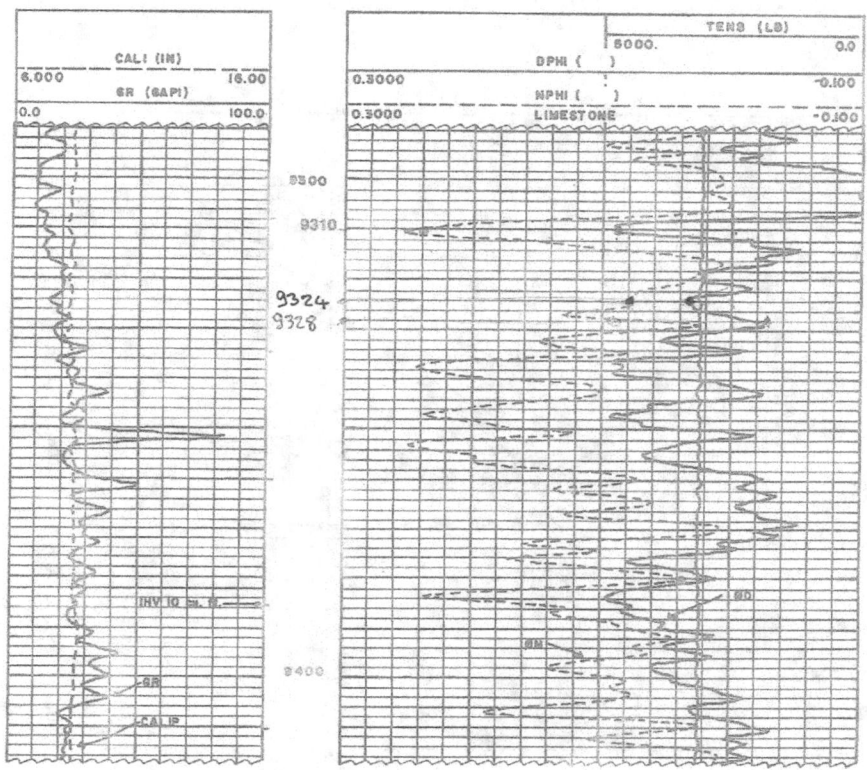

Fig 5-6 -A Combination Neutron-Density Log with Gamma ray log and calipar

Due to borehole irregularities, a given bulk density: $\rho_f = 1.0$ gm/cc (suggested fluid density of freshwater muds); and a given bulk density $\rho_s = 1.1$ gm/cc (suggested fluid density of saltwater muds). Porosity from a neutron-density Log can also be calculated mathematically, by using the root mean square formula:

$$\phi_{N-D} = \sqrt{\frac{\phi_N^2 + \phi_D^2}{2}}$$

where:

ϕ_{N-D} = neutron-density porosity
ϕ_N = neutron porosity (limestone units)
ϕ_D = density porosity (Limestone units)

If the neutron-density porosities from **Fig 5.6a at depth 9,324ft**, are entered into the root mean square formula, the calculated porosity value compares favorably with the value obtained from the crossplot method.

If the Neutron-Density log records a density porosity of less than 0.0- a common value in anhydritic dolomite reservoirs **(Fig 5.6a at depth 9328ft)**, the following formula should be used instead to determine neutron-density porosity (Schlumberger, AAPG, 1982).

$$\phi_{N-D} = (\phi_N + \phi_D)/2$$

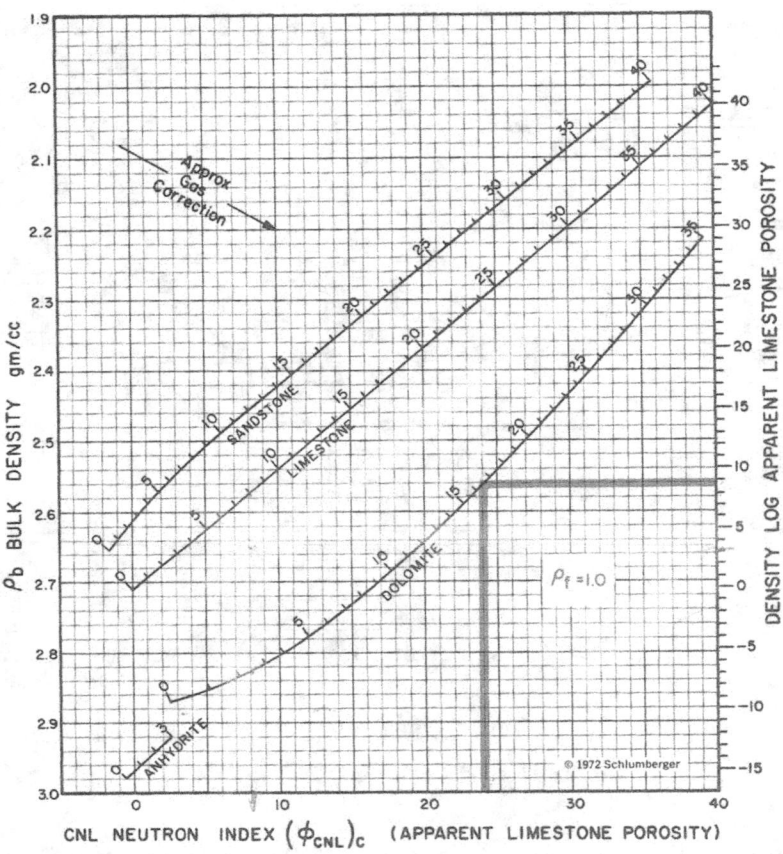

Example (2): Fig 5.7- is a generalized illustration of how lithology affects the combination Gamma Ray// Neutron-Density Log..Such relationship between log responses on the Gamma Ray-Neutron-Density Log and rock type provides a powerful tool for the subsurface geologist. It also shows changes in the log response from oil- or water-bearing rock units compared to gas-bearing sand. By identifying rock type from logs, a geologist can construct better facies maps.

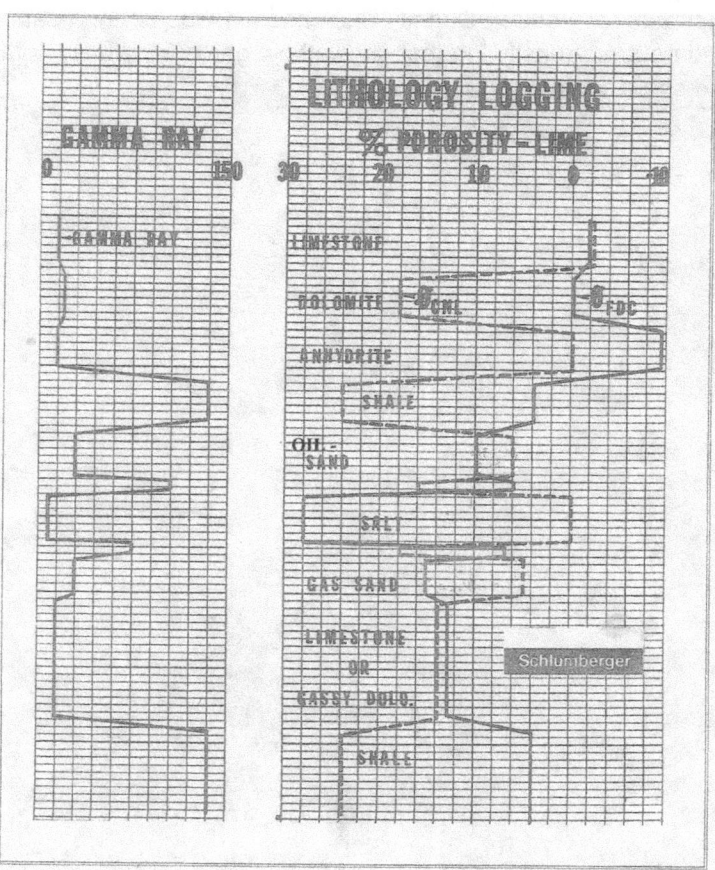

Fig 5.7.- A generalized lithology logging with combination Gama Ray Neutron-Density Log. It shows relationships between log responsees and the rock type as well as changes from oil-or water-bearing rock units compared to gas-bearing units.
(Schlumberger, AAPG, 1982)

The oil- or water-bearing sand has a density log reading of 4 porosity units more than the Neutron log (**NB:** each division on the Neutron density log equals to two porosity units); in contrast, the gas-bearing sand has a density reading of up to 10 porosity units more than the neutron log. The increase in Density-porosity that occurs along with a decrease in Neutron porosity in a gas-bearing zone, is known as "gas effect," which is created by gas in the pores. Gas in the pores causes the density log to record too high a porosity (i.e. gas is lighter than oil or water), and causes the Neutron log to record too low porosity (i.e. gas has a lower concentration of hydrogen atoms than oil or water). Therefore, the effect of gas on the Neutron-Density log is a very important log response to detect gas-bearing zones.

Fig 5.8- A schematic illustration of Neutron-Density responses in Gas bearing sandstone (modified after Truman et al, 1972); a generalized neutron-density log that shows how gas effect varies with depth of invasion , porosity , hydrocarbon, density, and shale content.

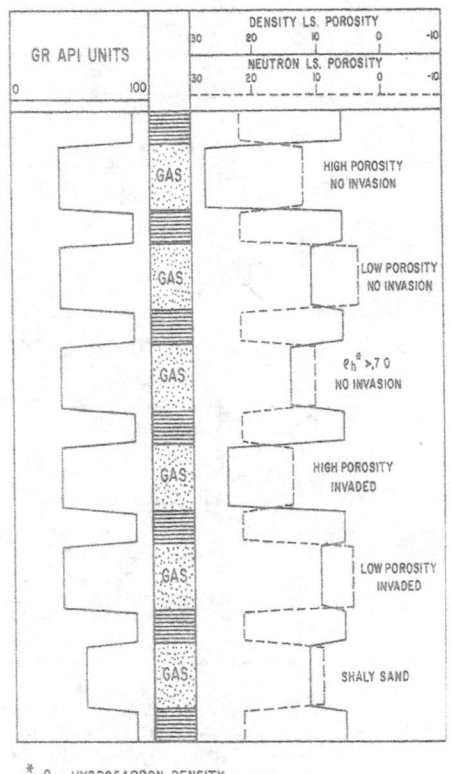

Fig 5.8 - Schematic illustration of neutron-density responses in gas-bearing sandstones (modified after Truman et al, 1972). Generalized neutron-density log responses show how gas effect varies with depth of invasion, porosity, hydrocarbon density, and shale content.

5.5.4 Gamma Ray Logs

Gamma Ray logs measure natural radioactivity in formations and can be used to identify lithologies, correlate zones, and provide information for calculating the volume of shale in a sandstone or carbonate. Shale-free sandstones and carbonates have low concentrations of radioactive material and therefore give low gamma ray readings. As shale content increases, the gamma ray log response increases because of the concentration of radioactive material in shale. However, potassium-feldspar, micas, and glauconite produce a high gamma ray response in clean or shale-free sandstones,, containing potassium feldspars, mica, glauconite, or uranium-rich waters, and in this case a "spectralog" can be run in addition to the gamma ray log, because it breaks the natural radioactivity of a formation into the different types of radioactive material such as potassium, and uranium .

The gamma ray log is recorded in Track #1, together with Caliper log. Tracks #2 & #3 often contain either a porosity log or a resistivity log **(Fig 5.9)**

Volume of Shale Calculation- Because of the radioactivity of shale, the gamma ray logs can be óóused to calculate the volume of shale in porous reservoirs by first determining thex gammax ray index (I_{GR}): (Schlumberger, 1972) :

$$I_{GR} = \frac{GR_{Log} - GR_{Log}}{GR_{Log} - GR_{Log}} \qquad (8)$$

where:

I_{GR} = gamma ray index,
GR_{Log} = gamma ray reading of formation,
GR_{min} = minimum gamma ray (clear sand or carbonate),
GR_{max} = maximum gamma ray (shale).

The volume of shale is also calculated from the gamma ray index (IGR) by the following Dresser Atlas (1979) formula.

For consolidated/older rocks:

$$V_{sh} = 0.33[2^{(2 \times IGR)} - 1.0] \qquad (9)$$

For unconsolidated/Tertiary rocks:

$$\text{Volume of shale } (V_{sh}) = 0.083[2^{(3.7 \times IGR)} - 1.0)] \qquad (10)$$

Fig 5.9- shows Gamma Ray/ /density logs. The Gamma Ray log in track #1 shows its scale increasing from left to right, and ranges from 0 to 150 API gamma ray units.

A quick review:

Bulk density (ρ_b) is shown by a solid line and ranges from 2.0 to 3.0 gm/cc., increasing from left-to-right. Density po`rosity (ϕ_D) is represented by a dashed line and ranges from -10% to +30% increasing from right to left. The correction curve is represented by a dotted- and dashed line and ranges from -0.25 to +0.25 gm/cc, increasing from left to right, but only uses the right half of the track. The tension curve is a log that measures how much weight is being pulled on the wireline during logging; it is represented by a broken line and ranges from 2,000 to 12,000 lbs., increasing from right-to-left, but it also increases the right half of the track.

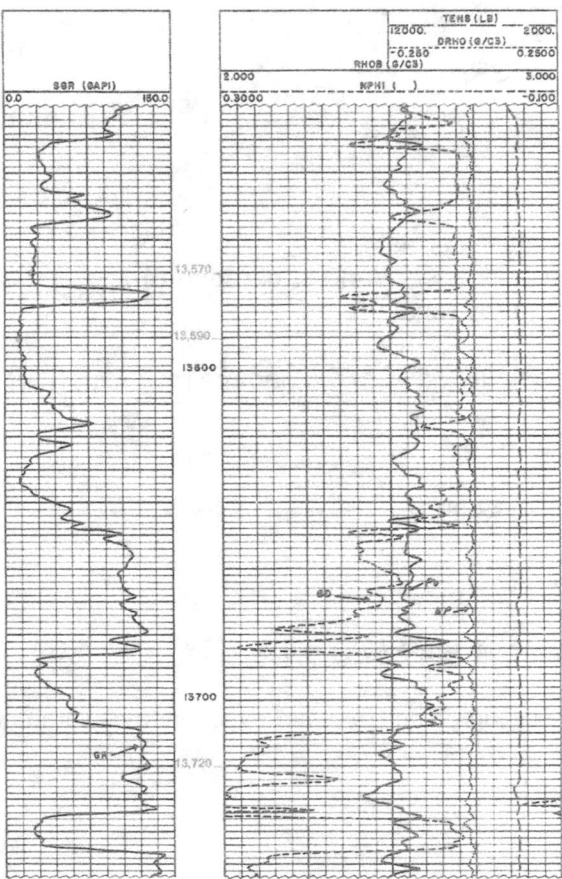

Fig 5.9 - Gamma Ray/Density Log shows curves and scales; Track #1 of Gamma Ray; Tracks #2 & #3 represent bulk density; density porosity and density correction and tension curves

At depth 13,570ft, the gamma ray of the formation reads 28 gamma ray units; noticing that the scale measures in increments of 15 units; the minimum gamma ray reading from the log is at depth 13,590 ft., GR_{min} =15 gamma ray units, and the maximum gamma ray reading is from depth 13,720 ft; GR_{max} = 128 gamma ray units.

5.6 Well Design and Well Type Completions

5.6.1 Scope

It is necessary to run casing at various depth intervals and cement it in place at every interval. The number and size of the casing strings vary with the location of the well, its depth, and the anticipated producing characteristics of the objective reservoir. The casing which includes several strings is the thin-walled steel pipe used to line the borehole, to prevent its caving and contamination of potable water reservoirs; to provide anchorage for the wellhead and blowout preventers; and to provide an adequate means of controlling well pressures.

Initial cementing jobs are performed in conjunction with setting various casing strings to afford additional support for the casing and to retard corrosion by minimizing the friction factor between the pipe and the corrosive formation waters. Besides the casings, a smaller diameter string known as tubing is added and is considered the actual flow conduit for the produced petroleum.

5.6.2 Open Hole Completions

The open hole completion is the only applicable method for the highly competent formations and is common in low-pressure limestone areas where cable tools are used for the drilling-in operation **(Fig. 5.10)**. Rotary tools are used until the oil string is set; at that time, the cable tool rig moves in, bails the mud from the hole, and drills the desired pay interval. This allows the zone in question to be tested as it is drilled, eliminating formation damage by drilling mud and cement, and allowing incremental deepening as necessary to avoid drilling into water; the latter factor is quite important in thin, water-drive pay sections consisting of a few feet of oil zone.

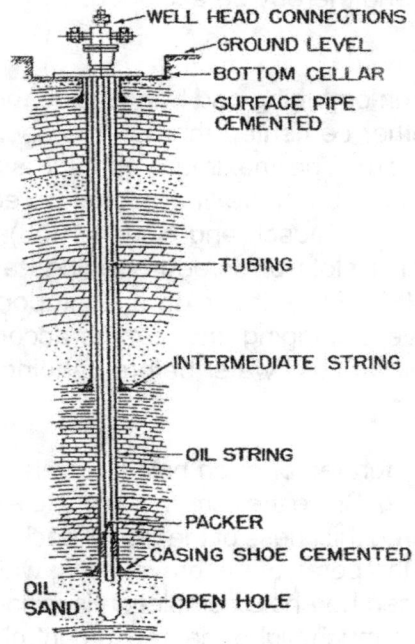

Fig 5.10 - A Sketch of open hole completion in oil-productive well
(Gatlin, 1960)

Fig. 5.10 -A sketch diagram of an open hole completion in a productive oil well

In open hole completion, there are three separate casing sizes: the surface pipe, the intermediate, and the oil strings; the oil string is set up at the top of the oil sand in open hole completion, whereas the packer isolates the tubing-oil string annulus from the pay sand. This category refers to wells that flow naturally and is only applicable to highly competent formations with no caving problems.

5.6.3 Perforated Completion

The perforated completion offers a much higher degree of control over the pay section and is only applicable on wells where the oil string is set through the pay section and cemented, to be perforated at the desired interval. **Mechanical perforators** were used prior to the early 1930s, in which casing can be perforated in place; each consists of either a single blade or wheel-type knife, opened at the desired interval, which cuts vertical slots in the casing. **Bullet perforators** are multi-barreled firearms designed for being lowered into the well, positioned at the desired interval, and electrically fired from surface controllers. A **jet perforator** uses a steam jet's high velocity impact, which is developed to penetrate the target, causing the liner's inward collapse and partial disintegration.

5.6.4 Screening Techniques

The completion of a well in unconsolidated sand is not simple, because of the problem of excluding sand produced together with the oil; unchecked sand can cause corrosion of equipment, and plugging of flow in the string to the extent that the well operation becomes uneconomical. Sand production is also sensitive to the rate of fluid production, especially when large quantities of sand are carried along in the production stream.

Screening techniques are employed for excluding sand by one of two methods: either using slotted (or screen) liners, or by packing the hole with aggregates of gravel. The basic requirement in both methods is that the openings through which the produced fluids flow should be of the proper size to cause the formation sand to form a stable bridge and thereby be excluded.

5.6.4.1 Screen Liners

The slotted or screen liner is normally run on tubing and hung inside the oil string opposite the producing zone. The oil string may have been either cemented through the section and perforated, or set on top of the pay zone (in open hole completion). The maximum opening size that will exclude a given sand is determined from screen analysis. The appropriate liner slot or screen width is taken as twice the 10 percentile size given from screen analysis (Tausch and Corley 1958). Highly variable sands may have the same 10 percentile size, and smaller slots or screen openings are advisable if an unusually high percentage of fine grains are present **(Fig.5.11)**. It is extremely important that the sand face be free of mud cake before the liner is set to prevent plugging; that can be accomplished either by using clay-free completion fluids or by washing the well with salt water before hanging the liner.

5.6.4.2 Gravel Packing

Gravel packing is performed in either perforated or open hole intervals. The gravel size for gravel packs is about five times the 10 percentile sand size. Since the screen is used to exclude gravel only, the slots may be slightly smaller than the gravel; the required thickness of the gravel pack is only four or five gravel diameters. The formation sand then bridges within the pores of the gravel pack, while gravel entry is prevented by the screened liner **(Fig. 5.12)**. In general, open borehole completion is initially cheaper, but the perforated well completion by hydraulic fracturing offers a much higher degree of control over the pay section.

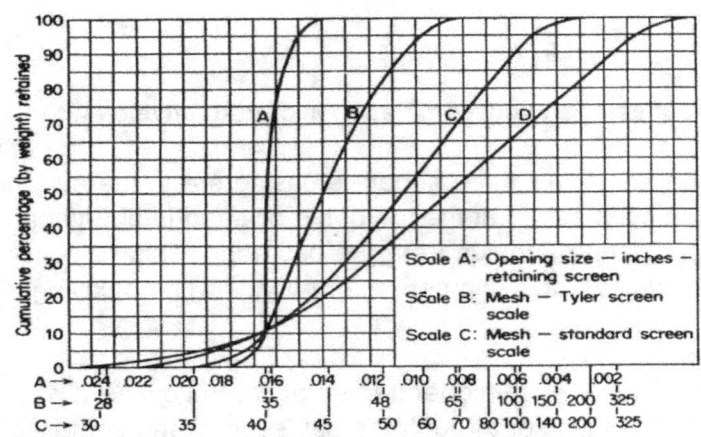

Fig 5.11 - Method of plotting screen analysis data. Note that four sands shown have essentially same 10 percentile point of 0.0165-in. despite size range variations. After Tausch and Corley,[25] courtesy *Petroleum Engineer*.

(Gatlin, 1960)

Fig. 5.11 Method of plotting screen analysis data. Note that four sands shown have ossentially same 10 percentile point of 0.0165-in. despite size range variations. After Tausch and Corley, courtesy *Petroleum Engineer*

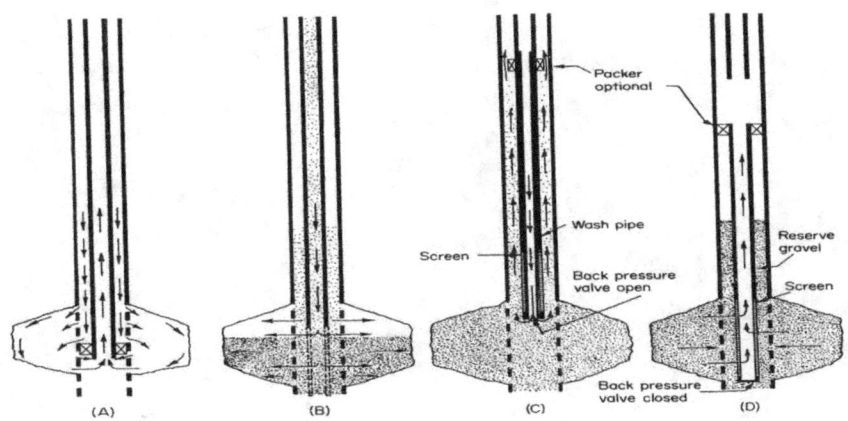

Fig. 5.12 Simplified method of gravel packing commonly used in the Gulf Coast: **(A)** Perforations are washed; **(B)** gravel is squeezed through perforations; **(C)** cavity is filled and screen is washed through gravel; **(D)** after screen placement, wash pipe is removed. After Tausch and Corley, courtesy *Petroleum Engineer (Gatlin 1960)*.

5.6.5 Other Well Type Completion

A permanent-type well completion is one in which the tubing is run and the wellhead is assembled only once in the life of the well. Perforating, swabbing, squeeze cementing, gravel packing, or other completion or remedial work is performed with special small diameter tools designed to run inside the tubing.

In multiple zone completions, there are one or more separate pay zones producing simultaneously from the same borehole, without commingling of the fluids; this segregation is co57-59mmonly required for purposes of reservoir control under state regulatory bodies. Dual completions are the most common, although triple and quadruple zone completions have been performed (Gatlin 1960).

References

Archie GE (1942) The electrical resistivity log as an aid in determining some reservoir characteristics. *J. Pet. Technol.*, 5:54–62

Asquith GB, Gibson CR (1982) Basic well log analysis for geologists. The American Association of Petroleum Geologists, Tulsa, Oklahoma, Fig 1/p7; Fig 5/p14; Fig 7A/p20; Fig 7B/p22; Fig 10B/p30.

Dresser Industries, Inc. (1975, 1979) Dresser Atlas Log interpretation fundamentals and charts Houston, Texas, 107 pp

Hilchie DW (1978) Applied openhole log interpretation. Golden, Colorado, D.W., Hilchie, AAPG (1979)

Schlumberger (1972) Log interpretation manual/principles. Schlumberger Well Series, Inc., Ridgefield, Vol. I

Schlumberger (1974) Log interpretation manual/applications. Schlumberger Well Series, Inc., Ridgefield, Vol. II.

Schlumberger (1977) Log Interpretation/Charts. Houston, Schlumberger, Well Sevices, Inc.

Schlumberger (1979) Log Interpretation/Charts. Houston, Schlumberger, Well Sevices, Inc.

US Environmental Protection Agency (1977) An introduction to the technology of subsurface waste injection. Environmental Protection Agency 600/2-77-240, 1977; Cincinnati, Ohio, pp 73–88

Wylie MRJ; Rose WD (1950) Some theoretical considerations related to the quantitative evaluation of electric log data. Jour. Petroleum Technology 189:105–110

Tausch GH and Corley CB Jr (1958) Sand exclusion in oil and gas wells The Petroleum Engineer, ppB-38, and B-58, respectively, in: Gatlin C (1960)Petroleum engineering, drilling and well completions. Dept. of Petroleum Engineering, Fig 15.9/p314 & Fig 15.12/p315, Univ. Texas, Prentice Hall. Inc., Englewood Cliffs, NJ USA.

PART II

A Guide to Computerized Lithostratigraphic Correlation Charts-Petroleum Resources- A Model Type /North Africa

Chapter 6

Mineral and Petroleum Resources-Petroleum Provinces, and Computerized Lithostratigrafic Correlation Charts of the sedimentary Section in North Africa

6.1 Introduction

The northern African countries include the Arabian Maghreb which comprises Morocco, Algeria, and Tunisia to the west; Libya and Egypt to the east (**Fig. 6.1;** Michael, 1969, AAPG 1970).

During the 18th century, the French colonialism spread widely all over the Arabian Maghreb countries, where Algeria, the latest French colony gained its independence in 1962, is bordered by Morocco and Mauritania in the west, Tunisia and Libya to the east; Nigeria and Mali to the south.

Algeria is a country of ~2.36 million km² (900,000 square miles), with a population of about 50 million; the Algerian territory covers an area between latitude 20°–36°N and longitude 10°E–08°W, which is approximately equal to that of its neighbors combined.

During the colonial Epoch, French oil companies, Sopefal and Becip carried out petroleum exploration studies in north Algeria; other French petroleum companies such as SN-Repal and Total were granted long-term contracts to carry out production operations in different basins of the southern Saharan Platform; where the assigned oil companies worked independently, and caused identical problems for geologists and research scientists, who joined later the government oil Company (Sonatrach), after independence, as they found it difficulty to review hundreds of documents and publications of different basins, with numerous geological terms and confusing abbreviations of the same lithological, and/or stratigraphical terms, by different Petroleum Companies.

Libya is a country of about 1.78 million km² (680,000 square miles), previously considered almost barren by an international mission after the Second World War; but prompted by discoveries in neighboring Algeria, when seventeen companies carried out oil exploration and invested more than $800 million by the beginning of 1963.

Egypt is a country of 1.21 million km² (460,000 square miles), with a population of about 100 millions; it has been called "the gift of the Nile," yet the Nile Valley is one of the mineralogical poorest regions in North Africa, though holding mineral resources in Sinai and the Eastern Desert, such as iron, phosphates, gypsum, salt, potash, and Talc ore deposits.

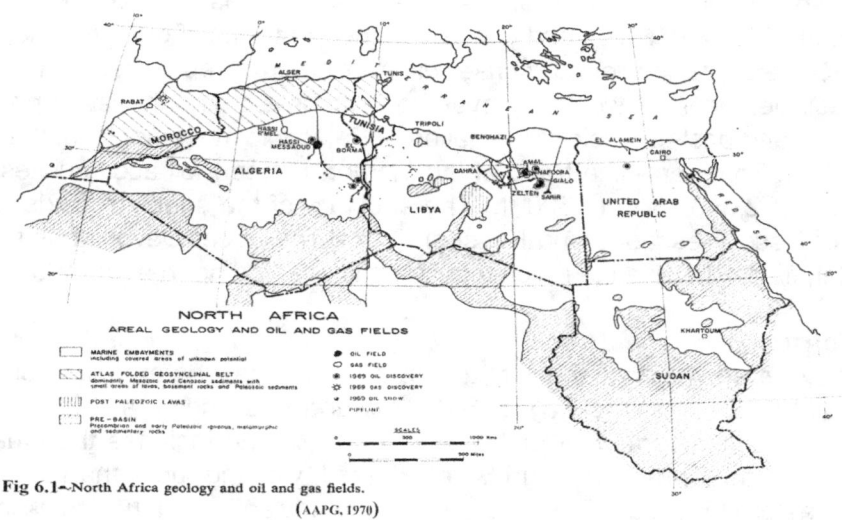

Fig. 6.1 North Africa geology and oil and gas fields. (Michael 1969, AAPG 1970)

6.2 Mineral and Petroleum Resources in North Africa

6.2.1 Scope

The sedimentary formation of the Arabian Maghreb accumulated from the Tethys Ocean, which flooded the area during the period between Cambrian and Eocene. From the Precambrian window of the Hoggar Mountains southeast of Algeria to the Mediterranean, successively younger sediments accumulated in the Epi-continental zones. Consequently, younger oil fields appear from the southwest towards the northeast. The traps, irrespective of age, were frequently controlled by Precambrian-Paleozoic structures. The Arabian Maghreb is well known as the continent's largest oil, gas, phosphate, and lead supplier. Egypt has limited mineral resources that are concentrated in the oases of the western and eastern deserts as well as in Sinai. After independence, the National Algerian Petroleum Society, "Sonatrach", increased its share with the French Sopefal, whereas Libya significantly increased its oil production after the military revolution in 1969. In the same period, Tunisia and Morocco awarded large blocks of their offshore acreage to French companies.

Recently the Russian and American oil companies, discovered gas in the Nile Delta of Egypt, as well as a new oil field in Sirte of Libya.

In 1965, crude oil reserves of the Arabian peninsula and North African countries represented almost two-thirds of the world total (300 BBO). The African rate of expansion was ten of times as high as the overall increase in world production in the past four decades.

6.2.2 The Algerian Territory

There are four main regions of mineral, oil and gas resources in the Algerian Sahara: a) The Precambrian mass of Hoggar on the southern border of Algeria, constitutes the anchor of the Sahara (rich in platinum, asbestos, tin, zinc, etc.); b) The Paleozoic oil fields to the south of the Saharan platform; c) The Triassic Province of oil and gas to the north of the Sahara; and d) The mineral resources of the Eocene to the east of the Sahara (Kun 1965; **Fig. 6.2a**).

The Algerian Sahara can be classified into four groups of structures and basins: (1) The Hassi er R'Mel, the world's largest gas field, of 320 km2 (200 miles) south of Algeria, was assigned to SN Repal and CFP, which had been expanding rapidly since the late 1960s; (2) The oil field of Hassi Messaoud, 300 miles south of Constantine, owned by SN Repal; (3) the Salah Basin (480 km or 300 miles south of Hassi er R'Mel), where oil was struck at Edjelah; and (4) the Fort Polignac-Tinrhert Basin, the largest oil producer in Africa until 1963, owned by CREPS (Kun 1965; **Fig. 6.2b**), and consists of the Zarzaitine, Tiguen-tourine, Edjeleh fields and a number of other fields. The Zarzaitine field comprises different oil stratigraphic beds in the Devonian formation. The Fort Polignac Basin lies near the western Libyan border where the northeastern part has been intensely faulted by the Mesozoic and post-Jurassic rejuvenation of Paleozoic movements; at the Fort Polignac basin, the faulted Ohanet North/South anticline trend of the Tinrhert Plateau, oil accumulates in Lower Devonian sandstones at a depth of 2,343.75 m (7,500 ft.) The structure of the southern fields is very different from the flat-lying basins of Hassi Messaoud and Hassi er R'Mel. The upper Devonian is the most favorable oil reservoir in the Edjeleh field, whereas the Edjeleh anticline is oriented north-south (Sonatrach 1972).

6.2.3 The Libyan Territory

The Paleozoic Sea covered most of the Libyan territory; in the eastern part, oil fields are oriented east-southeast parallel to Es-Sidr, the Bay of Sirte. Oil and gas shows are found in a broad zone of northwestern Libya, extending from west of Tripoli to Edjeleh, and includes the Oued Chebbi gas field in the northwestern corner of Tripolitania and Bir el-Rhezeil, yielding more than 2 million ft^3 of gas **(Fig. 6.2c; Kun 1965)**. In eastern Libya, the Paleocene is particularly saline; though thick interbeds of dolomite and anhydrite were deposited during the Lower Eocene. In Middle Eocene, the subsidence continued, as shown by limestone and thin seams of interceded lignite and coal. In the Upper Eocene another emergent phase, which reached its climax in the Oligocene was initiated, as indicated by conglomerates and sands.

The basins of Tripolitania and northern Fezzan extend along a south-southwest trend for 640 km (400 miles) as a long broad arc. However, the Oued Chebbi gas field, at the northernmost part of the alignment, belongs to the northern Triassic basin.

As in Algeria, the pre-basin relief of Libya strongly affects the accumulation of hydrocarbons. Sediments were first deposited in the troughs, and then over the whole pre-Cretaceous land surface. The differential compaction of sediments along with tectonics, further impressed the imprint of the Paleozoic-Precambrian topography. Upper Cretaceous carbonates form more favorable traps than Lower Cretaceous sandstone, although basal sandstone and partly dolomitized calcarenites of the coast constitute the principal reservoirs. As in the northern Sahara, many of the reservoirs are located on uplifts, structural heights, or on their edges. The Samah and Waha fields are good anomalies of Upper Cretaceous accumulation of oil (Kun 1965).

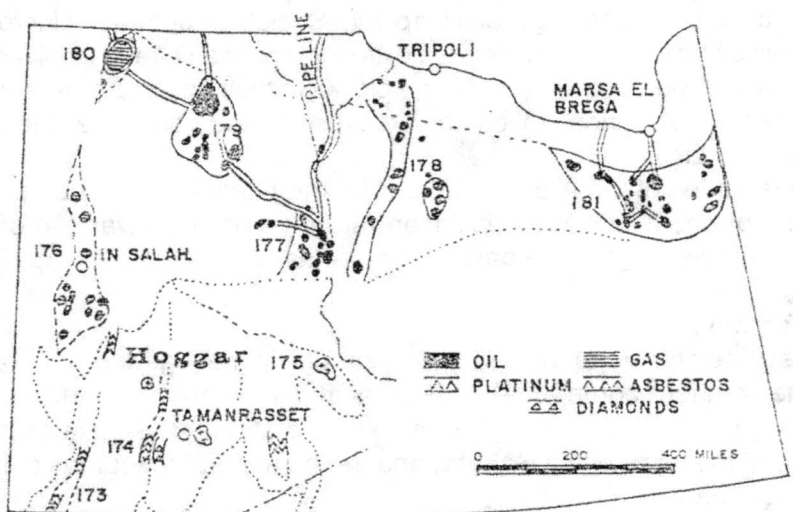

Fig 6.2a - The Mineral resources of the Sahara (Kun, 1965)

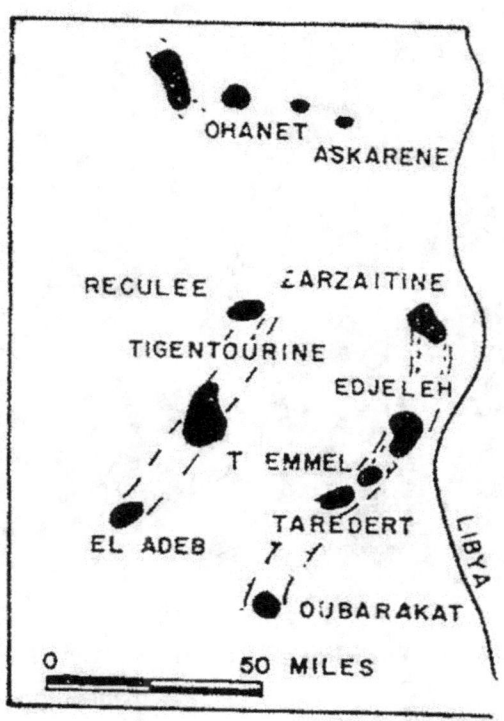

Fig 6.2b - Fort Polignac Oil Field
Kun, 1965)

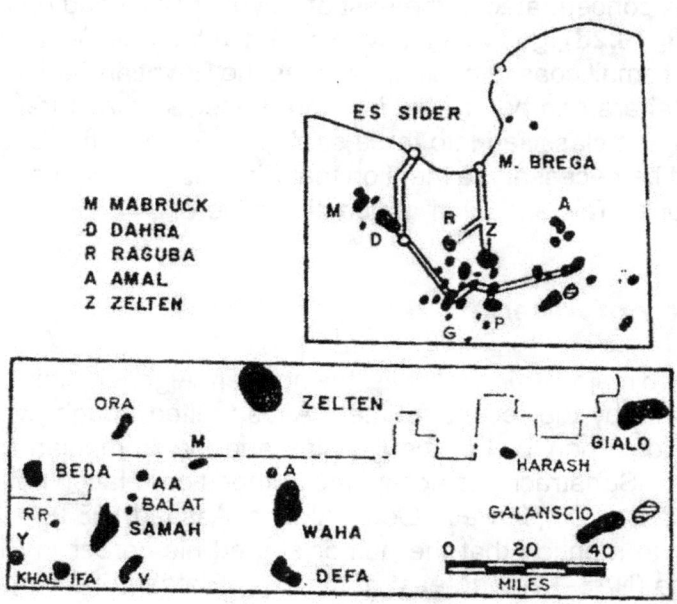

Fig 6.2c - The East Libyan Oil Field
Kun, 1965)

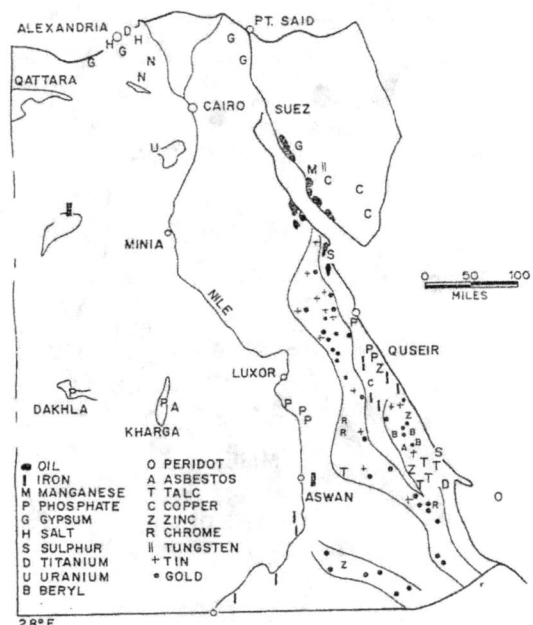

Fig 6.2d - The Mineral Resources of Egypt
Kun, 1965)

Fig. 6.2 (**a**) Mineral resources of the Sahara (Kun, 1965); (**b**) Fort Polignac oil field (Kun, 1965); (**c**) The East Libyan oil field (Kun, 1965); (**d**) The mineral resources of Egypt (Kun, 1965).

6.2.4 The Egyptian Territory

In 1963, oil production was concentrated in the east of Egypt at Sinai and on the southern coastal areas of the Gulf of Suez **(see Fig. 6.2d)**; gas was lately explored in the Nile Delta of Egypt. The Miocene Gulf of Suez oil fields are in fact small coastal basins, whereas the Egyptian Sahara has yielded little oil in the present. The rest of the Sahara can be divided into three unequal structural units, the east, the center, and the west, and can be also classified into three stratigraphic units, the Eocene, the Triassic, and the Paleocene basins. It might be necessary to mention that Sinai as well as, the eastern Desert of Egypt are rich with different iron, copper, Talc and other radioactive minerals.

6.3 Structural Geology of Algeria

Algeria is subdivided into two major structural units, the northern Algerian Alpine and the southern Saharan Algerian Platform, separated by the recent southern Atlas faulted trough, which caused a problem for the French scientists, besides the Chief of the Russian and the Romanian contractors working for the National Algerian Company "Sonatrach", to accept the author's correlation between the Triassic deposits of both units (See Ch. 10 & Ch. 11- (Jr. Petr.. Geol., Algeria, Assaad,1983).

It might be necessary to mention that the author started his career in Algeria by citing a key well location at Nador Sud area (NAS-1), north-east of the High Plateau of Algeria, and carried out regional electric well log correlation of the Triassic deposits with that of the nearest well in the Saharan Platform, inspite of the denial of the French scientists, the Russian and the Romanian Delegates working for the State oil company "Sonatrach", due to the presence of very much younger deposits of the Saharan Atlas Trough in between both the High Plateau and the Saharan Platform, that could be only caused by a *"huge Sink Hole"* several millions years ago, followed by younger deposits that confused the delegates working for the State petroleum Co. (Sonatrach, 1972).

Actually, it was a surprise for the author to find the electric log correlation of NAS-1 well was a mirror image of the chosen well at the Saharan Platform, together with the palaentological and palynological correlation of both structures; the author then defined a new Triassic basin and its trends to the NE/SW of the High Plateau, therefore disclosed the old issue of the *"missing cycle of the Triassic reservoir"*.

NB: It will be very appreciated to carry on a special project by the UAB-Geology Dept.(Birmingham, Alabama), to estimate the different eras that took place starting from the era of the Sink Hole of the great trough, and that of its filling up of much younger deposits.

6.3.1 The Algerian Alpine
The northern Algerian region, mainly affected by Alpine orogeny, is formed of young mountains of Tertiary times and consists of a number of structural sedimentary units: From north to south as follows:

(**a**) **The off-shore area** -is a reduced continental shelf of Tertiary and Quaternary sediments overlying a metamorphic basement. (**b**) The **Tellian Atlas** is a northern complex area comprising of a chain of a set of Nappes in place, during the Lower Miocene. (**c**) The **Hodna Basin,** a fore-deep basin of continental Eocene and Oligocene deposits, is overlain by marine Miocene sediments. (**d**) The **High Plateaus** of delineated features in the center with large plains to the west and high structures to the east, are the foreland of the Alpine range of thin sedimentary cover (Liassic objective). (**e**) The **Saharan Atlas,** is formed from an elongated trough pinched between the high plateaus and the Saharan platform, and was filled in by thick Mesozoic sediments (7,000–9,000 m); the Tertiary compressive tectonic stresses, modified the elongated trough into a number of reverse faulted structures that led to the creation of the mountain range.

Both the high plateaus and the Saharan Atlas comprise the first major tectonic unit of the Epihercynian platform, which was active during the Alpine stage and caused the separation of the southern portion of the Saharan Atlas from the Saharan Platform, that constitutes the second major tectonic unit in northeast central Algeria. Both the Epihercynian and Precambrian platforms attained their main tectonic features after the Early Alpine orogenic movements, whereas the successive southerly migrations of the equator during Late -Early Liassic Time led to the formation of several Lagoons.

6.3.2 The Algerian Saharan Platform
The Algerian Saharan Platform is located to the south of the Alpine domain and is part of the North African Craton that yields most of Algeria's hydrocarbon resources. It has been a relatively stable area of low tectonic activity over the course of the geologic time. Numerousythe xbasement uplifts accompanied by faults, have affected the Paleozoic sediments; but the tectonics never reached the violence of the Alpine folds that disturbed the northern portion of the shield and produced folding sheets in the Saharan Atlas Mountains, disengaging the Mesozoic sediments from the plastic saliferous beds of the Triassic-Liassic Formation. No Triassic/Liassic diapirs and no Triassic outcrops were known in the Saharan platform because of the slow effect of the epeirogenic movements on the relatively thin sediments overlying the solid and hard Precambrian basement **(Fig 6.3)**.

The Precambrian basement is unconformably overlain by thick sediments of the Paleozoic, structured into a number of separated basins by high zones, followed by Mesozoic and Cenozoic sediments. The basins, from west to east, are: the **Tindouf/Reggane Basins**; **Bechar Basin**; the **Ahnet-Timimoun Basin**; and the **Mouydir** and **Aguemour-Oued M'ya basins**, where major fields are located within the Cambrian of Hassi Messaoud and the Triassic formations of Hassi R'Mel. To the far east, the **Illizi-Ghadames syncline** yields hydrocarbons both in Paleozoic and Triassic sediment **"http://www.memalgeria.org/hydrocar-bons/ geology"**.

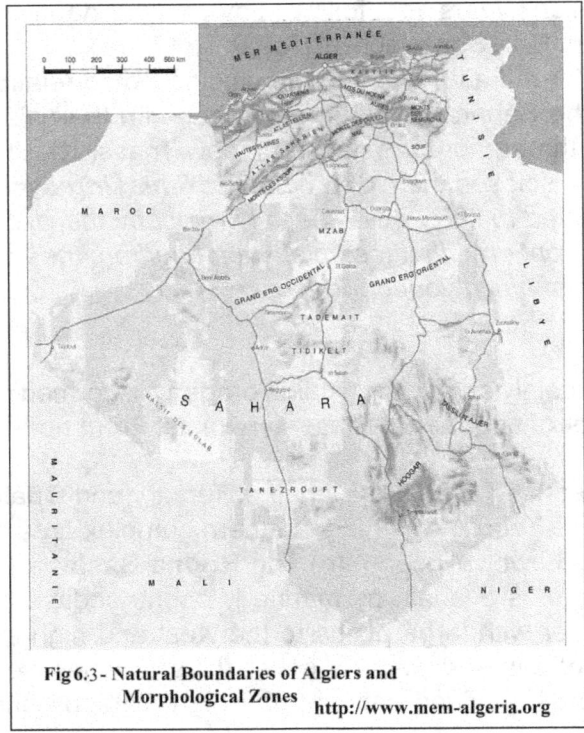

Fig 6.3 - Natural Boundaries of Algiers and Morphological Zones http://www.mem-algeria.org

Fig. 6.3 Natural boundaries of Algeria and the morphological zones http://www.mem-algeria.org http://www.mem-algeria.org/hydrocarbons/geology.htm

The Paleozoic structural basins south of the Algerian Sahara comprise the following:

1 – **Tendouf Basin** is characterized by an east/west synclinal axis and divides the basin into two unequal parts; a northern structural zone and a southern monocline dipping to the north. Stratigraphically, the basin is represented by a series of Paleozoic formations that thickened in the northern part, pinching to the south, and ranging from Cambro-Ordovician to Upper Carboniferous. The Ordovician and/or Devonian formations played an important rôle as gas reservoirs.

2 – **Bechar Basin** shows complicated tectonics and stratigraphic aspects of Paleozoic formations (Cambrian to Carboniferous). The granular and the carbonate reservoirs attained poor permeability and fairly good porosity, though secondary fissures especially in the highly tectonic zones might improve the petrophysical characteristics.

3 – **Reggane and the Timmimoun Basins** show gas reservoirs in the Upper Ordovician (Ashgillian), Upper Devonian (Strumian and Frasinian), Lower Devonian (Emisian and Siegenian), and Lower Carboniferous (Visean) sandstones; to the South, the Ahnet and Teguentour basins are gas producers in the lower Ordovician, Devonian, and Carboniferous reservoirs. The Ahnet Basin includes three regional north-south trending faults, composed of local isolated highs, and is productive in the Ordovician; whereas the Tinguentour field is productive in the Tournasian formation.

4 – **The Illizi Basin** to the southeast of the Algerian Saharan platform consists of the Tinfoyé Tabenkurt oil and gas fields that are productive from Gothlandian (Unit A) and Upper Ordovician (Unit IV) reservoirs. The Illizi basin is mainly oil and gas productive in the Upper Shaly sandstone series (T1 + T2), in the lower sandstone series (SI-series) of the Triassic reservoir, and in the Devonian (Unit F_3+F_6) reservoir. To the south, the Illizi central region is productive from Lower Devonian and Gothlandian reservoir rocks **(Chart 6.1)**.

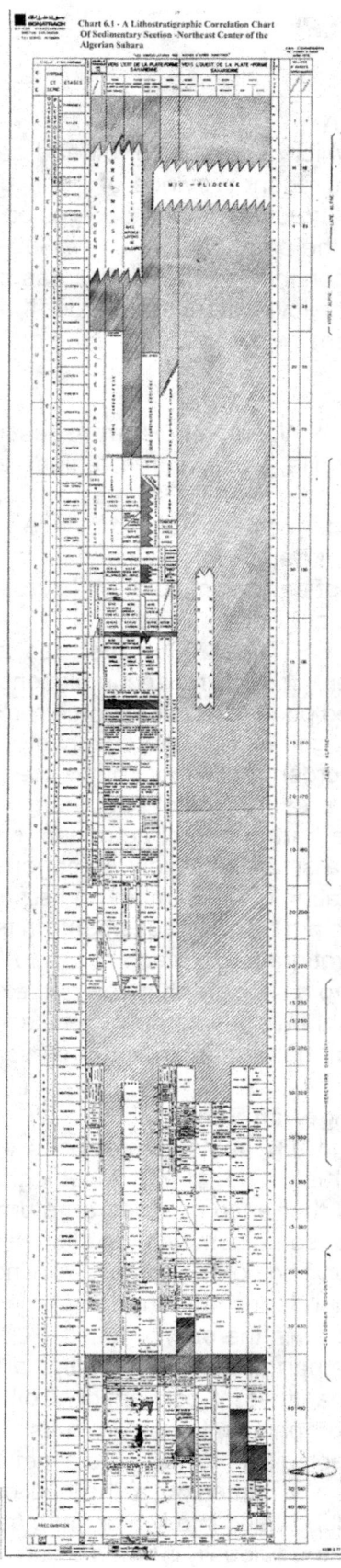

6.4 Petroleum Classification of Provinces

6.4.1 Petroleum Province—Definition
A petroleum province is a region in which a number of oil and gas pools and fields occur in a similar geologic environment, e.g., the mid-continent province of the south central United States has definite regional characteristics of stratigraphy, structure, and oil and gas occurrence; thus the term "province" has a specific meaning for geologists and the petroleum industry. Sub-provinces may occur within prvinces. The occurrence of petroleum is not confined to the land areas of the world; there are vast potential sources of petroleum underlying the shallow waters that border the continents. These submerged lands, sloping down to a depth of 600 ft (191m) below sea level, are known as continental shelves.

6.4.2 Total Petroleum System
The Assessment Methodology Team of World Petroleum Assessment 2000, developed a hierarchical Scheme of geographic and geologic units, and covered regions and geologic provinces that comprise total petroleum systems, and assessment units (USGS, Klett and others, 1997). Total petroleum systems and assessment units were delineated for each geologic province considered for assessment; the boundaries of both types are not necessarily being entirely contained within a geologic province. Particular emphasis is placed on the similarities of petroleum fluids within total petroleum systems, unlike geologic provinces in which similarities of rocks are emphasized.

The Total Petroleum System (USGS-WPA 2000), includes all genetically related petroleum that occurs in petroleum hydrocarbon, where shows and accumulations (discovered and undiscovered) were generated by a pod or by closely related pods of mature source rock. Total petroleum systems exist within a limited mappable geologic space, together with the essential mappable geologic elements, such as source, reservoir, cap rock, and overburden rocks that control the fundamental processes of generation, expulsion, migration, entrapment, and preservation of petroleum within the total petroleum system.

An assessment unit is a mappable part of a total petroleum system in which discovered and undiscovered oil and gas fields constitute a single relatively homogeneous unit so that the methodology of resource assessment, based on estimation of the number and sizes of undiscovered fields, is applicable. (Pubs.usgs.gov/bul/b2202-a/b2202aso.pdf). The assessment units are named after the total petroleum system with a suffix of "Structural/ Stratigraphic", which refers to the progression from a structural and combination trap exploration strategy to a stratigraphic trap and exploration strategy.

The Total Petroleum system should be considerably more effective than the traditional methods in performing certain geological tasks, normally spent in non-professional work (Assaad, 1981).

6.4.3 Geological Provinces of the Algerian Sahara
The world is divided into 8 regions and 937 geologic provinces, which have been ranked according to the discovered oil and gas volumes (Klett et al. 1997). A geologic province is an area of characteristic dimensions of hundreds of kilometers that encompasses a natural geologic entity (e.g., a sedimentary basin, thrust belt, or accreted terrain) or some combination of contiguous geologic entities. Each geologic province is a spatial entity with common geologic attributes. Province boundaries were drawn as logically as possible along natural geologic boundaries, although in some regions, they were located arbitrarily (e.g., along specific water depth contours in the open oceans).

Klett (2006) prepared a report as a part of the U.S. Geological Survey World Petroleum Assessment (USGS-WPA 2000), to assess the quantities of conventional oil, gas, and natural gas-liquids, and to develop a hierarchical scheme of geographic and geologic units.

The scheme consists of regions, geological provinces, total petroleum systems, and assessment units. Assessment of undiscovered resources was carried out at the level of total petroleum system or assessment unit. A total petroleum system may be subdivided into two or more assessment units, and each assessment unit is sufficiently homogeneous to assess individually in terms of geology, exploration considerations, and risk.

A number of factors such as formation thickness (1,000–8,000 m), lithology, tectonic deformations, and subsidence, were primarily used to define different sedimentary basins which can be grouped into three regions: ***the western*** (Tindouf, Reggane, Bechar basins); ***the northeast central*** (Triassic of Hassi R'Mel and Oued M'ya basins); ***and the eastern*** (Ghadames and Iliizi basins. Actually, the geologic provinces encompass several basins and can be considered as "multi-basin provinces;" on the other hand, some of the basins share with more than one province **(Fig 6.4)**.

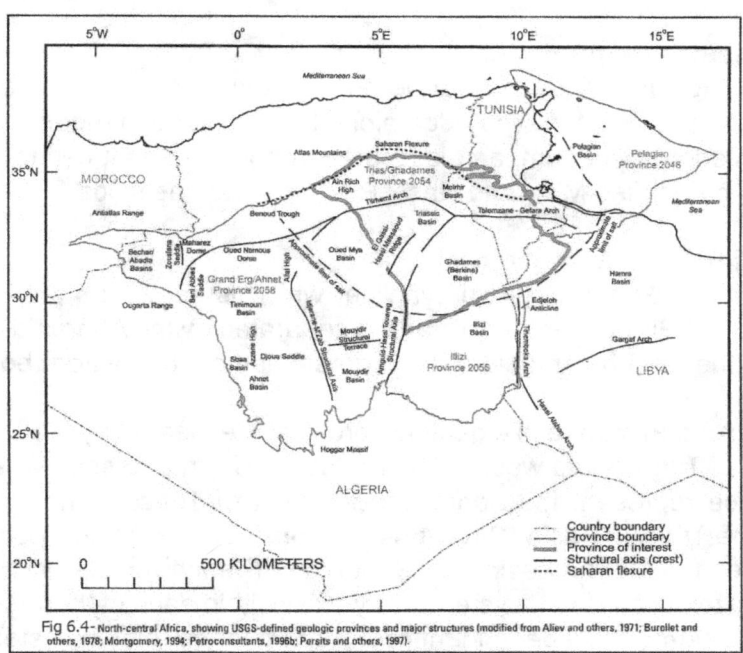

Fig. 6.4 North-Central Africa, showing USGS-defined geologic provinces and major structures (modified from Aliev and others, 1971; Burollet and others, 1978; Montgomery, 1994; Petro-consultants, 1996b; Persits and others, 1997)

In general, it might be best if we define the geological provinces in relation to the main regions of the Algerian Saharan Platform; the Western, the Northeastern, and the Eastern Regions as follow**s::**

6.5 Regions of the Saharan Platform

6.5.1 The Western Saharan Region
The Western Saharan region mainly covers the Grand Erg/Ahnet Province that includes Tindouf, Reggane, and Bechar basins to the west of Algeria, whereas the Triassic basin encompasses the northeastern portion of the Grand Erg/Ahnet province as well as the northwestern portion of the Trias-Ghadames province.

The Grand Erg/Ahnet Province, is a gas producing province but remained practically unexplored. The northwest tip of the province extends slightly into Morocco and the province area encompasses approximately 700,000 km^2; includes the Timmimoun, Ahnet, Sbaa, Mouydir, Benoud, Bechar, Abadla basins and part of the Oued M'ya basin; as well as the giant Hassi R'Mel gas field.

The Grand Erg/Ahnet province is mainly a Precambrian basement unconformably overlain by thick Paleozoic sediments, structurally defined by four basins (reservoir rocks of Cambro-Ordovician, Silurian, Devonian, and Carboniferous age), ranging from Cambrian to Namurian (upper Carboniferous), separated by high structures, followed by insignificant thicknesses of Mesozoic and Cenozoic sediments.

More than one total petroleum system may exist within each of the basins in the Grand Erg/Ahnet Province. The *composite total* petroleum systems are coincident with the basins in which they exist and are known as Tanezzuft-Timimoun, Tanezzuft- Ahnet, Tanezzuft-Sbaa, Tanezzuft-Mouydir, Tanezzuft-Benoud, Tanezzuft-Bechar-Abadla, and Tanezzuft-Oued M'ya total petroleum systems. The latter extend across both the Grand Erg/Ahnet and the neighboring Trias/Ghadames Province. Tanezzuft refers to the Tanezzuft formation (Silurian), which is the oldest major source rock of the total petroleum system, Tanezzuft is then followed by the basin name in which the total petroleum system exists.

6.5.2 The Northeast Triassic Region

The **Triassic Region of Algeria** divides the Northern Sahara into eastern and western basins where the Paleozoic strata form a north-south trending ridge along the longitude of Algeria. The Triassic region is a large an east/west oriented anticlinorium, and is located on the northeast central portion of the Saharan Platform that comprises both Oued M'ya and Hassi R'Mel oil and gas fields.

6.5.3 The Eastern Region

The Eastern Region of the Ghadames-Illizi syncline, which is known as the East Algerian Syncline, consists of both the Illizi and Ghadames basins, and is separated by the Ahara ridge. The Ghadames-Illizi Syncline is bounded on the west by Amguid-El Biod dorsal and by the eastern border of Algeria:

(a) **TheTrias/Ghadames Province** is a geologic province delineated by the USGS, located in central east Algeria, southwest of Tunisia and west of Libya. The southern and southwestern boundaries of the Trias/Ghadames province represent the approximate extent of Triassic and Jurassic evaporates that were deposited within the Triassic Basin (**Pub.usgs.gov/bul/b2202-c/b2202-bso.pdf**).

The province includes the Melrhir basin, the Ghadames (Berkine) basin, and part of the Oued M'ya basins. Although several total petroleum systems may exist within each of these basins, only three "composite" total petroleum systems have been identified. Each total petroleum system occurs in a separate basin, and comprises a single assessment unit. The composite systems are known as **(1)** the Tanezzuft-Oued M'ya, **(2)** the Tanezzuft-Melrhir, and **(3)** the Tanezzuft-Ghadames total petroleum systems; the last two systems are located almost entirely within the **Trias/Ghadames Province**.

The Tanezzuft-Oued M'ya petroleum system extends into the neighboring Grand Erg/Ahnet Province, whereas the Trias/Ghadames Province contains the giant Hassi Messaoud oil field.

(b) The Illizi Province is located southeast of Algeria and a small portion of western Libya. The province and its total petroleum system coincides with the Illizi Basin. The Illizi province contains both the Tin Foyé-Tabenkort and Zarzaitine oil fields. More than one total petroleum system may exist within the Illizi Basin. One "composite" total petroleum system is described as the Tanezzuft-Illizi Total Petroleum System. In the Illizi basin, the Tanezzuft is sometimes referred to as Argillaceous (or Argileux). The Tanezzuft Formation is the principal source rock, deposited during a major regional flooding event.

6.6 An Approach to Computerized Lithostratigraphic Charts in North Africa

6.6.1 Discussion

Geological and geophysical activities greatly increased in northwest African countries over the past four centuries. A precise interpretation of the regional geological framework of the sedimentary formation in different regions of the Saharan Platform of the Arabian Maghreb was carried out, and rock units were better correlated on a regional scale and were cited by the author in their proper stratigraphic levels. Also, a regional updated terminology of rocks had been replaced to avoid misleading abbreviations submitted by previous foreign contractors.

A lithostratigraphic chart was built up to define the stratigraphic level of each rock unit among different structural regions of the Algerian Saharan Platform by correlating the rock units among different regions of the eastern Sahara; such regions comprise the Oued M'ya-M'Zab region to the northwest; the Erg Oriental-Agreb el-Gassi and Hassi Messaoud region to the southeast; and the Tinrhemt-Illizi region to the south. **(NB: Chart 6.1/300MB; charts 6.1a, 6.1b &6.1c /each ~66MB).**

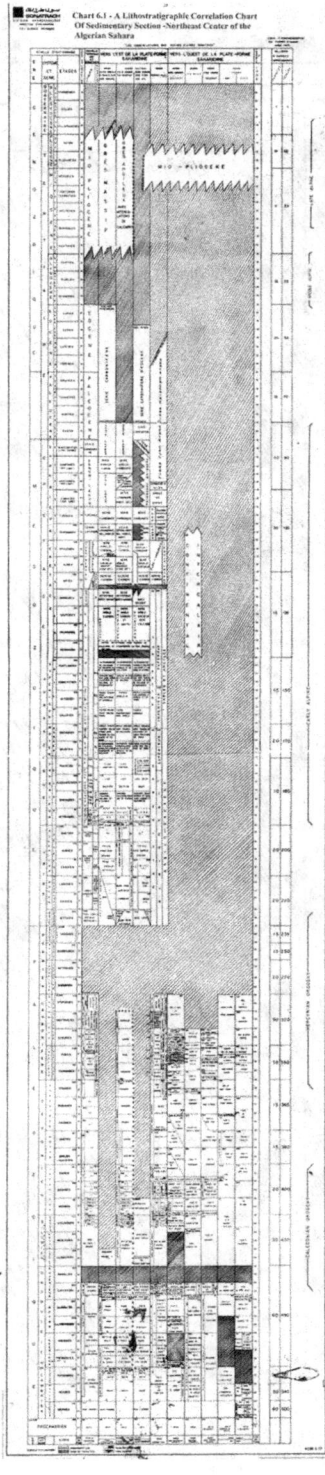

Chart 6.1a - A Lithostratigraphic Correlation of the Cenozoic Sedimentary section of the Algerian Saharan Platform

6.6.2 Numeric Coding of Different Elements of the Geologic Provinces of the Algerian Sahara

Klett (2000) designed a numeric code to identify each region, province, total petroleum system, and assessment unit. A graphical depiction that places the elements of the total petroleum system into the context of geological time is provided in the form of an events-chart, the items of which include the major rock unit names; the temporal extent of source rock deposition, reservoir rock deposition, seal rock deposition, and overburden rock deposition; and the generation, migration, accumulation, and preservation of petroleum, defined in a petroleum system (Magoon and Dow 1994).

A numeric code is given to identify each region, province, total petroleum system, and assessment unit of the Erg/Ahnet province, Illizi Province, and Ghadames Province:

Table 6.1 A numeric code applied for the Grand Erg/Ahnet Province, Algeria, and Morocco and their total petroleum systems (Klett 2000) /[pubs.usgs.gov/bul/b2202-b/b2202-bso.pdf]

Unit	Name	Code
Region	Middle East and North Africa	2
Province	Trias/Ghadames	2058
Total petroleum systems	a- Tanezzuft-Timimoun	205801
	b-Tanezzuft-Ahnet	205802
	c-Tanezzuft-Sabaa	205803
	d- Tanezzuft-Mouydir	205804
	e- Tanezzuft-Benoud	205805
	f – Tanezzuft-Bechar/Abadla	205806
Assessment units	a- Tanezzuft-Timimoun Struc/Stratig	20580101
	b- Tanezzuft-Melrhir Struc/Stratig	20580201
	c- Tanezzuft-Ahnet Struc/Stratig	20580301
	d. Tanezzuft-Mouydir Struc/Stratig	20580401
	e- Tanezzuft-Benoud Struc/Stratig	20580501
	f- Tanezzuft-Benoud Struc/Stratig	20580601

Table 6.2- A numeric code applied for the Illizi province [pubs.usgs.gov/bul/b2202-a/b2202-bso.pdf]

Unit	Name	Code
Region	Middle East and North Africa	2
Province	Illizi	2056
Total petroleum systems	Tanezzuft-Illizi	205601
Assessment units	Tanezzuft-Illizi Structural/Stratigraphic	20560101

Table 6.3 A numeric code for the region, province, total petroleum systems, and assessment units of the Trias/Ghadames Province (Klett 2000)

Unit	Name	Code
Region	Middle East and North Africa	2
Province	Trias/Ghadames	2054
Total petroleum systems	a- Tanezzuft-Oued M'ya	205401
	b-Tanezzuft-Melrhir	205402
	c-Tanezzuft-Ghadames	205403
Assessment units	a- Tanezzuft-Oued M'ya Struct/Stratig	20540101
	b-Tanezzuft-Melrhir Struct/Stratig	20540201
	c-Tanezzuft-Ghadames Struct/Stratig	20540301
	d. Tanezzuft-Mouydir Struc/Stratig	20580401
	e- Tanezzuft-Benoud Struc/Stratig	20580501
	f- Tanezzuft-Benoud Struc/Stratig	20580601

Table 6.3- is a numeric code applied for the Trias/Ghadames province in Algeria, Tunisia, and Libya, including the Tanezzuft-Oued M'ya, Tanezzuft-Melhir, and Tanezzuft-Ghadames **{pubs.usgs.gov/bul/b2202-c/b2202-bso.pdf}**.

The criteria for assigning codes are uniform throughout all publications of the USGS-WPA 2000. The author considers such numeric codes useful for further application of the proposed computerized lithostratigraphic correlation chart of the Arabian Maghreb in northwestern Africa, to be extended and cover North Africa as well.

6.6.3 Structural Settings of the Algerian Sahara

The primary sedimentation of the Saharan platform has been disturbed by deep-seated movements that separated the platform into more or less well defined basins, each with its own specific geologic history and each posing different problems for petroleum exploration.

The Algerian Saharan platform is generally classified into three major stages; the Precambrian, the Paleozoic, and the Mesozoic. The **Precambrian formation** constitutes the basement rock, and is formed of strongly dislocated metamorphosed and granitized rocks outcropping at the Hoggar massif to the far southeast of Algeria. The **Paleozoic** and **Mesozoic formations** include the main reservoir rocks in the Algerian Saharan platform and can be classified into three types of structures **(See Fig 6.4a)**: (1) High structures comprise Hassi Er R'Mel Mole, Hassi Messaoud, Tilrhemt, Rhourd El- Begoul, Tinfoye', etc. (2) Low structures comprise Bechar, Tindouf, Reggane, Oued M'ya, and Rhadames Basins. (3) Structural terraces include the central structure of Illizi, and the structural terrace of Mouydir (**Fig 6.5**). The Paleozoic-Mesozoic reservoir rocks are oil and gas productive in three main regions: the Triassic Province of Oued M'ya-M'zab and Hassi Messaoud to the northwest, the Illizi basin to the southeast, and the Paleozoic basins in isolated regions to the southwest (Tindouf, Bechar, Regane, and Timmimoune.

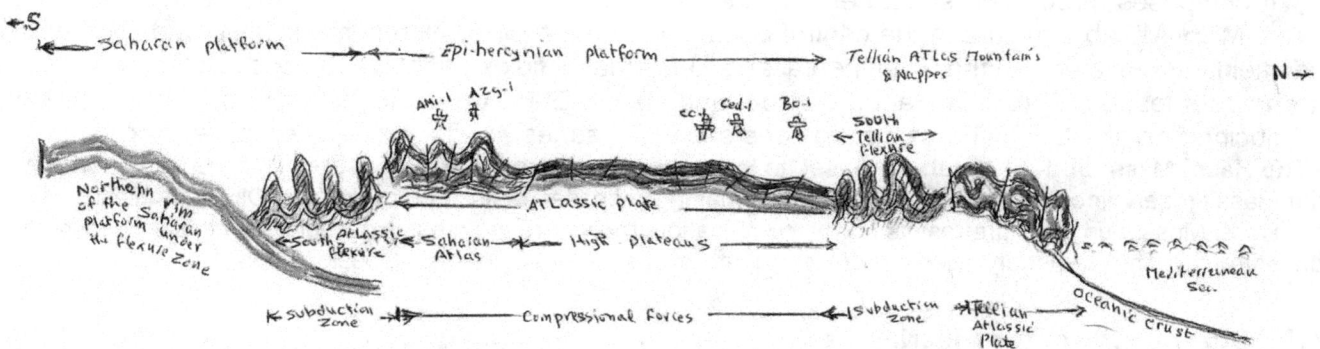

Fig 6.4a - A postulated sketch Diagram showing the Tectonic structures in Algeria

Fig 6.4a -A Postulated Skretch Diagram showing the Tectnic structures in Algeria (Fakhry A. assaasd

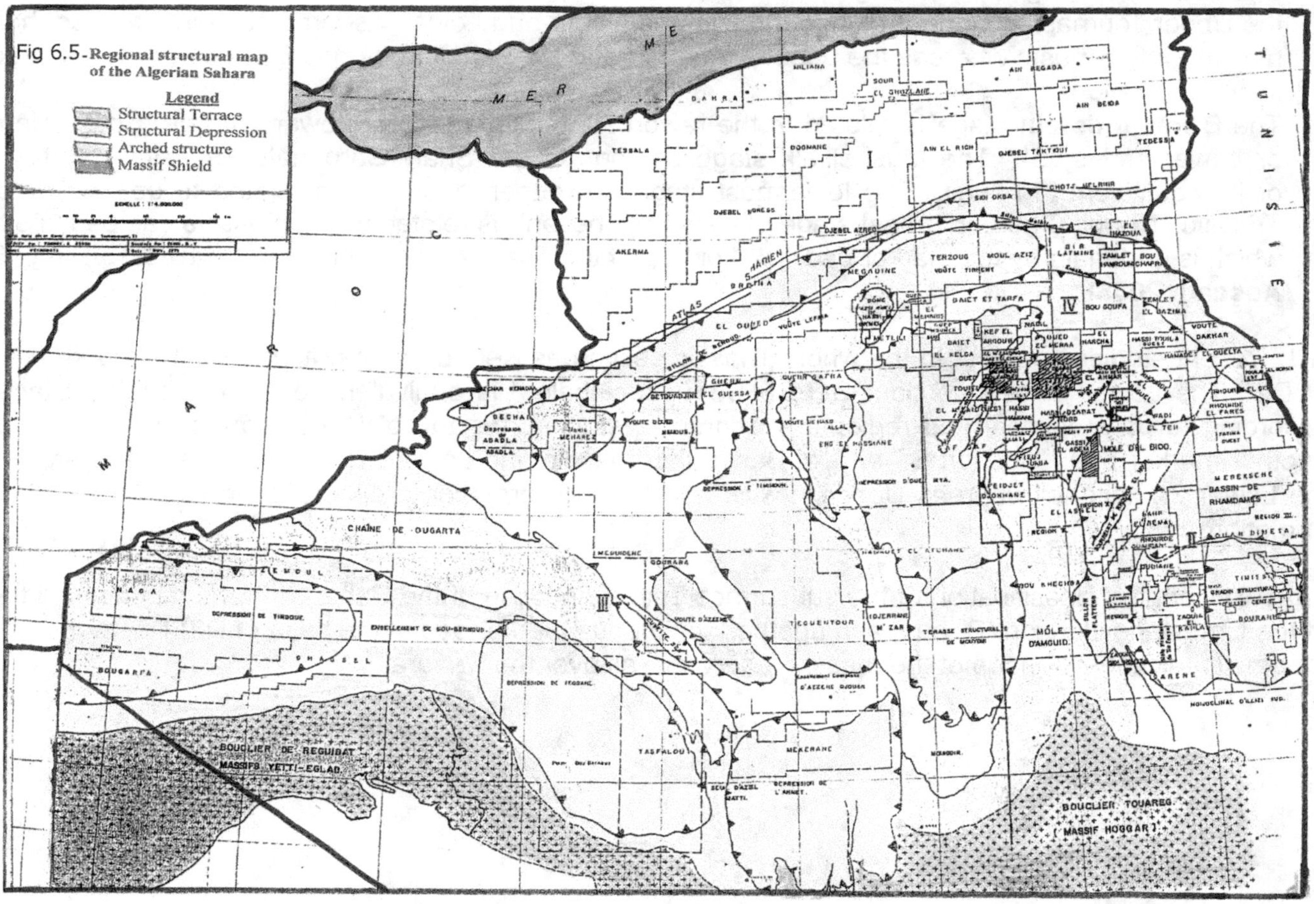

Fig 6.4b- Regional Structural Map of the Algerian Sahara.

-The Triassic Province of Oued M'ya-M'Zab- It was so named because of the thick Liassic evaporite deposits that overlie the Triassic reservoir (according to the latest classification (Stoica and Assaad, 1981).

It comprises three main structures: (1) the Hassi R'Mel Domal structure to the northwest, and the Oued M'ya–M'Zab structure in the central portion, where the Oued Noumer, Ait Kheir, and Djorf oil and gas fields are located; (b) the Erg-Oriental and El-Borma oil fields capped by Triassic limestone, located to the southeast; and (c) the Haoud Berkoui and Haniet El-Mokta oil fields to the east of Oued M'ya, producing from the Lower Triassic sandstone series (SI-series), and capped by eruptive rocks.

- The Hassi Messaoud–El Agreb–El Gassi, to the southeast of Hassi R'Mel, consists of different fields where the massif reservoir of the Cambrian formation is mainly composed of fissured quartzites. Oil accumulation of the Hassi Messaoud structure results from a combination trap where the crest is associated with the Hercynian unconformity. The Cambrian reservoirs ($R_1 + R_2$), are overlain by Late Triassic-Liassic deposits.

6.6.4 Stratigraphy of the Algerian Sahara*

A brief summary of different stratigraphic stages, systems, and series is given in chronological sequence:

1. The **Aschgillian** stage of the upper Cambro-Ordovician formations is not represented; the uppermost rock units are related to the middle and upper Caradocian **(Chart 6.1c)**.
2. The **Carboniferous** section at Illizi Basin (Lower Tournasian stage to Upper Stephanian stage) was classified according to palynological studies; the petroleum reservoirs D6–D8 series are related to the Upper Tournasian stage. At Tindouf basin, the Upper and Lower Visean reservoirs attained their maximum thickness; whereas the upper Numerian and Westphalian formations are not represented in some locations.
3. The **Buntsandstein** stage is related to the reworked detrital series of Lower Triassic (Scythenian) or Lower series "SI." The Mushelkalk stage belongs to the Shaly Carbonate Triassic formations or its equivalent "T1- unit." The lowermost stage of Keuper, "Carnian" is related to upper detrital Triassic "T2-unit. The uppermost stage of Keuper "Rhetian", is related to the massive salt bed "TS4" which is underlain by the marker bed "D2," on top of the Triassic formation **(See Chart 6.1b-Fig 6.5; Assaad, 2006)**.

NB: Recent studies considered the whole Triassic section as Keuper in age, its top is "the marker bed of D2". The Lower Liassic section "S1+S2+S3+upper shales" is equivalent to the Hettangian Stage, according to palynological and regional stratigraphic studies; the top of Hettangian is at the base of another marker bed: "Horizon B." At Illizi Basin, the Triassic and Lower Liassic sections are related to the Lower Zarzaitene, whereas Upper Liassic and Dogger formations represent the Upper Zarzaitine formation.

Fig 6.6- Updated classification of the subsurface Triassic section of the Saharan Platform, showing top of the Limestone Horizon "B" bed, and another continuous marker bed at the base of D2"-limestone bed (of 5ms in thickness) on top of theTriassic reservoirs, all over the Sahara .

Fig 6.5 - OLD AND RECENT CLASSIFICATIONS OF LOWER LIASSIC/TRIASSIC FORMATIONS IN NE CENTRAL ZONE OF ALGERIA

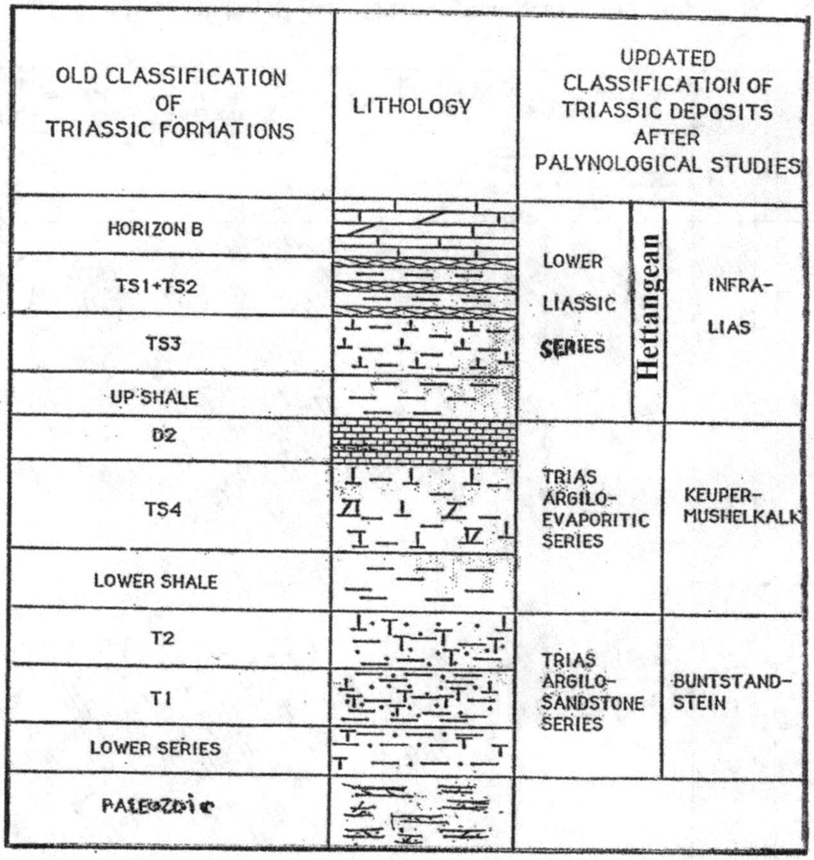

OLD CLASSIFICATION OF TRIASSIC FORMATIONS	LITHOLOGY	UPDATED CLASSIFICATION OF TRIASSIC DEPOSITS AFTER PALYNOLOGICAL STUDIES		
HORIZON B		LOWER LIASSIC SERIES	Hettangean	INFRA-LIAS
TS1+TS2				
TS3				
UP-SHALE				
D2		TRIAS ARGILO-EVAPORITIC SERIES		KEUPER-MUSHELKALK
TS4				
LOWER SHALE				
T2		TRIAS ARGILO-SANDSTONE SERIES		BUNTSTAND-STEIN
T1				
LOWER SERIES				
PALEOZOIC				

1. **Upper Toarcian-Top Barremian (Upper Liassic to Base Aptian)** formations are classified according to regional electric log correlation into six rock units (Units I–VI). The lower evaporitic Liassic section "**Pliensbachian**" is considered as the Upper Evaporite unit. Each of the rock units of the Jurassic-Lower Cretaceous formations shows a complete cycle of sedimentation to the north where the transgression and regression from and to the north occurred, and ended with sea regressions where continental deposits then prevailed.
2. **The Middle and Upper Cretaceous formations "Aptian-Maestrichtean"** are studied to define stratigraphic levels of different rock units, regional structural trends, the evolution and thickness of facies, the environment of deposition and paleo-geographic evolution. The carbonate barrier base of the Aptian formation, overlying the Austrichtean discordance, separates the continental sandstone masses of the Albian and Barremian formations
3. **Miopliocene** eroded surface (Aquitanian to Astian) outcrops in some portions of the Algerian Sahara (**Chart 6.1a**).

(NB: **Chart 6.1**-a complete Lithostartigraphic sedimentary section of the Saharan Platform; available electronically (of 310MB) from the original personal drafted size (~ 3ms x 0.40m) /"**Stratigraphy**".

Chart 6.1a - A Lithostratigraphic Correlation of the Cenozoic Sedimentary section of the Algerian Saharan Platform

Chart 6.1b - A Lithostratigraphic Columnar Section of the Mesozoic "Algerian Saharan Platform"

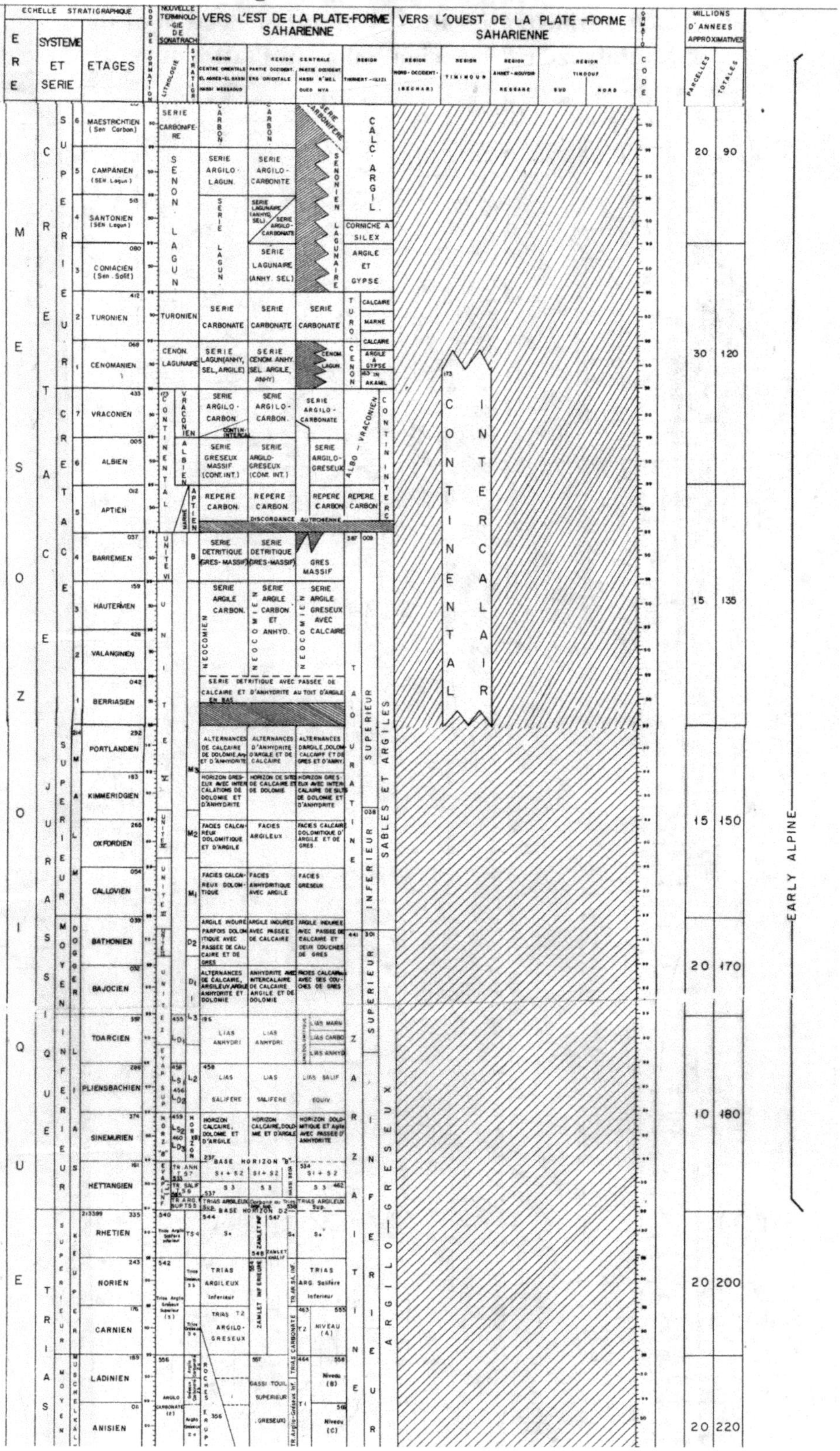

Chart 6.1c - A Lithostratigraphic Columnar Section of the Paleozoic "Algerian Saharan Platform"

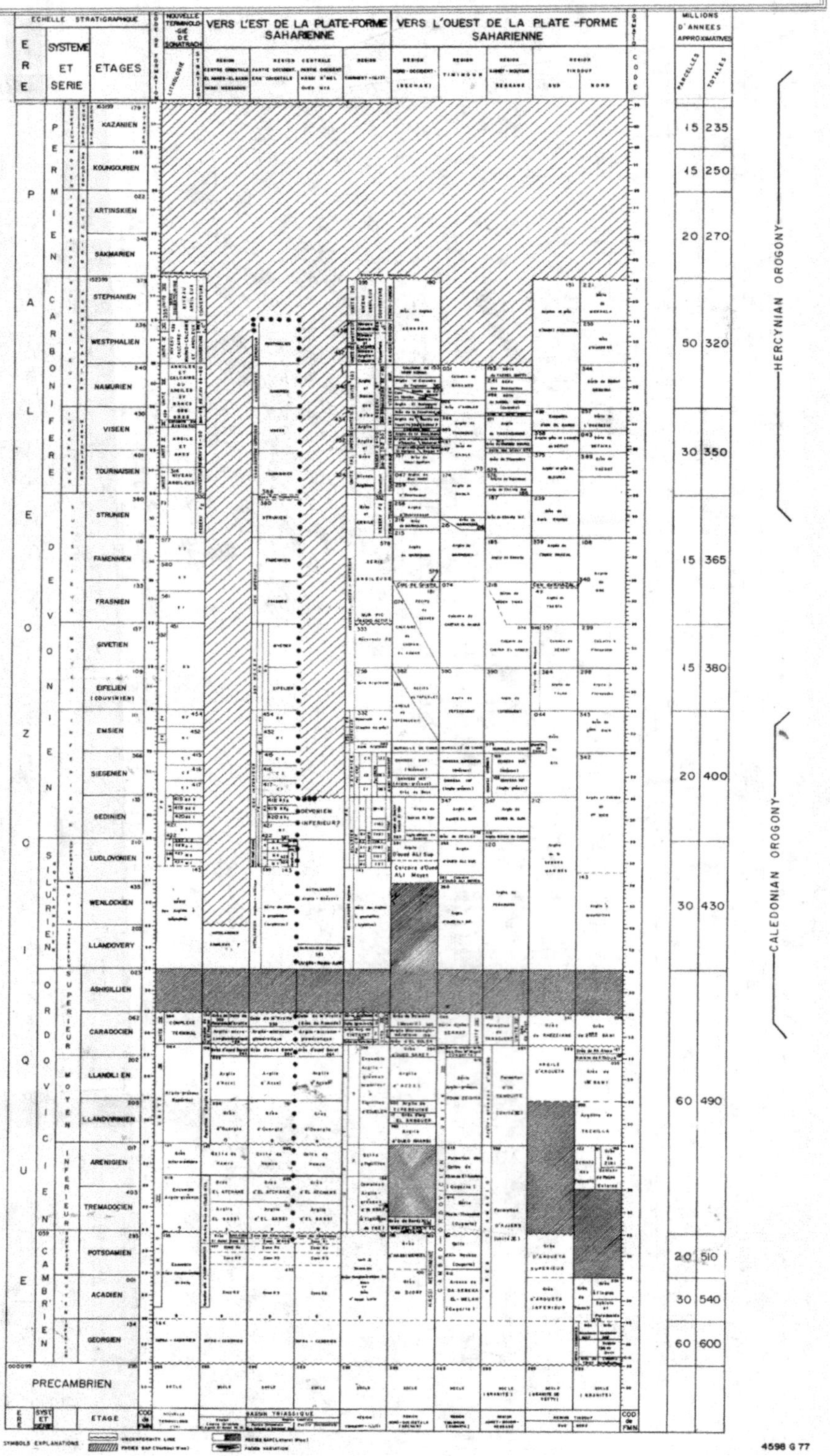

6.7 A Geological Study of an Exploratory well (Ry-1), at the Eastern Border of the Algerian Sahara*

The Rhourde Yacoub petroleum exploration well (Ry-1), drilled in February 1972, is located at the eastern border of the Algerian Saharan platform, and extends to the Tunisian Sahara. The Miopliocene outcrops at the well location, and was completed in the Lower Devonian at a total depth of 4,340m. The Triassic sandstone reservoir yields salt water, whereas the Middle Devonian produces poor gas and gasoline; the Lower Devonian yields saltwater. **Table 6.4** shows the general well data of the exploratory well Ry-1. **Figures 6.7 and 6.8** show the location map of the Ry-1well, and a cross section among nearby wells, (Chebourou, H., 32008; Personal communication) respectively; both figures were the Miopliocene outcrops at the well location, and was completed in the Lower Devonian at a total depth of 4,340ms. The Triassic reservoir yields saltwater; whereas, the Middle Devonian produces poor gas and gasoline; the Lower Devonian yields saltwater. **Table 6.4**-shows the general well data of the exploratory well Ry-1, **figures 6.7 & 6.8** show the location map, and a cross section among nearby wells, respectively; both figures were submitted by the Directorae of Studies and Syntheeses of Sonatrach

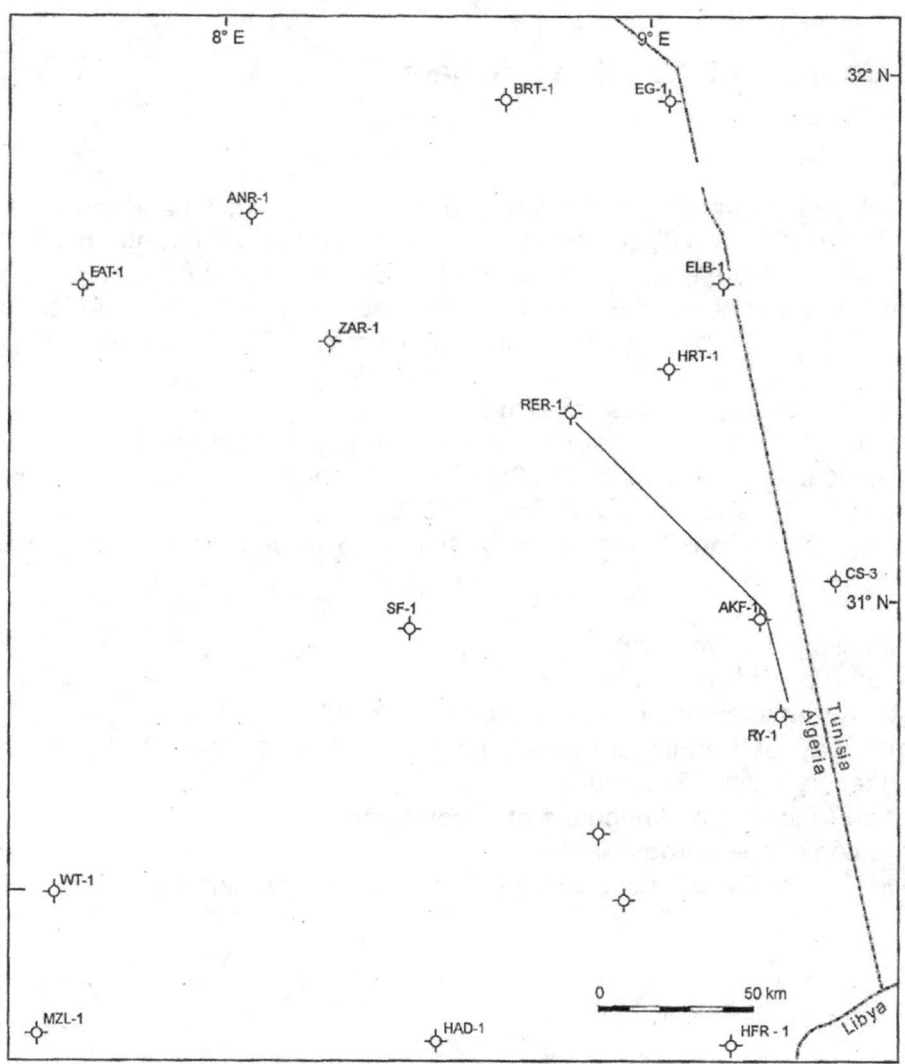

Fig 6.7 – LOCATION MAP OF WELL RY-1 AND CROSS SECTION

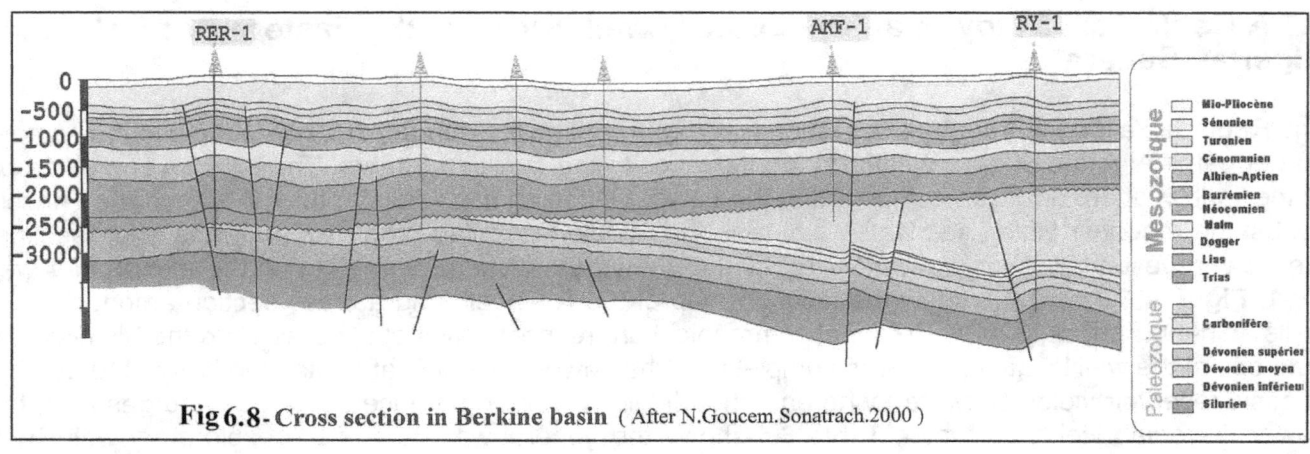

Fig. 6.8 - Cross section in Berkine basin (After N. Goucem. Sonatrach. 2000)

6.8 Petroleum Geology of the Libyan Sahara

6.8.1 Scope
North Africa contains two outstanding provinces: the Sirte basin of Libya and the Triassic/Illizi basins of Algeria; both the Sirte and Triassic/Illizi provinces contain reserves in the range of 30–35 BBOE, placing them amongst the world's largest; together they contain 85% of oil and 80% of gas discovered in North Africa. The African Sahara contains the largest oil and gas fields in the North African region: the Hassi Messaoud and Hassi R'Mel, which were producing approximately 9000 BBO and 60 TCFG, respectively.

6.8.2 Stratigraphy of Sebha Area West of Libya
The Regwa Company, of Cairo, Egypt, carried out a stratigraphical study on the Sebha area. The surface geology of the area, located between latitudes 25°–27°30′ N and longitudes 12°30′–15°E, includes three main Wadis—El-Ajal, Shatti, and Traghen (Werwer* 1973)

The **outcropping formations** range from Cambrian to Recent and are generally interrupted by unconformities:

1. Quaternary deposits (Early Tertiary)
2. Murzuk formation (Lower Cretaceous)
3. Nubian sandstone (undeveloped because of unconformity)
4. M'rar formation-Uwaynat Wannin formation (Upper Devonian to Lower Carboniferous)
5. Memouniat formation (Upper Devonian)
6. Haouaz formation (undeveloped because of unconformity)
7. Hassouna formation (Upper Ordovician)
8. Cambrian-Lower Middle Ordovician rocks (unidentified) and overlain by unconformity.

Table 6.4 General well data of the well Rhourde Yacoub (Ry-1)

Formation tops	Thickness	From	To	Remarks
Mio-Pliocene	(30 m)	00 (279)	35 (+244) m	Friable Sandstone
Senonian Carbonate	(333 m)	35 (244)	368 (−89) m	Calcareous
Senonian Lagunaire	(153 m)	368 (−89)	521 (−242) m	Salt & Anhydrites
Turonian	(89 m)	521 (−242)	610 (−331) m	Limestone
Cenomanian	(174 m)	610 (−331)	784 (−505) m	Brown Shaley ss
Varconian	(26 m)	784 (−505)	810 (−351) m	Dolo. Brown clay
Albian	(95 m)	810 (−351)	905 (−626) m	Sandstone and silt
Aptian	(30 m)	905 (−626)	935 (−656) m	Compact dolomite
Barremian	(141 m)	935 (−656)	1,076 (−797) m	Shaley sandstone
Neocomian	(219 m)	1,076 (−797)	1,295 (−1,096) m	Silty shale and ss
Malm	(197 m)	1,295 (−1,096)	1,492 (−1,213) m	Silt, shale and ss
Dogger Argileux	(74 m)	1,492 (−1,213)	1,566 (−1,287) m	Shale and silt
Dogger Lagunaire	(186 m)	1,566 (−1,287)	1,752 (−1,473) m	Anhydritic Sh. & silt
Lias Anhydrite	(135 m)	1,752 (−1,473)	1,887 (−1,608) m	Anhydrite and shale
Lias Salifere	(49 m)	1,887 (−1,608)	1,936 (−1,657) m	Shaley salt
Horizon (B)	(20 m)	1,936 (−1,657)	1,956 (−1,677) m	Limestone marker bed
Trias evaporates	(115 m)	1,956 (−1,677)	2,071 (−1,792) m	Shaley anhydrite & salt
Trias Carbonate	(66 m)	2,071 (−1,792)	2,137 (−1,858) m	Shaley carbonate
Lower Trs. Sh. ss	(103 m)	2,137 (−1,858)	2,240 (−1,961) m	shaly sandstone
Lower Carbonifere	(679 m)	2,240 (−1,961)	2,919 (−2,640) m	Sandstone, silt & shale
Up. Devonian	(604 m)	2,919 (−2,640)	3,523 (−3,244) m	Sandy Shale
Middle Devonian	(468 m)	3,523 (−3,244)	3,991 (−3,712) m	Calcareous shaly ss
Lower Devonian	(>394 m)	3,991 (−3,712)	TD = 4,340 m	sandy shale

Ry-1 well location:
Ground surface (Zs) = 274.34 m
Rotary table (Zt) = 279.08 m
Coordinates: X = 9° 20' 33" E
Y = 30° 46' 08" N

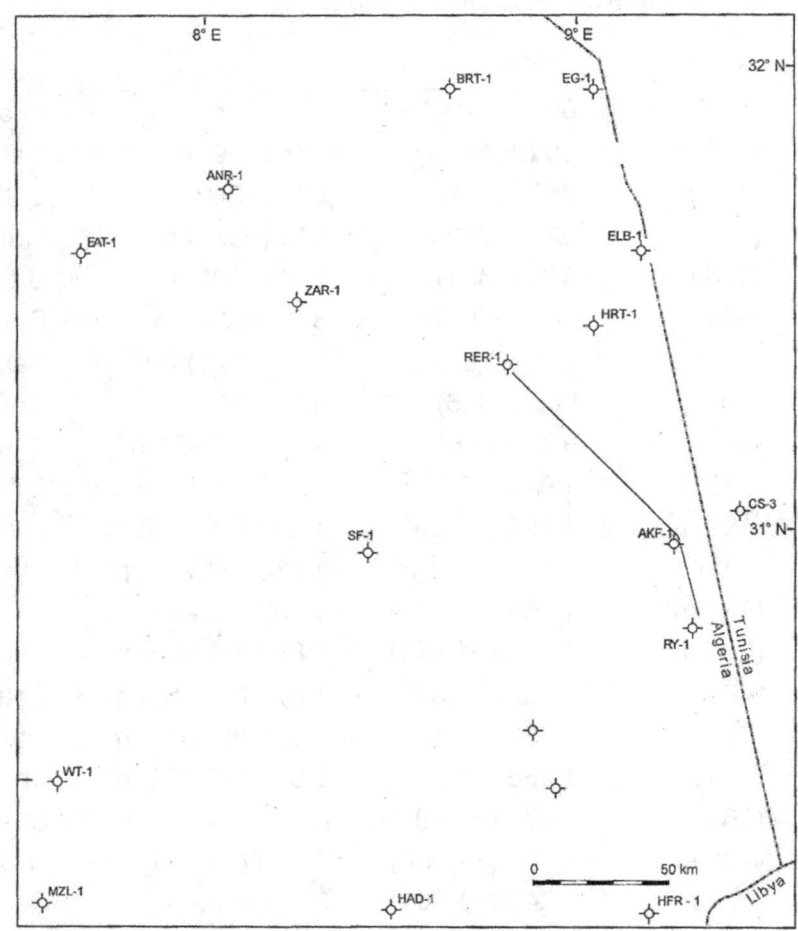

Fig 6.7 – LOCATION MAP OF WELL RY-1 AND CROSS SECTION

Fig 6.7- A location of RY-1 and cross section.

A detailed lithologic description of formations was prepared from driloss ling 39 test wells; vertical and lateral facies changes of rock units were given below:

1. Murzuk formation unconformity
2. Nubian Sandstone ~unconformity
3. Zarzaitine post-Tassilian
   ~Unconformity
       (b)- El-Regeba green clay member
4. Continental post-Tassilian group
5. Tournaisian post-Tassilian group
   ~Unconformity
6. Tiguentourine post-Tassilian group
       (a) Temenrhent red clay member
7. Dembaba formation (Upper Carboniferous)
8. Assedi Jefar formation-M'rar formation (Lower Carboniferous)
9. Uwaynat Wennin formation (Middle Upper Devonian)
10. Cambro-Ordovician formation

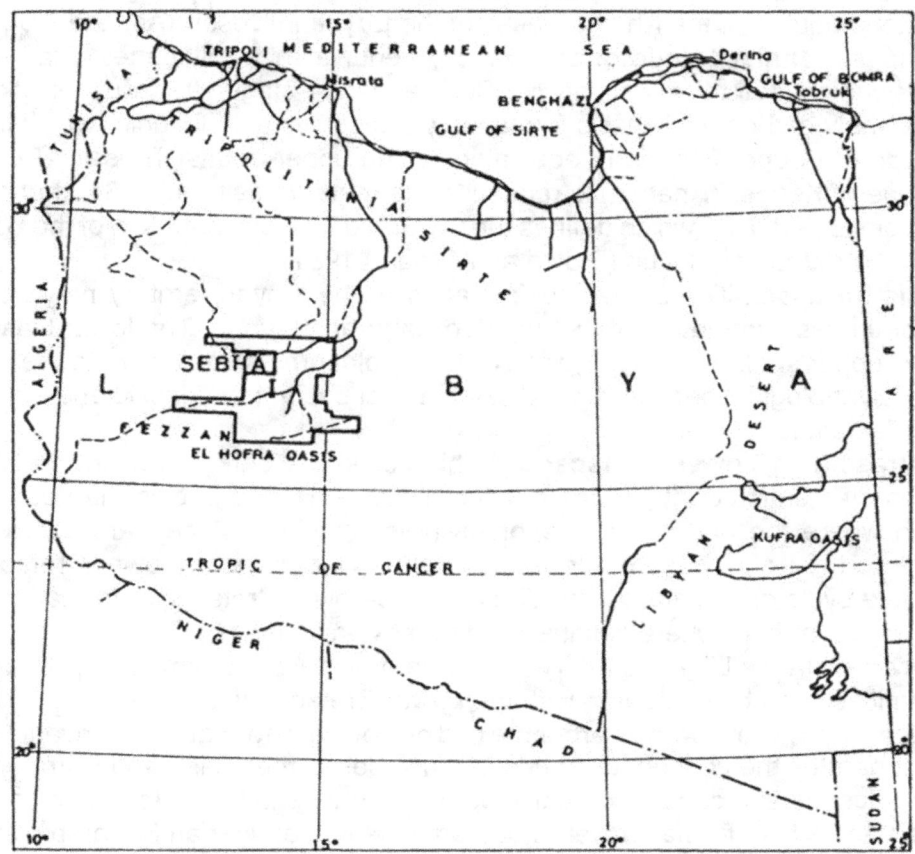

Fig 6.9 - Location map of Sebha Area, Libya

6.8.3 A General Lithostratigraphic Comparison of Sedimentary Section, West of Libya and Northeast Algerian Sahara

The sedimentary section of the Libyan Sahara can be summarized in chronological sequence:

1. **Caradocian of Upper Ordovician**: The Hassouna formation of Libya may be compared with Caradocian of the Upper Ordovician in the Saharan platform.
2. **Frasinian -Givitian of Middle-Upper Devonian**: The Awenat Wennin formation (Middle-Upper Devonian) of the Libyan territory can be compared with the Frasinian-Givitian of the Saharan platform.
3. **Strunian-Tournasian**: The Mrar and Assed Jefar formation of the Libyan territory may be compared with the Upper Devonian-Lower Carboniferous formations (Strunian-Tournasian) of the Algerian Saharan platform.

    The Carboniferous section at the Illizi basin of the Algerian Sahara was classified according to palynological studies as Upper Tournasian. At Tindouf basin, the Upper and Lower Visean reservoirs attained their maximum thickness, whereas the Upper Numerian and Westphanian formations are not represented in some locations.
4. **Stephanian of the Upper Carboniferous**: The Dembaba formation of the Libyan territory can be compared with the Upper Carboniferous (Stephanian) of the Algerian Saharan platform, whereas the Tiguintourine formation is represented in both territories as Upper Carboniferous and is differentiated in the Libyan Territory by two clay members: a lower El-Regeba green clay overlain by Temenhert red

clay—"compare with the variegated shales (Maestrichtian of the Upper Cretaceous?) overlying the Nubian sandstone aquifer of the Kharga oases of the Egyptian Desert (Assaad, 1988).
5. **The Lower Triassic Shaley Sandstone** (Scythian)—equivalent to "SI" series of the Algerian Saharan platform as reworked detrital series of the Buntsandstein stage; the Mushelkalk stage followed upward by the shaly carbonate Triassic formation or its equivalent "T1-unit," which is overlain by the lowermost stage of Keuper (Carnian), equivalent to the upper detrital Triassic "T2-unit" series. The uppermost stage of Keuper (Rhetian) is related to the massive salt bed "TS4" that directly underlies the marker dolomite bed "D2," which defines the top of the Triassic formation of the Saharan Platform, according to an updated correlation (Stoica and Assaad 1981).
6. **Liassic-Middle Jurassic**: The Zarzaitine formation of the Libyan territory may be compared with the Lias-Upper Jurassic formation of the Algerian Saharan platform. The lower Liassic section "TS1 + TS2 + TS3 + upper shales" of the Algerian Saharan platform is equivalent to the Hettangian stage according to Palyanological bed "Horizon B." At the Illizi basin, the Triassic and Liassic sections are related to the Zarzaitine formation.
7. **From Top Jurassic to Lower Cretaceous**: This section is only known in the Algerian Saharan platform and is differentiated into six rock units, "Units I-VI," and is classified according to electric log correlation where the underlying evaporites (Pliensbachian stage), are considered the Upper evaporite unit. To the north, each of these rock units of the Jurassic-Lower Cretaceous formations show a complete cycle of sedimentation. The Jurassic-Lower Cretaceous formation ended with sea regression toward the north where continental deposits prevailed.
8. **Lower Cretaceous**: In the Libyan territory, Touratine and Marzouk formations, separated by successive unconformities, may be compared with the Lower Cretaceous of the Algerian Saharan platform.
9. **Middle Cretaceous**: The lowermost carbonate bed of the Aptian formation that overlies the Austrichtian discordance separates the continental sandstone masses of the Albian and Barremian formations of the Middle Cretaceous and can be compared with the continental post-Tassillian group of the Libyan territory. Aptian and Albian formations are good aquifers in the Algerian Saharan platform.
10. **Mio-Pliocene Outcrops**: The eroded surface of the Mio-pliocene outcrops in some parts of the Algerian Sahara. (**N.B**: A precise stratigraphic comparison is highly recommended).

6.9 The Egyptian Territory (Northeast Africa)

There are three main areas of petroleum interest in Egypt: the northern part of the Western Desert, The Nile Delta, and the Gulf of Suez province.

6.9.1 The Western Desert of Egypt *

The geology of the Western Desert of Egypt is greatly different from that of the Western Arabian Maghreb region. The Western Desert of Egypt includes the most fertile desert land and embraces an area of 681,000 km^2, or more than two-thirds of the Egyptian territory. It is one of the most arid regions and hottest places in the world; the sources of water are found hundreds of km apart; it is essentially a plateau with extensive areas of rubbly rock surface, covered in some places by long swaths of sand dunes.

Active oil exploration in the vast area of the Western Desert started in the mid 1950s by the Sahara Petroleum Company (SAPETCO).

The detailed subsurface structural synthesis in the northern part of the Western Desert of Egypt became of great interest for its petroleum potential as it comprises the second most promising locality in Egypt after the Gulf of Suez province where the first drilled well was in 1908

Table 6.5 -Formation tops from micro-palaentological results in some wells at Sabh area

Well No.	Depth	Age
12	600–606	Lower Cretaceous
12	669-672	Lower Jurassic
12	686-692	Jurassic
9	400–406	Jurassic
9	501–507	Lower Carboniferous
23	218-224	Upper lower
23	358-364	Carboniferous
23	403–409	Carboniferous

A Review –Oil and Gas fields in Egypt*

Oil and gas fields in Egypt were found in the following three areas: Two oil and gas fields were studied in El-Razzak and the Abu Gharadig area, and the third oilfield in the Meleiha area; the two petroleum areas are located between latitudes 29°30' –30°40' N and longitudes 28°22' –28°38' E, covering about 180 km² and 240 km², respectively; the stratigraphic section in the northwestern desert of Egypt ranges in age from the Cambro-Ordovician to Recent. The Quaternary, Pliocene, and Miocene sediments are exposed on the surface and form the blanket over the subsurface section that has been penetrated by 14 wells in the El-Razzak field area and 18 wells in the Abu-El-Gharadig field area **(Fig. 6.10-** (Metwalli et al. 1979).

Most of the wells drilled in the area were completed in the Lower Cretaceous deposits. The reservoir rocks in both fields, of Aptian, Cenomenian, Turonian clastics and carbonates, generally include a relatively marine and non-marine series of sedimentary rocks that overlie the basement rocks, and vary in thickness from one locality to another. A detailed study of the structural modeling of the Abu-Gharadig and El-Razzak oil and gas fields has shown that both structures have changed in shape, size, or amplitude and shifted their position laterally as a result of repeated folding stages that affected the area during the Cretaceous age.

*M. Hamed Metwalli, George Philip, A.M.A.Wali (1979)

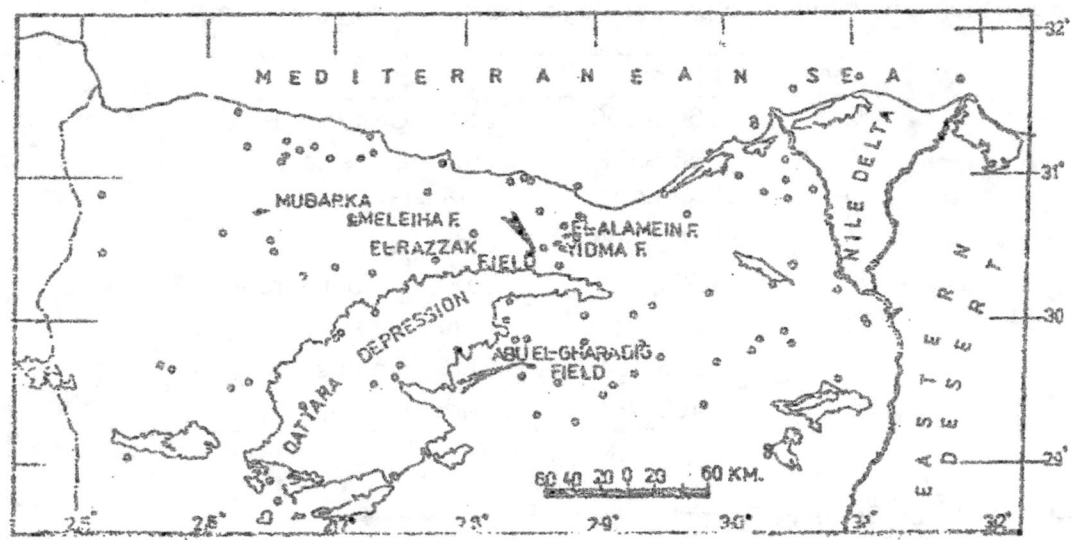

Fig 6.10- Location of El-Razak and Abu Gharadig Oil and Gas Fields, northwest of the Western Desert of Egypt

A third field at the Meleiha oil fields, produces oil from the Upper Bahariya formation (Upper Cretaceous/ Lower Cenomanian). The field area covers 700 km², at latitude 30°36' –30°54' N and longitude 27°00' –27°18' E. The Lower Cenomanian (the upper part of the Bahariya formation) is the main producing zone in the Meleiha oil fields. The study of lithofacies, structural features, and electric logging indicates that the sediments were deposited in tidal-flat, near-shore, shallow marine environments.

The main depressions of the Western Desert are: Kharga-Dakhla, the Bahariya-Farafra oases, the Quattara-Siwa, and the Fayium and Wadi El Natrun depressions; the total area of the oases is 75,165 km², and extend between latitudes 24°30' –30°00' N and longitudes 28°30' –30°35' E; each has its own tradition and dialect.

6.9.1.1 The Qattara-Siwa Depression

The Qattara-Siwa Depression occupies about 19,000 km² in the northwestern desert hewed into Miocene rocks. The Qattara depression is the deepest and largest, where the floor descends at its lowest point to 134 m below sea level, and is very suitable for establishing an electric power system from a waterfall that could be generated from the Mediterranean Sea, expecting therefore a great successful project in future (during the Nowadays Era of the so called Middle East Spring).

To the north of Qattara rises a third plateau, averaging 200 m above sea level, which slopes northward toward the coast of the Mediterranean between Alexandria and Sallum. The Qattara Depression contains only salt lakes and salt marshes and is consequently uninhabitable.

There is a high plateau of Nubian sandstone in the south, extending from the mountains of Uwaynat (over 1,800 m), which descends slowly till it reaches the depression of both the Kharga and Dakhla oases. To the north of the depression, a limestone plateau (500 m above sea level) extends to the depressions of Farafra and Bahariya. The surface of the plateau slopes northward and ends in a great depression, some parts of which are below sea level (e.g. Siwa and Qattara).

There are various places in the depressions of the Western Desert where long parallel lines of high sand dunes extend in a north-south trend, sometimes for great distances. The most famous sand dunes are the eastern series on the line of the oases, extending about 700 km to the south of Kharga.

6.9.1.2 The Bahariya Oases

The Bahariya oasis is a large natural excavation. It is entirely surrounded by escarpments and has a large number of isolated hills within the depression. The general shape of the depression is oval, with its major axis running northeast, with a narrow blunt extension at each end. It is situated on the stable-unstable shelf hinge; its principal structural feature is a very broad anticlinal fold, with its axis trending southeasterly from Gebel Ghorabi in the north and seeming to continue south to include the Farafra structure. The topography of the Bahariya oasis is characterized by a large number of hills within the depression. The most strongly marked group of hills extends in a straight northeast direction, cutting the oasis in two. The largest hill of the range is Gebel Hefhuf, a narrow, ridge-like hill of limestone **(Fig. 6.11-Balland Beadnell, 1903)**.

The basement in the Bahariya oases was reached at a depth of 1,825 m, whereas the Paleozoic sediments are unconformably overlain by the Mesozoic and by later sediments of shallow marine deposits that are relatively thin.

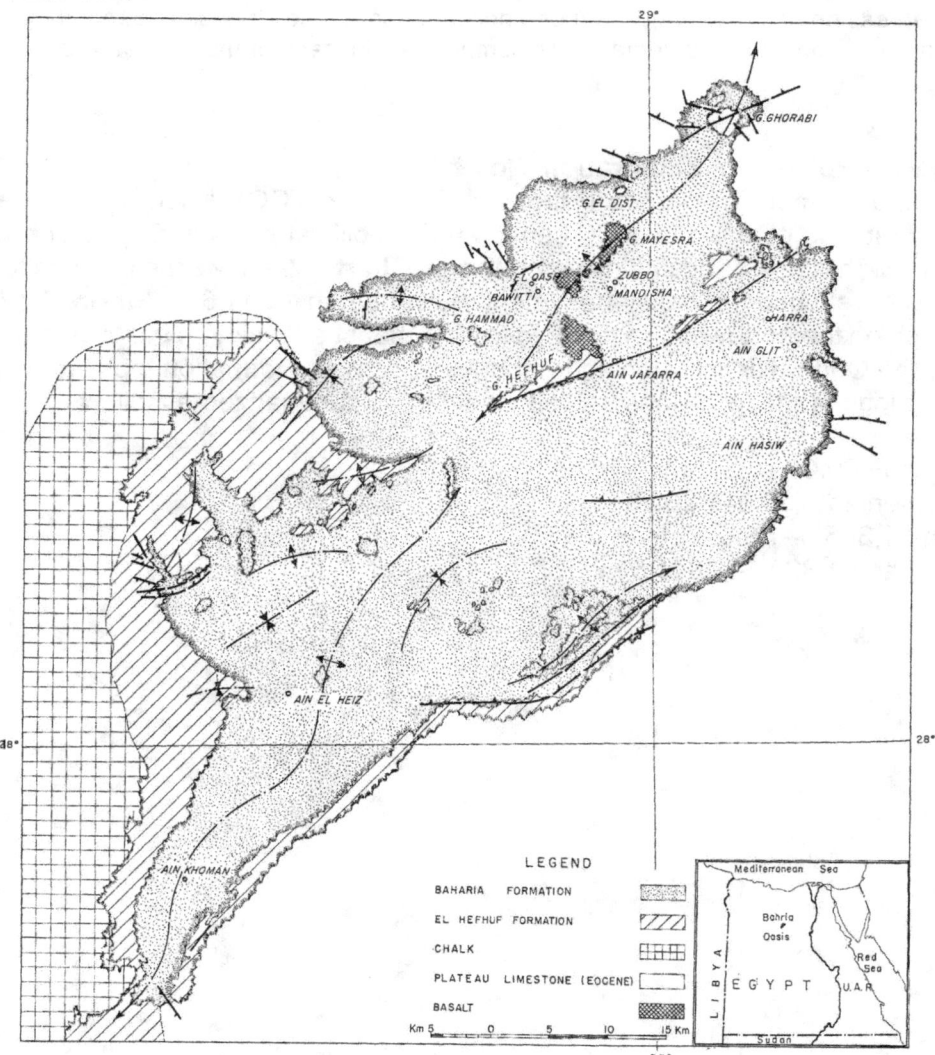

Fig6.11 - A Geological Map of Bahariya Oases, Western Desert, Egypt
(Ball and Beadnell (1903)

6.9.1.2.1 The Sedimentary Section

The stratigraphic rock units that make up the geology of the Bahariya oasis of the Western Desert of Egypt are described in chronological sequence:

1. **Dolerite intrusions**—mainly of laccolithic dolerites, which cap Gebel Mendisha and Gebel Maysera and are intruded as sill bodies in the Cretaceous rocks of Gebel Hefhuf.
2. **The limestone plateau**—mainly of nummulitic limestones of the Lower Middle Eocene age, is unconformably overlying the Bahariya formation both in the wall of the escarpment and in the isolated hills within the depression.
3. **Chalk formation**—comprised of chalky limestones that belong to the chalk rock unit and are unconformably overlain by the Hafhuf Formation.
4. **El-Hafhuf formation**—mainly brown to gray crystalline limestones in its lower part and variegated shales and sandstones in its upper part. It occurs in series of hills at the center of the depression, situated more or less along a fault line that crosses the oasis diagonally.
5. **Bahariya Formation**—mainly of sandstones and variegated shales, forming the floor of the oasis and parts of the wall of the escarpment. In the northern part, the sandstones are overlain immediately by the Eocene plateau limestone. In the south they are followed by El-Hafhuf limestones and variegated shales. In the isolated hills within the depression of the Bahariya, sandstones and variegated shales are capped either by the Eocene plateau limestone or by basalts or dolerites. The composite thickness of the exposed part of the unit is about 300 m.

6.9.1.2.2 El-Bahariya Petroleum Exploration Well # 51

In the early 1950s, the American Sahara Petroleum Co. (SAPETCO) drilled an exploration well in the Bahariya oasis, but it was plugged in 1956 because of a political crisis in Egypt. The well was mainly drilled in sandstone with little silt and shale intercalations. **Chart 6.2-** shows the composite log of Bahariya exploration well-51; it reached the basalt basement rock at depth 1,841.6 m. **Figure 6.12** shows a cross section from Matruh on the northwestern coast to the southeast at Assiut on the Western Bank of the Nile River, passing by the Qattara depression and Bahariya oases. The formation tops are defined by gamma rays and resistivity logs at the following intervals and can be summarized as follows:

Cenomenian : 00.00–705.6 m
Pre-Cenomenian: 705.6–1,365.5 m
Cambrian : 1,365.5–1,822.7 m
Pre-Cambrian : 1,822.7–1,841.5 m (TD)

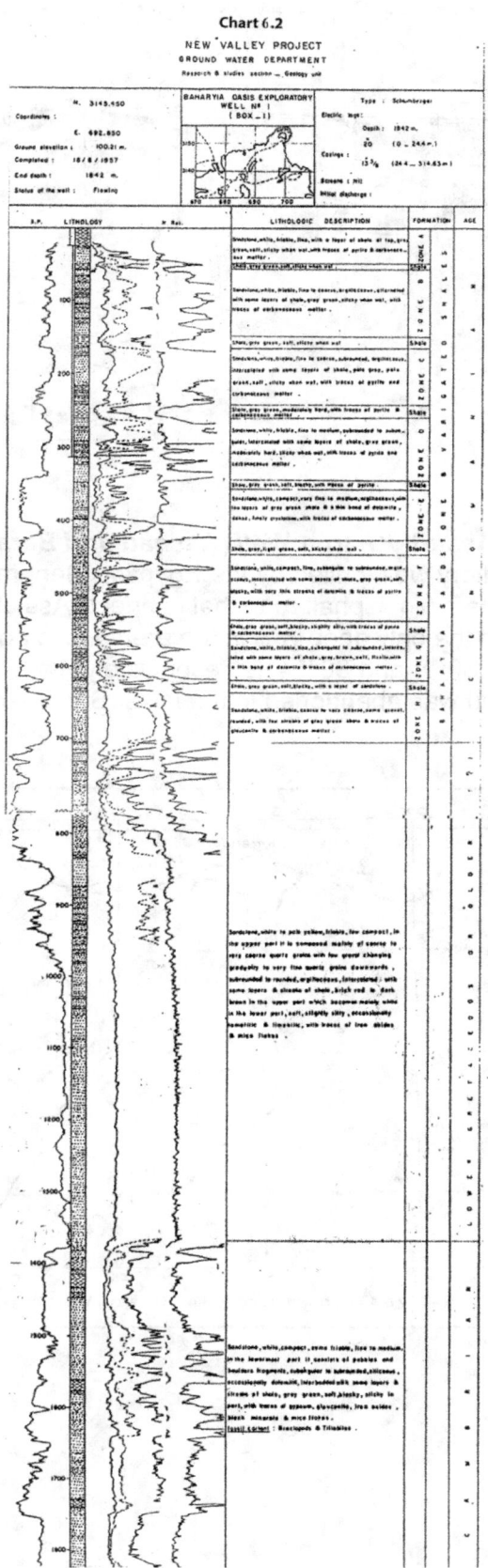

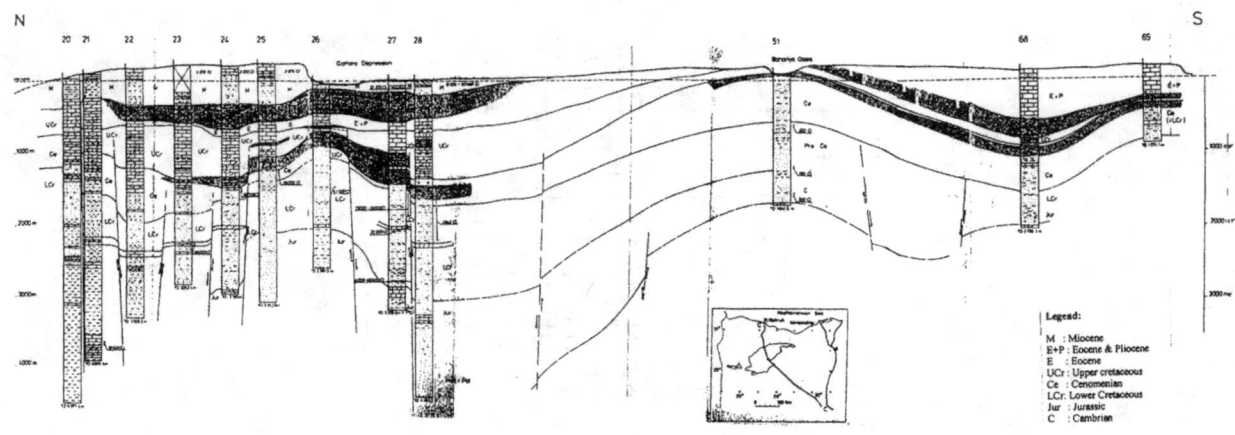

Fig 6.12- Generalized Cross Section : Mersa Matruh - Qattara Depression – Bahariya – Assuit

6.9.1.3 Geological Results of Assiut- Kharga Well (Assaad and Barakat, 1965)

The Assiut-Kharga deep exploratory well, drilled by the Egyptian General Desert Development Authority (EGDDA), is nearly midway on the main asphalt road that connects Assiut city on the River Nile valley and the Kharga oasis **(Fig. 6.13)**. It is the only deep well in the area and is about 120 km southwest of Assiut. Its latitude and longitude are 26°30' N and 30°54' E, respectively. The total depth reached at 1,140.5 m spudded at the lower Eocene and was abandoned in the lower Senonian due to pipe stuck **(Chart 6.3)**

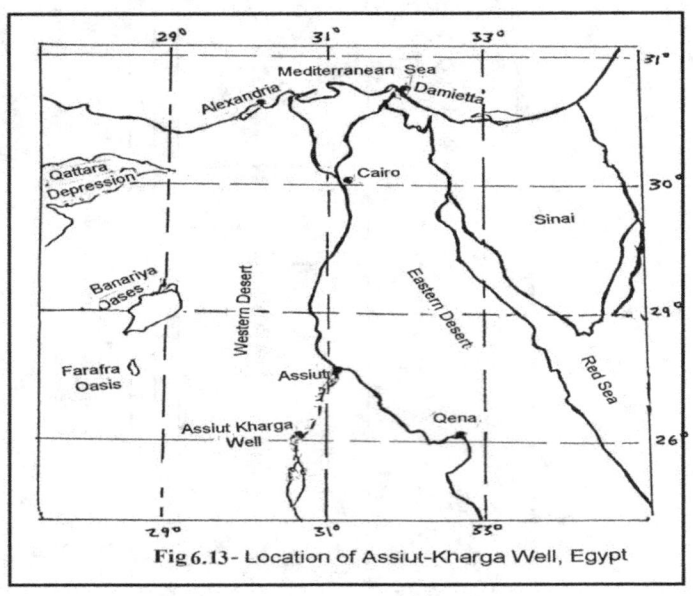

Fig 6.13- Location of Assiut-Kharga Well, Egypt

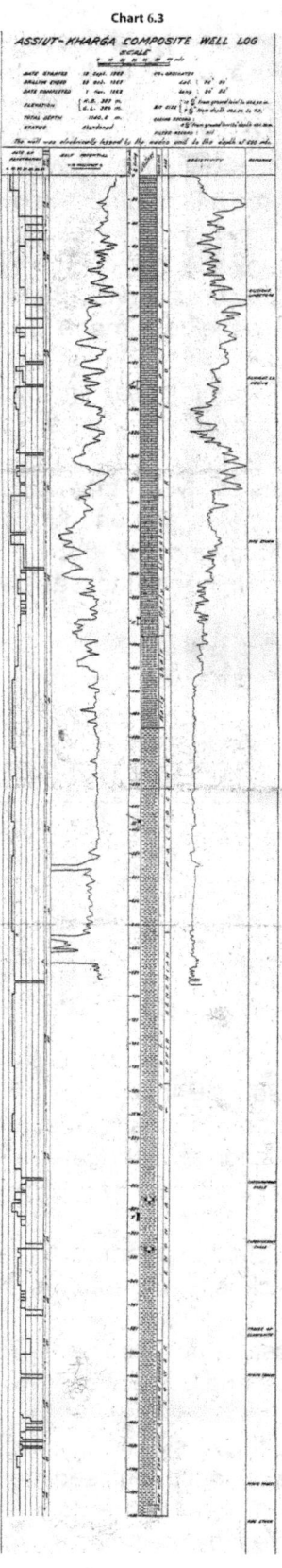

Chart 6.3

6.9.1.3.1 Discussion

Eocene beds outcrop at the well site area and are mainly non- clastics, hard limestone, partially dolomitized, gray to yellowish, and are intercalated by flint nodules. The beds are similar to those exposed in the Gebel Drunka section to the southwest of Assiut. It was estimated that the thickness of the whole Eocene section (particularly the Lower Eocene) in the vicinity of the well site ranges from 300 to 400 m.

Bishay and Tawfik (General Petroleum Co. 1965) described similar beds exposed to the east of Assiut on the other side of the Nile which are differentiated by two formations: a lower non-fossiliferous unit of 180 m of chalk marly limestone with intercalations of flint known as the "Assiuty chalk formation." The upper unit is nummulitic crystalline limestone of 350 m thick, known as the "Manfalout formation." To the northwestern side of the Kharga depression, a section of Gebel El-Tir made up of upper Cretaceous beds is overlain by 10 m of lower Eocene, and is estimated of 138 m in thickness.

No detailed geophysical investigations have been made to cover the area between Assiut and Kharga. A magnetic survey was only carried out in the Wadi El-Assiuty area northeast of Assiut; it was concluded that the basement complex is a NE-SW trend of a local high or a dome structure complicated by faulting. The thickness of the sedimentary section overlying the basement complex ranges from 500 m in the highs to 3,000 m in the lows.

The stratigraphic description and facies analysis of the subsurface section are mainly based on ditch samples which were collected every one meter, and on four core samples recovered from the following depths: 378–384 m, 551–557 m, 799–800 m, and 882–888 m.

The stratigraphic boundaries were determined at the General Petroleum Company (**Table 6.6**; Tawfik, personal communication).

Table 6.6 Stratigraphical units in Assiut-Kharga well

Stratigraphic unit	Depth from Derrick From (m)	To (m)	Top to sea level (m)	Thickness (m)
Lower Eocene	000	430	+320	430
Paleocene	430	648	−110	218
Upper Senonian	648	775	−328	127
Lower Senonian	773	1,140.5	−455	365.5

6.9.1.3.2 Local Correlation

The stratigraphic formations encountered in Bahariya well #51 are correlated with those of the Um El-Kosour deep borehole located in the northernmost part of the Kharga depression.

The uppermost beds of the Upper Senonian variegated shales ae about 40 m thick, overlying the Nubia sandstone of 715 m thick, and are bottomed by weathered basement complex, drilled in granite wash for 20 m.

The Nubia sandstone at Gebel Um El-Ghanayem is 40 m thick and is overlain by the Paleocene shales and chalk that reach 106 m in thickness; the topmost Nummulitic limestone is 94 m thick of the Lower Eocene.

The Nakb Assiut escarpment (or Assiut pass) to the far east of the Kharga oases chronologically comprises of Upper Senonian, Paleocene, and Lower Eocene beds of 107.5 m, 143 m, and 36 m, respectively **(Fig 6.14, Assaad, et al.,1965)**.

From the stratigraphic point of view, the Assiut well occupies a site on the fringe of a sedimentary basin that developed under the influence of continuous sedimentation during Eocene–Upper Cretaceous times. Increase in thickness is expected further to the northeastwards. The possibility of encountering marine Jurassic, Triassic, and upper Paleozoic sediments like those recorded by the deep exploration wells in the Western Desert, is still realistic. The basin is the extension of the Nile Valley basin from Beni-Suef in the north to "nearby" Assiut in the south. It seems to offer good oil prospects, because black shales are well developed in the Lower Eocene, Paleocene, and Upper Cretaceous sediments, and could act as the mother rock of petroleum hydrocarbons, just as in the case of the Paleocene shales of Libya. Oil could be trapped in the structural highs of the fissured limestone of the Lower Eocene. Thus, further drilling is strongly recommended to delineate clearly the limits of the basin and define its facies changes both laterally and vertically.

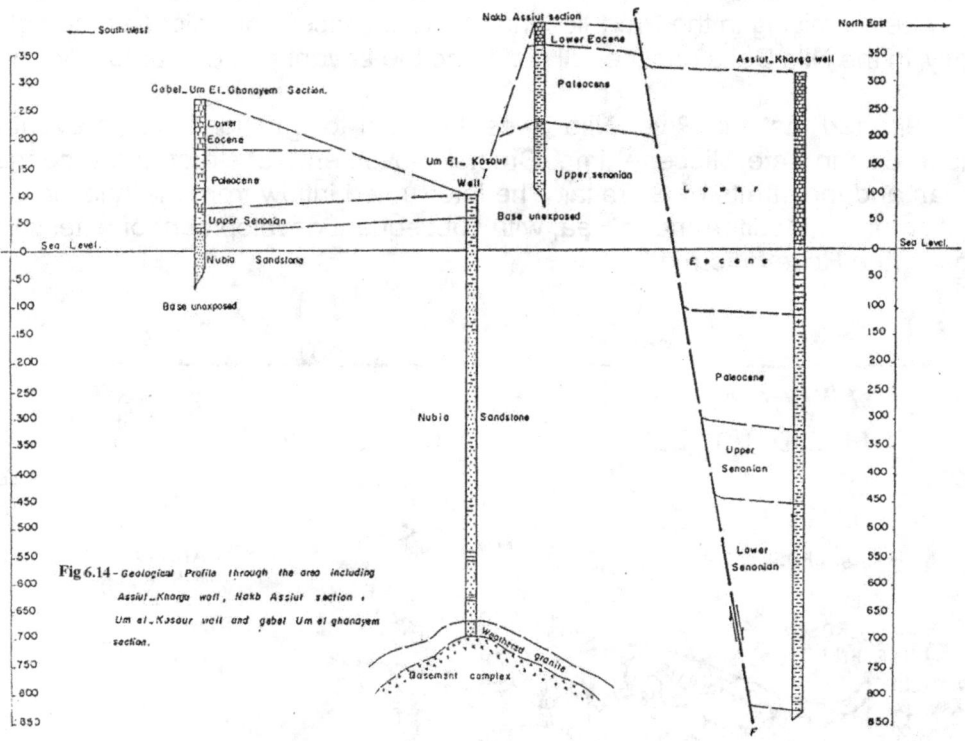

Fig 6.14 - Geological Profile through the area including Assiut-Kharga well, Nakb Assiut section, Um el-Kosour well and gebel Um el ghanayem section.

6.10 The Nile Delta of Egypt*

The Nile Delta stretches 170 km in a north-south trend, attaining a base of 220 km along the Mediterranean Sea; it covers an area of 22,000 km² onshore, and about an equal area in the offshore continental shelf and slope (Metwalli et al., 1978; **Fig. 6.15**). The Nile Delta represents a consequence of the conflict between the River Nile and the Mediterranean Sea. An understanding of the interaction of the Nile Delta together with the southern Mediterranean, provides the key that unlocks the possibilities and vision of significant petroleum reserves hiding in the delta area (on-shore and off-shore).

The Nile Delta, like the Niger Delta, is a product of approximately equal contributions from riverine and wave forces (fluvial-marine interaction). Continuous beach and beach-ridge formations fringe the coastlines of both deltas but are most extensively developed along the coast of the Nile Delta.

The episodical fracturing of the overlying salt deposits of the Miocene clastics permitted the hydrocarbons to migrate towards the continental shelf and the Nile Delta body mass. The high pressure source areas might represent abnormally pressured hydrocarbon compartments that are the source of lateral and vertical migration of hydrocarbon gases into the normally pressured Pliocene clastics, which form significant reservoir-traps in the northern Delta offshore area.

*By M.H. Metwalli, G. Philip, El Sayed, and A. Yousef

The Tertiary Nile Delta is not a true basin but later overprints onto other basin types which represent the interaction of the coastal setting of the Mediterranean Sea. It is temping to visualize the Nile Delta as an echo of the rifting of the Red Sea and the Gulf of Suez. It might be a result of a triple tectonic junction (rifting) from different tectonic sources in time and space, reflecting: (1) the Red Sea, the Gulf of Suez, and Gulf of Aqaba rifting trends;

(2) The Western Desert tectonics echo; and (3) the acquired post-rifting Atlantic margin from the north, the "Tythes and the Mediterranean ridge" (Metwalli 2000).

The Nile Delta area produces very high hydrocarbon gases (biogenic and thermogenic), yet crude oil prospects may be promising in the Pre-Miocene offshore concessions close to the delta's depocenter and consequently, in the Nile Delta continental cone and the Levant platform areas (Metwalli et al., 1978; **Fig. 6.16**).

Saiid (1981) reported that the River Nile owes its origin to a major tectonic event which affected the Mediterranean Sea in Late Miocene time. Crustal movements obstructed the connection between the Atlantic Ocean and the straits of Gibraltar. The interrupted inflow from the Atlantic Ocean led to the gradual desiccation of the Mediterranean Sea, with subsequent development of extensive thick salt and evaporitic deposits (the Rosetta formation).

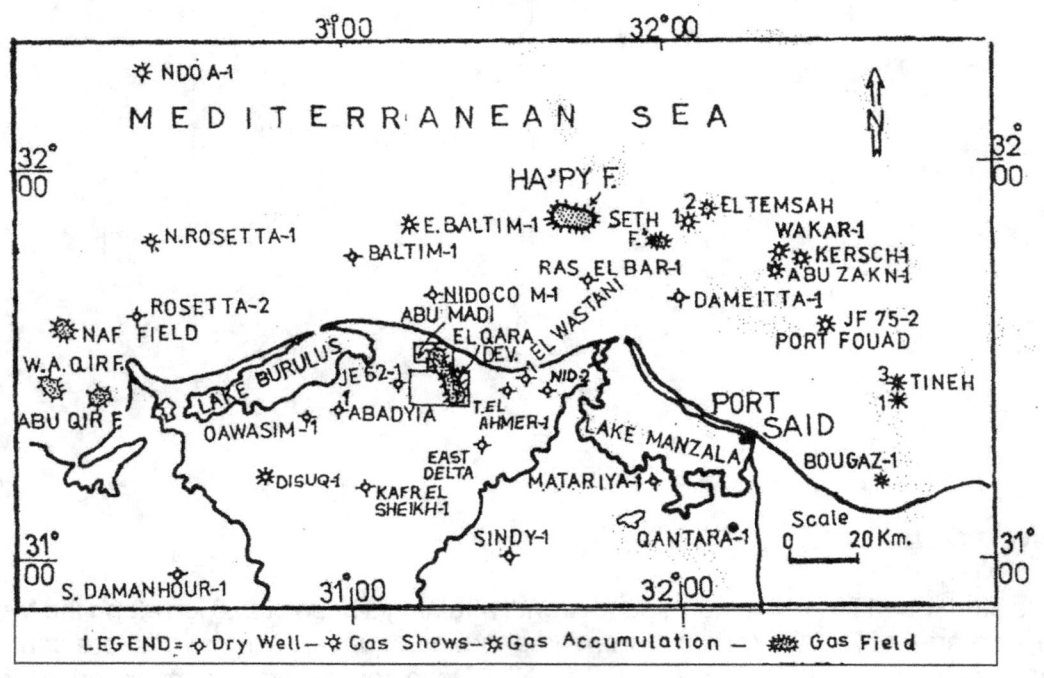

Fig 6.15-A Location Map of Northern Nile Delta, Egypt

(Metwalli, 1978)

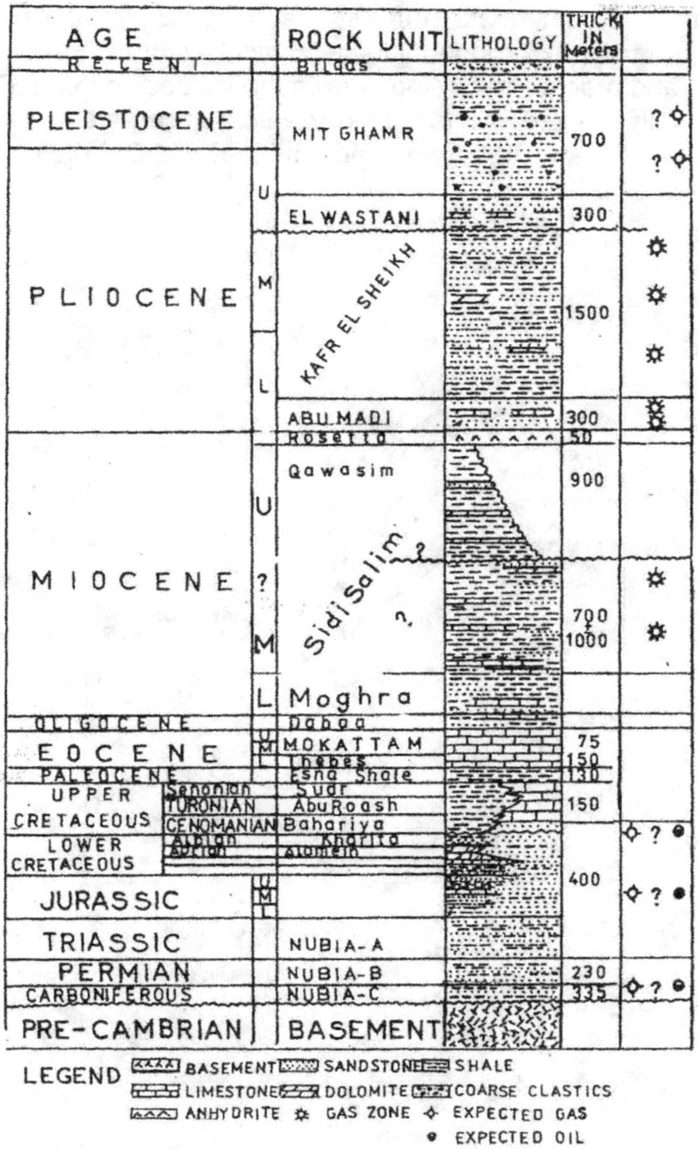

AGE		ROCK UNIT	LITHOLOGY	THICK IN Meters	
RECENT		Bilqas			
PLEISTOCENE		MIT GHAMR		700	? ◇ ? ◇
PLIOCENE	U	EL WASTANI		300	✿
	M	KAFR EL SHEIKH		1500	✿ ✿
	L	ABU MADI		300	✿
		Rosetta		50	
MIOCENE	U	Qawasim		900	
	?	Sidi Salim			✿
	M			700 1000	✿
	L	Moghra			
OLIGOCENE	U	Dabaa			
EOCENE	M	MOKATTAM		75	
		Thebes		150	
PALEOCENE		Esna Shale		130	
UPPER CRETACEOUS	Senonian	Sudr		150	
	TURONIAN	Abu Roash			
	CENOMANIAN	Bahariya			◇ ? ●
LOWER CRETACEOUS	Albian Aptian	Kharita Alamein		400	◇ ? ●
JURASSIC	U M				
TRIASSIC		NUBIA-A			
PERMIAN		NUBIA-B		230	◇ ?
CARBONIFEROUS		NUBIA-C		335	
PRE-CAMBRIAN		BASEMENT			

LEGEND: BASEMENT, SANDSTONE, SHALE, LIMESTONE, DOLOMITE, COARSE CLASTICS, ANHYDRITE, ✿ GAS ZONE, ◇ EXPECTED GAS, ● EXPECTED OIL

Fig 6.16 - GENERALIZED HYDROCARBON-BEARING LITOSTRATIGRAPIC COLUMN OF THE NILE DELTA WITH INFERRED OLD TERTIARY AND PRE-TERTIARY SEQUENCES (MODIFIED AFTER SCHLUMBERGER, 1984)

6.11 The Gulf of Suez, Egypt

The Gulf of Suez, though significantly smaller in terms of total reserves, yet it is considered the third most significant province. It lies within the stable belt of Egypt. It runs in a NW-SE direction, following the Erythrian trend of faulting, and forms an elongated depression separating the massifs of central Sinai from those of the backbone of the Eastern Desert (Fig. 6.17 & 6.18- Metwalli et al., 1978), Fig 6.18).

The coastal strip along are the Gulf of Suez region that lies between the waters of the Gulf and trment the major African trending faults, represents one of the most promising areas for the finding of future oilfields. Saiid (1962) stated that bordering of the Gulf of Suez depression on both sides, are two marrginal faulting zones.. Sittuated within the stable belt of Egypt, the Gulf of Suez has been an active zone of subsidence (or trough) throughout its geological history . Great thicknesses of sedimernts accumulated in the trough and practically all stages from the Paleozoic to the Recent are reprsented by regional faulting zones, usually marked by lines of high vertical Escarpments on the upthrown sides. The sucessive Paleoozoic --Mesozoic and Tertiary movements that affected the Gulf od Suez have resulted in themost inensively faulted area

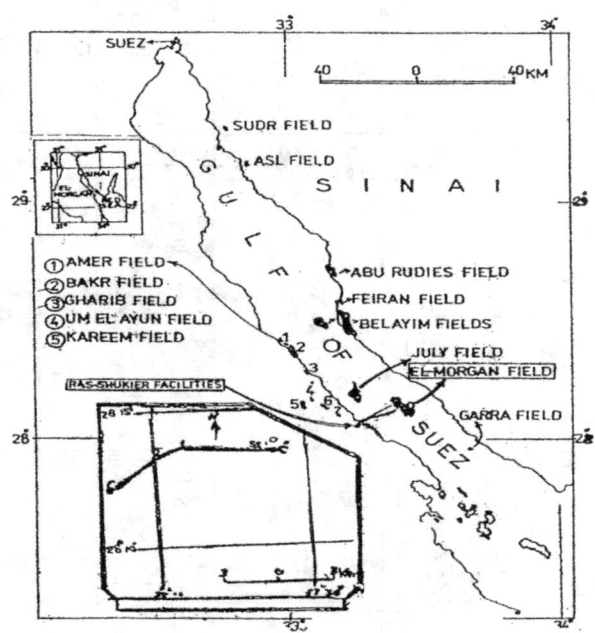

Fig 6.17 – A Location Map of Gulf of Suez showing the Morgan Oil Field and the position of cross line C-C'

(Metwalli, 1978)

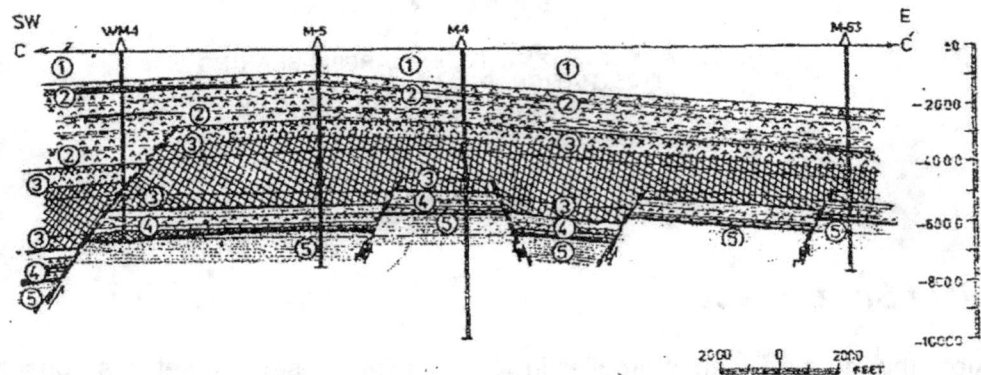

Fig 6.18 - A structural cross section across the line C-C'

6.12 The Miocene Reservoir of El-Morgan Offshore Oil

The startigraphic section of the Gulf of Suez region ranges in age from Paleozoic to Recent, but most of the drilled wells ended in Tertiary rocks; whereas the Miocene evaporates and the underlying Miocene clastics are the main sequence penetrated. Several inshore and offshore oil fields were drilled, of which the Morgan offshore oil field is presented as a type of producing well.

The oil field of Morgan is an offshore oil field, SW of the Gulf of Suez petroleum province of Egypt; it embraces an area of 46 km² and is situated in faulted blocks of reservoir rocks. It was a traditional concept that hydrocarbon accumulations in the Gulf of Suez region were confined within the Tertiary sediments, especially Miocene. However, the pre-Miocene deposits were not fully penetrated, although detailed studies on the geologic, tectonic setting and the mode of salt movement, together with detailed structural analysis of the nearby areas, might help in exploring unknown hydrocarbon reserves underlying the Miocene and in deeper reservoirs that were normally capped by thick Miocene salt deposits (Metwalli et al. 1978). The Morgan field lies on the eastern bank rich of coral reefs that give its Arabic name, "El-Morgan".

Metwalli et al. (1976) referred to the tectonic setting of the field, with special reference to salt structures, and to its movements that affect to some extent the hydrocarbon accumulations as well as the source-reservoir relations. **Figure 6.19** (Metwalli et al. 1978) shows the litho-stratigraphic section of the Morgan oil field where the Miocene evaporite and shale deposits overlie different reservoir rocks, the most important of which are:

(a) **Zeit formation** (Late Miocene-Pliocene)—Lies on top of Miocene deposits, mainly of evaporites and shale intercalations, directly underlying the post-Miocene clastic deposits. Hydrocarbons in the Zeit formation at El-Morgan oil field are not recorded; however, it is oil bearing at the Belayim onshore oil field.

(b) **South Gharib formation**— It is oil producing in many fields of the Gulf of Suez petroleum province, characterized by a remarkable increase in the thickness of evaporites (mainly rock salt) of proper lagoon facies, where salt bulges are defined in some locations and cause many problems in interpreting seismic data. No traces of hydrocarbons are known in the South Gharib formation at the El-Morgan oil field; however, it is oil producing at the Belayim onshore and Bakr oil fields. The formation acts as a sealing unit that prevents the oil migration of the underlying Belayim and Karim formations updip at the El-Morgan oil field.

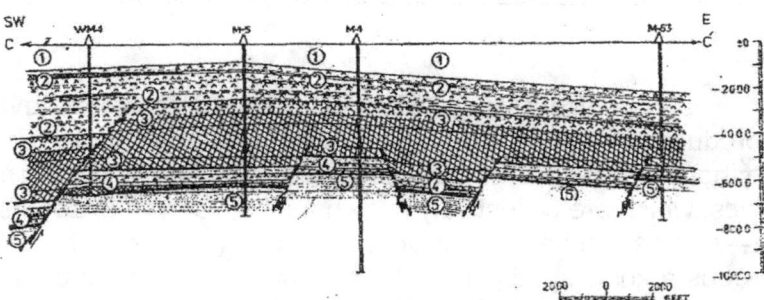

Fig 6.18 - A structural cross section across the line C-C'

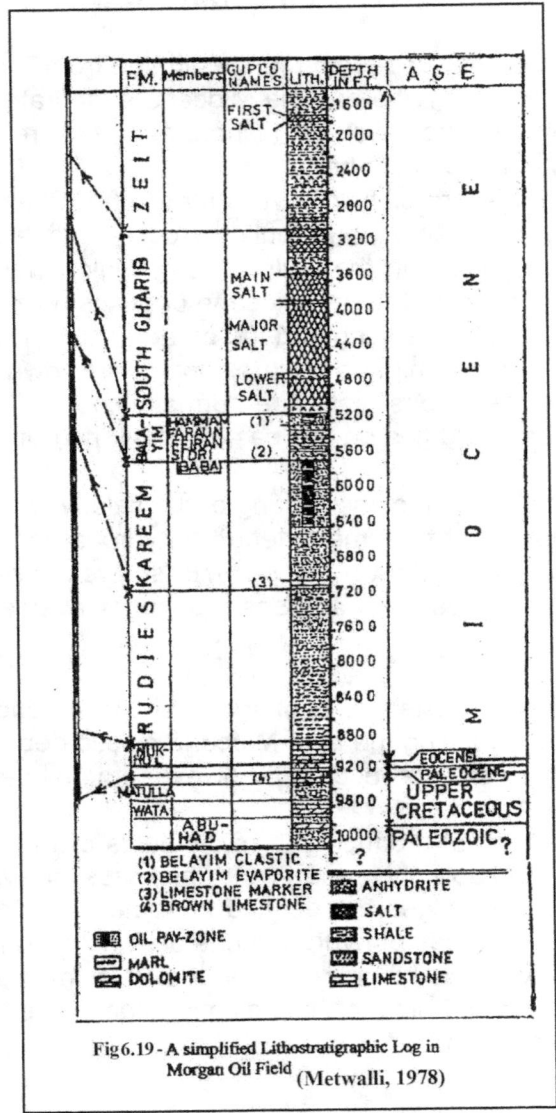

Fig 6.19 - A simplified Lithostratigraphic Log in Morgan Oil Field (Metwalli, 1978)

(c) **Belayim formation**—It is s about 1,000 ft (~333 m) in thickness, consists of evaporites and interevaporite marls. It is oil producing in many fields of the Gulf of Suez petroleum province, e.g., El-Morgan, Bakr, and Gharib offshore, as well as the Belayim onshore oil fields. The topmost member is known as the Belayim clastics, which are collectively known as the Belayim evaporites.

(d) **Kareem formation**—Attains a thickness of about 1,400 ft (~400 m) at the El-Morgan field and is composed of calcareous arkosic sandstone with thin shale interbeds and minor evaporite intercalations. The sand and sandstone of the Kareem formation form the pay zone, producing oil of about 30° API; it is also productive in both the Kareem and the Belayim onshore oil fields. At the Gharib field, it reaches about 800 ft (~250 m) in thickness of clastics, and is interbedded by anhydrite deposits and occasional limestone.

(e) **Rudies formation**—Attains a thickness of about 2,550 ft (~830m) at the Suez Gulf region and consists mainly of sandy clays. The clays are highly calcareous with abundant planktonic foraminifera (Globigerina marl). The Rudies formation is oil producing at the Balayim onshore and offshore oil fields, and the Feiran and Kareem oil fields.

References

Aliev, M., Ait Laussine, Said, A. et al. (Sonatrach, 1971) Geological Structures and Estimation of Oil and Gas in the Sahara of Algeria: Spain, Altamira Rotopress, S.A., 255 pp.

Assaad, F. (1972) Contribution to the study of the Triassic formations of the Sersou-Megress Region (High Plateaux) and the area of Daia M'Zab (Saharan Platform), No. 84(B3), Eighth Arab Petroleum Congress, Algeria.

Assaad, F., (1981) A further geologic study on the Triassic formations of North-Central Algeria with special emphasis on Halokinesis, Journal of Petroleum Geology, Vol. 4, pp 163–176.

Assaad, F.A. (1988) Hydrogeological aspects and environmental concerns of the New Valley Region, western desert, Egypt, with special emphasis on the Southern Area,. Environmental Geology Vol. 12, No. 3, 141–161.

Assaad, F.A, and Barakat, M.G. (1965) Geological results of the Assiut Kharga well, Egypt, theJournal of Geology, UAR, Vol. 9, No. 2, pp. 81–87.

Bishay, F. and Tawfik, A.E. (1965) in Journal of Geology, UAR, Vol. 9, No. 2, pp. 81–87.

Klett, T.R., Ahlbrandt, T.S. Schmoker, J.W., and Dolton, G.L., (1997) Ranking of the world's oil and gas provinces by known petroleum volumes: U.S. Geological Survey Open-File Report 97–463.

Kun, N.D. (1965) The Mineral Resources of Africa, Columbia University in the City of New York, N.Y., Elsevier Publishing CO., 52 Vanderbilt Av., New York, N.Y. 10017, (Figs.2, 58, 123 & 125).

Metwalli, M.H. (2000) A New Concept for the Petroleum Geology of the Nile Delta, A.R. Egypt, Geol. Dept. Fac. Science, Cairo, Egypt, 5th., International Conference on the Geol. Of he Arab World, Cairo, Univ., pp. 713–734.

Metwalli, M.H., Philip, G., Wali, A.M.A. (1979) Repeated Folding and its-significance in northern Western Desert, Petroleum Province, Egypt, Vol. 29, No. 1, pp. 133–150 Acta Geologica, Polonica, Poland, Warsawa.

Metwalli, M.H., Philip G., Youssef, A.A.E. (1978) El Morgan Oil Field as a Major-Blocks Reservoir masked by thick Miocene Salt; Vol. 28, No. 3; Acta Geologica Polonica, Poland, Warsawa.

---------------- (1982) Petrographic characteristics of the oil-bearing formations in El-Morgan oil field "Gulf of Suez Petroleum Province", A.R.Egypt. Acta Geologica Academiae Scientiiarum Hungaricae, Vol. 25 (3–4), pp. 275–295.

---------------- (1981) El-Morgan oil field crude oil and cycles of oil generation and accumulation in the Gulf of Suez petroleum province, A.E.R., Egypt; Acta Geologica Academiae Scientiarum Hungaricae, Vol. 24 (2-4), pp. 369-387.

Michel R. (1969) - Petroleum developments in North Africa. Am. Assoc Petrol Geol, AAPG 1970; Bull 54(8): Fig. 1.1

Montgomery, S., (1993) Ghadames Basin of north central Africa; Stratigraphy, geologic history, and drilling summary, Petroleum Frontiers, Vol. 10, No. 3, 51 p.

Persits, F. and others (1997) Maps showing geology, oil and gas fields and geologic provinces of Africa: U.S. Geological Survey Open-File Reopoport 97-470A, CD-ROM.

PetroConsultants (1996b) PetroWorld 21: Houston, Tex., PetroConsultants, Inc. [database available from Petroconsultants, Inc., P.O. Box 740619, Houston, TX 77274-0619].

SN RÉPAL (1961) Les séries Permo-Triassiques dans le Nord Sahara. Études Pétro-graphique du cycle Détritique. Etudes du Cycle Salifére (texte et planches).

Said, R. (1962) The Geology of Egypt, Cairo University, Egypt, UAR, Fig. 13, Elsevier Amsterdam, Publishing CO.

Stoica, I., and Assaad, F.A. (1981) Geological Studies on the Triassic Reservoir of Condensate and Gas at the Hasssi R'Mel Field (Internal Report, Sonatrach).

Werwer, A.M., (1973) Hydrogeological Study of the Wadis El-Ajal, Shatti, Traghen, Libya, by REGWA, Cairo, Egypt.

Websites
– Map of Algeria/AAPG: [mem-algeria.org/hydrocarbons/geology.htm]
– Klett, T.R. (9/23/2006): [http://greenwood.cr.usgs.gov/pub/bulletins/b2202.bl/

(1) Total Petroleum Systems of the Illizi Province, Algeria and Libya: [Pubs.usgs.gov/bul/b2202-a/b2202-bso.pdf]

(2) Total Petroleum Systems of the Grand Erg/Ahnet Province, Algeria and Morocco: [Pubs.usgs.gov/bul/b2202-b/b2202-bso.pdf]

(3) Total Petroleum Systems of the Trias/Ghadames Province, Algeria, Tunisia, and Libya: [Pubs.usgs.gov/bul/b2202-c/b2202-bso.pdf]

Cited References
Burollet, P.F. et al. (1978) The geology of the Pelagian Block: the margins and basins off southern Tunisian and Trilopitania, in Nari, A.E.M., Kanes, W.H., and Stehli, F.G., eds., The Ocean Basins and Margins: New York, Plenum Press., v. 4B, pp. 331–359.

Magoon, L.B., and Dow, W.G. (1994) The petroleum system, in Magoon, L.B. and Dow,

W.G Eds, The Petroleum System – From Source to Trap: American Association of petroleum Geologists, Memoir 6-**Appendix 6.A1- Geological Time Scalehere**

Appen·6A-1

GEOLOGIC TIME SCALE

PHANEROZOIC			PRECAMBRIAN

CENOZOIC

AGE (Ma)	Period	Epoch	Stage	AGE (Ma)
0	Quaternary	Holocene		
		Pleistocene		1.81
		Pliocene L	Gelasian	2.59
		E	Piacenzian	3.60
5			Zanclean	5.33
	Neogene	Miocene L	Messinian	7.25
10			Tortonian	11.61
		M	Serravallian	13.65
15			Langhian	15.97
		E	Burdigalian	
20				20.43
			Aquitanian	23.03
25	Paleogene	Oligocene L	Chattian	28.4
30		E	Rupelian	33.9
35		Eocene L	Priabonian	37.2
40			Bartonian	40.4
45		M	Lutetian	
				48.6
50		E	Ypresian	
55				55.8
		Paleocene L	Thanetian	58.7
60		M	Selandian	61.7
		E	Danian	
65				65.5

MESOZOIC

AGE (Ma)	Period	Epoch	Stage	AGE (Ma)
70	Cretaceous	Late	Maastrichtian	70.6
75			Campanian	
80				
85			Santonian	83.5 / 85.8
90			Coniacian	89.3
95			Turonian	93.5
100			Cenomanian	99.6
105		Early	Albian	
110				112.0
115			Aptian	
120				
125			Barremian	125.0
130				130.0
135			Hauterivian	136.4
140			Valanginian	140.2
145			Berriasian	145.5
150	Jurassic	Late	Tithonian	150.8
155			Kimmeridgian	155.7
160			Oxfordian	161.2
165		Middle	Callovian	164.7
			Bathonian	167.7
170			Bajocian	171.6
175			Aalenian	175.6
180		Early	Toarcian	183.0
185			Pliensbachian	189.6
190			Sinemurian	
195			Hettangian	196.5
200				199.6
205	Triassic	Late	Rhaetian	203.6
210			Norian	
215				216.5
220			Carnian	
225				228.0
230		Middle	Ladinian	
235				237.0
240			Anisian	
245		Early	Olenekian	245.0
250			Induan	249.7 / 251.0

PALEOZOIC

AGE (Ma)	Period	Epoch	Stage	AGE (Ma)
255	Permian	Lopingian	Changhsingian	253.8
			Wuchiapingian	260.4
260		Guadalupian	Capitanian	265.8
265			Wordian	268.0
270			Roadian	270.6
275		Cisuralian	Kungurian	275.6
280			Artinskian	284.4
285			Sakmarian	
290				294.6
295			Asselian	299.0
300	Carboniferous	Pennsylvanian Late	Gzhelian	303.9
305			Kasimovian	306.5
310			Moscovian	311.7
315			Bashkirian	318.1
320		Mississippian Late	Serpukhovian	326.4
325				
330		Middle	Visean	
335				
340		Early	Tournaisian	345.3
345				359.2
350				
355	Devonian	Late	Famennian	
360				
365				374.5
370			Frasnian	
375				385.3
380		Middle	Givetian	391.8
385			Eifelian	397.5
390		Early	Emsian	407.0
395			Pragian	411.2
400			Lochkovian	416.0
405	Silurian	Pridoli		418.7
410		Ludlow	Ludfordian	421.3
415		Wenlock	Homerian	426.2
420			Sheinwoodian	428.2
425		Llandovery	Telychian	436.0
430			Aeronian	439.0
435			Rhuddanian	443.7
440	Ordovician	Late	Hirnantian	445.6
445				455.8
450		Middle	Darriwilian	460.9
455				468.1
460		Early		471.8
465			Tremadocian	478.6
470				488.3
475	Cambrian	Furongian	Paibian	501
480		Middle		513
485		Early		
490				542.0

PRECAMBRIAN

AGE (Ma)	Eon	Era	Period	AGE (Ma)
600	Proterozoic	Neoproterozoic	Ediacaran	542 / 630
700			Cryogenian	850
800			Tonian	
900				1000
1000		Mesoproterozoic	Stenian	
1100				1200
1200			Ectasian	
1300				1400
1400			Calymmian	
1500				1600
1600		Paleoproterozoic	Statherian	
1700				1800
1800			Orosirian	
1900				2050
2000			Rhyacian	
2100				2300
2200			Siderian	
2300				2500
2400	Archean	Neoarchean		2800
2500		Mesoarchean		
2600				
2700				3200
2800		Paleoarchean		
2900				3600
3000		Eoarchean		
...			Lower limit is not defined	
4600				

PART III
Several Article Papers of Chapter 7-ABC on petroleum activiies in North Africa

Chapter 7
ABC – Several Articles Scientific Papers on the Petroleum Activities in Northeast Africa.

7A- Potential Hydrocarbon in North Africa

7A.1-Introduction

Africa is the second largest continent (after Asia), covering about one-fifth of the total land surface of the earth. The continent is bordered on the west by the Atlantic Ocean, on the north by the Mediterranean Sea, on the east by the Red Sea, and the Indian Ocean, and on the south by the merging waters of the Atlantic and Indian oceans; in otherwords, the northern margin of the African plate is bounded by the Atlantic to the northwest, the Mediterranean to the north and the Arabian plate to the east (Klett, USGS, 1997).

The whole of Africa can be considered as a vast plateau rising steeply from narrow coastal strips and consisting of ancient crystalline rocks. In general, the plateau may be divided into a southeastern portion and a northwestern portion. The northwestern part, which includes the Sahara (desert) and that part of North Africa known as Maghrib, has two mountainous regions: The Atlas Mountains in northwestern Africa, which are believed to be part of system that extends into southern Europe; and the Ahaggar (or Hoggar) Mountains of the Sahara; the southeastern plateau includes the Ethiopian plateau. Africa contains an enormous wealth of mineral resources, including some of the World's largest reserves of fossil fuels, metallic ores, Gems and precious metals (Article from the Encyclopaedia Britannica).

African natural gas resources are surpassed only by those of North America, which market 75 % of the world's 20,000M. ft^3. Crude oil reserves of the Arabian Peninsula and North African countries represent almost two-thirds of the world total of 300BBO in 1965. The African rate of expansion was tens of times as high as the increase of world production in the past four decades.

Fig 7.A.1- **Hydrocarbon resources in North Africa (Macrgregor, 1994).**
Northeast of Africa was joined to Asia by Sinai Peninsula until the construction of the Suez Canal. Because of the bulge formed by the western Africa, the greater part of Africa's territory lies north of the Equator, while it is crossed from north to south by the prime meridian (0° Longitude), which passes a short distance to the east of Accra, Ghana. The Romans, who for a time ruled the Northern African coast, are also said to have called the area south of their settlements Afriga, or the land of the Afrigs- the name of a Berber community south of Carthage.

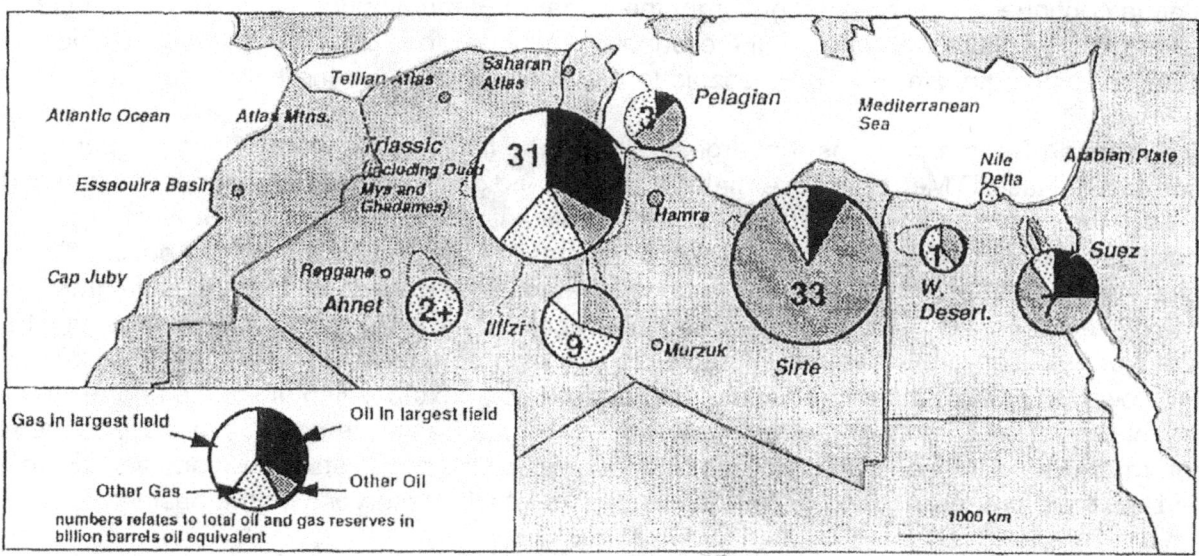

Fig 7A.4 - Hydrocarbon resources in North Africa. Condensate reserves are included in gas figures. Note the concentration of rese particularly in Eastern Algeria and that, predominantly of oil, in the Sirte Basin of Libya. Note also the high proportion of l reserves in two major Algerian fields (Hassi R'Mel and Hassi Messaoud), which contrasts with the more dispersed nat reserves in the rift provinces of Sirte and Suez (Macgregor, 1994)

7A.2 Hydrocarbon Reserves in North Africa

North Africa's hydrocarbon reserves are spread over a wide range of reservoirs. The petroleum systems of the North African countries bordering the Mediterranean (Morocco, Algeria, Tunisia, Libya and Egypt) actually form a geographically contiguous unit as well as a common geological history, as a part of a single geological plate through Phanerozoic history with the exception of the Atlassic domains that were accreted in Hercynian (Late Carboniferous) and the Alpine (Tertiary) events.

The northwest Africa or the Arabian Maghreb (Mauritania, Morocco, Algeria, and Tunis) covers regions of different tectonic and sedimentation facets. The total thickness of sedimentary formations of the Mesozoic regions of the Saharan Atlas to the northwest, undergone the maximum sinking that exceeds 12,000ms.

Algeria actually repesented the northwest African coountries (or the so called the Arabian Maghreb), as the Mediterranian Sea border it from the north, Mauritania and Morocco from the west, Tunisia and Libya from the east and Nigeria and Niger from the south.

Algeria, of an area 900,000 square miles, is the second largest country in Africa and the Arab world after Sudan. Libya is a country of 680,000 square miles and was considered by an international mission after the Second World War, as almost barren but prompted by discoveries in neighboring Algeria, when 17 companies carried out exploration and invested more than $800 millions by the beginning of 1963.

During the early time of independence, Algeria had new legislation regarding concession contracts and royalties that had disscouraged exploration by foreign companies; since late 1980's, however, Algeria after revised its ligislation, encouraged petroleum Companies to explore and develop oil and gas resources (Sonatrach, 1972).

In northwest Africa, several French publications were issued on the Moroccan and the Algerian Atlas dealing with particular facets of stratigraphy, tectonics, lithology, and petrography; but despite the large number of geological and geophysical studies carried out during the French colonialism, the National Society of Transport and petroleum of Algeria "Sonatrach", started after the independence on 1962, a precise survey of the existing materials for establishing prospects of petroleum hydrocarbons, and setting up a work program for research in diverse regions.

Algeria continued to flow oil and gas from the Saharan Platform since 1962. The Precambrian mass of the Hoggar constitutes the anchor of the Sahara. There are four groups of provinces: Hoggar (rich in platinum, asbestos, tin, zinc, etc.), Paleozoic to the South, Triassic to the North, and Eocene province to the East.

The Algerian Sahara contains the largest oil and gas fields in the North African region, e.g. Hassi Messaoud and Hassi R'Mel; both cover half the total oil and gas resources and about one fifth of the total reserves in the region as a whole. The third most significant province, the gulf of Suez of Egypt deserves a special attention in terms of its unusual high productability. The offshore Pelagian basin of Tunisia and Libya ranks one forth of reserves that were concentrated in the Eocene carbonate sediments.

The dispersal of petroleum reserves through a wide strtatigraphic range of reservoirs might suggest that the North African provinces are characterized by high degrees of vertical migration. Closer examination, however, reveals the importance of lateral migration (juxtaposition of source and reservoir beds) on unconformities, e.g. most of the reserves within the Sirte Basin are developed close to the main rift unconformity (Shardanov and Shumliova, 1983), over which the mid Cretaceous sand acts as an efficient carrier bed, therefore, distributing oil from source rocks into the reservoirs on the basinal margins.

7A-3 Stratigraphy

The regional stratigraphy is continuous across North Africa, but petroleum generation, migration, and entrapment within each total petroleum system, were controlled by the tectonic history of each individual basins. Throughout most of the Paleozoic, North Africa was a single depositional basin on the northern shelf of the African craton. The basin generally deepened northward where deposition and marine influence were greater. Some gentle but large structures existed in the area throughout the Paleozoic and affected the thickness of the sedimentary cover. In Late Silurian and Early Devonian, Laurasia separated from Gondwanaland resulting in minor deformation, uplift, and local erosion. Later, in the Middle to Late Devonian, the initial collision of Laurasia and Gondwanaland began, resulting in erosion and further modification of pre-existing structures **(USGS, Bulletin 2202-A; klett, 1997). Fig 7A.2- Paleogeographic Map of Gondwana Land, Klott 1997.**

The Hercynian event marks the collision between Laurasia and Gondwanaland and caused regional uplift, folding, and erosion. Paleozoic basins, delineated by earlier tectonic events, were modified, resulting in the development of several intracratonic sag and foreland basins.

Several transgressive-regressive cycles occurred throughout the Paleozoic. Two major flooding periods, one in Silurian and the other in Late Devonian, were responsible for the deposition of the mudstone source rocks. Many of the prograding fluvial, estuarine, deltaic, and shallow marine sands were deposited during these cycles, and became reservoirs.

During the early Mesozoic, extensional movements caused by the opening of the Tethys and Atlantic oceans developed a cratonic sag basin called the Triassic basin. Triassic fluvial sands followed by a thick Triassic to Jurassic evaporite section were deposited within the sag basin (Sonatrach, Aliev and others, 1971; Boudjema, 1987). Sandstones resulting from the fluvial deposition are major reservoirs where evaporate deposits provide a regional seal for the Mesozoic and Paleozoic reservoirs.

In general, the sedimentary formation in the Arabian Maghreb, accumulated from the Tethys that flooded the area during the period between Cambrian and Eocene. From the Precambrian window of the Hoggar southeast of Algeria towards the Mediterranean, successively younger sediments accumulated in Epicontinental zones. Consequently, younger oil fields appear from the southwest towards the northeast. Traps, of different times, were frequently controlled by Precambrian-Paleozoic structure.

The initial stages of the Africa-Arabia and Eurasia collision during Late Cretaceous to Middle Tertiary (Pyrenean) caused compressional movements, deformation and uplift (Petterson, 1985), that tilted the Triassic Basin to its present configuration. Basins that existed where the present-day Atlas Mountains overlie, were inverted by these events. **(Pubs.usgs.gov/bul/b2202-a/b2202aso.pdf).**

Compressional movements during the Late Cretaceous and Pyrenean deformation tilted the Triassic Basin to its present configuration, whereas rhe existing basins on which the present-day Atlas Mountains overlie, were invered by these events (Klett, USGS 2000).

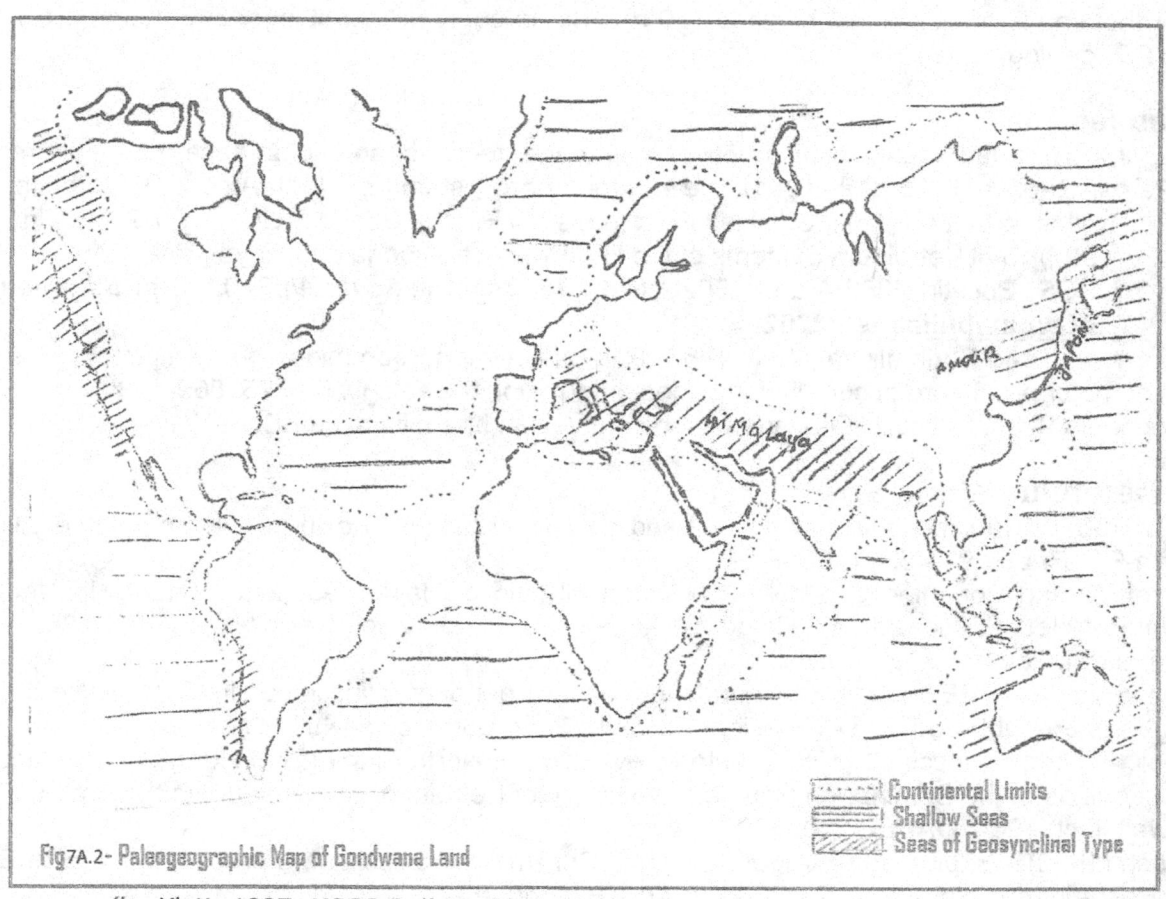

Fig7A.2- Paleogeographic Map of Gondwana Land

(by Klett, 1997; USGS-Bulletin 2202-A)

7A.4- Geological History and the Mesozoic Era of North Africa

Davidson S.H.W. Nicol- in his article on the geologic history of North Africa, mentioned that the Mesozoic era (about 252 to 66 million years ago), divided into three periods: The Triassic, Jurassic, and Cretaceous, is remarkable for the transgression of ancient seas and for the emergence of massive land formations containing interesting fossil remains;

The above author further divided the Mesozoic Era into Marine and Continental Formations:

(A) Marine Formations-During the Triassic period (~252-201 M years ago), the ancient seas left deposits of marine formations in North Africa; the southern Sahara, Egypt, and northern Madagascar. Deposits from Jurassic Period (~201-145 M years ago) extend to the Atlantic basins of Western Sahara along the northwest coast of the continent.

(B) Continental Formations- In North Africa and in Arabia, the Mesozoic continental formations covered large areas. During the Triassic, the Saharan Zarzaitene series, was depposited containing Dinosaur and other reptilian fossil remains. The Saharan Touratine series, containing fossils of vegetation and of great repltiles, was laid down during the Jurassic. The final stages of the Triassc and the early

Jurassic periods, were characterized by the terminal folding of the Cape mountain chain, creating formations some 4,000 ft (1200 ms) thick.

(C) The Technoexport (1971) of Sonatrach publioshed two volumes on the geological structures, stratigraphy of the Saharan Atlas, and the Saharan platform; both volumes presented a tremendous effort that last for 15 years to systemiaze and generalize a great deal of geological and geophysical data over thousands of diagraphs, besides 750 reports on composite well logs, 300 geophysical reports and 250 geological reports.

References

Macgregor, D.S. (1993) Relationships between seepage, tectonic, and subsurface petroleum reserves; in Petrol. Geol. 10, 606-619 (1994) The Hydrocarbon systems of North Africa; Conference, Malta, B.Exploration, 4/5 Long Walk, Stockley Park Industrial Estate, Ubridge, UK. (Fig 1.2 and Fig 1.3).

Klett, T.R. (1997) Total Petroleum Systems of the Illizi Province, Algeria and Libya-Tanezzuft-Illizi, version 1.0., , USGS, Bulletin 2202-A.and 2202C (http://pubs.usgs.gov/bul/b2202-a) http://greenwood.cr.usgs.gov/pub/bulletins/b2202-c/

Shardanov, A.N. and Shumuliova, M.B. (1983)-Regional hydrocarbon migration as a factor in the formation of major petroleum accumulation zones. International Geology. Rev.25, 569-573.

Encyclopaedia Bitannica-http://O-academic.eb.com.vulcan.bham.lib.al.us/EBchecked /topic/792..

Cited References

Klett, T.R.1997 – Ranking of the world's oil and gas provinces by known petroleum volumes: U.S.G.S Open-File Report 97-463, 1, CD-ROM.

Balbucchi, A. and Pommier, G. (1970) Cambrian oil field of Hass Messaoud, Algeria. In Geology of Giant Petroleum Field (ed. M.T Halbouty) Am. Assoc. Petrol. Geol. Mem. No. 14, 477-488; Figs 2&3 Marine Petroleum

Boudjema, A. (1987) Evolution Structurale du Bassin petrolièr Triassique du Sahara Nord-Oriental (Algerie), un-published Ph.D., Thesis, Universitè Paris-Sud, pp477-488 . Centre d'Orsay, 290 p.

Mengnoli, S., and Spinicci, G. (1985) Tectonic evolution of North Africa (from North Sinai to Algeria).in: Proceedings of the Seminar on Source and Habitat of Petroleum of petroleum in the Arab Countries, Kuwait, Oct. 1984, OAPEC, 119-174

Sonatrach (1992) – Exploration in Algeria: Algeria, Sur Presses Speciales U.A.F.A., 36p. - A Review of the East Algerian Sahara oil and gas (2000) – Historique (http://www.mem-algeria.org.hydrocarbons/geology.htm)

7B. Petroleum Geology and Potential Hydrocabnon Plays of Heavy Oil

Reservoirs: Examples from Egypt and Libya (N.A. Nabih, A. Youssef &W.M.Abd Raboh) -N. A. Nabih, A. Youssef & W. M. Abd Raboh,

Ganoub Elwadi Pertoluem Holding Company

The paper was presented at the 10th Offshore Mediterranean Conference and Exhibition in Ravenna, Italy, March 23-25, 2011. It was selected for presentation by OMC 2011 Programme Committee following review of information contained in the **abstract submitted by the author(s).**

ABSTRACT

Unconventional crude oil production will expand as the production of conventional oil will decline after 2020 (Odell, 1994). Production from the richest portion of the largest unconventional oil resources is forecasted to increase as escalating prices will allow profitable operations. The paper **stresses** the need to understand & re-evaluate the heavy oil reservoirs in Egypt and Libya as a clue for ascertaining the potentialities of heavy oil reserves at the edges of the Gulf of Suez in Egypt and Sirte basin in Libya. Heavy oil occurs in shallow and deep structure geologic settings, which are unconventional by accepted standards, yet they are economically interesting prospects in the light of modern method of production.

The encouraging oil potentiality (proven oil of AP1 gravity ranging from 11.4° to 20.5 ° AP1) in both carbonate and sand deposits of Miocene and pre Miocene will add more oil reserves for future exploration phases. Due to the present high oil prices a second look at old unprofitable wells, using new technology to eke every last drop out of reservoirs.

Introduction

The International Energy Administration, the U.S.Department of Energy, the National Defence Council, and industry experts are warning the world of an oil crisis when demand overtakes the capacity of the world to produce (Moody, 1995). Edwards (1997) forecasted that "world crude oil production will peak at 90 million barrels per day (33 billion barrels per year) in 2020". This peak may be shifted beyond 2020; thanks to the technology advances and the realization that unconventional resources will provide continuous opportunities for extraction (Blanchard, 2011). Crude oil will be able to supply the increasing demand until the peak of world production is reached. The energy gap caused by declining conventional oil production must then be filled by expanding production of coal, natural gas, unconventional oil from " tar sands"; the term bituminous sand is more technically correct. Bituminous sand is often applied to sand containing petroleum that has an API gravity of less than10°. (Speight, 1991)], heavy oil [a generic term which "is often applied to a petroleum that has an API gravity of less than 20° and usually, but not always, a sulfur content higher than 2% by weight. Furthermore, in contrast to the conventional crude oils, heavy oils are darker in colour and may even be black (Speight, 1981). The term "heavy oil" has also been arbitrarily used to describe both the heavy oils that require thermal stimulation for recovery from the reservoir and the bitumen in bituminous sand (or tarsand-) formations from which the heavy bituminous material is recovered by mining operation and oil shales (A kerogene-bearing), finely laminated brown or black sedimentary rock that will yield liquid or gaseous hydrocarbons on distillation], nuclear and hydrostatic power, and renewable energy sources (solar, wind and geothermal). Unconventional crude oil production will expand as the production of conventional oil declines after 2020 (Odell, 1994). Production from the richest portions of the largest unconventional oil resources are forecast to increase as escalating oil prices permit profitable operations. Unconventional crude oil production could reach 20 million barrels per day by 2100 (Edwards, 1997).

The shortfall of conventional crude oil supply in Egypt should be filled partly by expanding production of all other fuels (natural gas, coal, heavy oil, and possibly oil shales). The encouraging oil potentiality (proven oil of AP1 gravity ranging from 11.4° to 20.5 ° AP1) in both carbonate and sand deposits of Miocene and pre Miocene will add more oil reserves for future exploration phases. Production from the richest portions of the largest unconventional oil resources is forecasted to increase as escalating prices will allow profitable operations.

The paper stresses the need to understand and re-evaluate the heavy oil reservoirs in Egypt and Libya, especially after estimating the heavy oil reserves in the four study areas as examples (Abu Durba, Feiran and El Nezzazat in the Gulf of Suez half graben in Egypt and a field in the western side of the Sirt basin in Libya, **Fig7B.1**). Heavy oil occurs in shallow and deep structure geologic settings, which are unconventional by accepted standards, yet they are economically interesting prospects in the light of modern method of production.

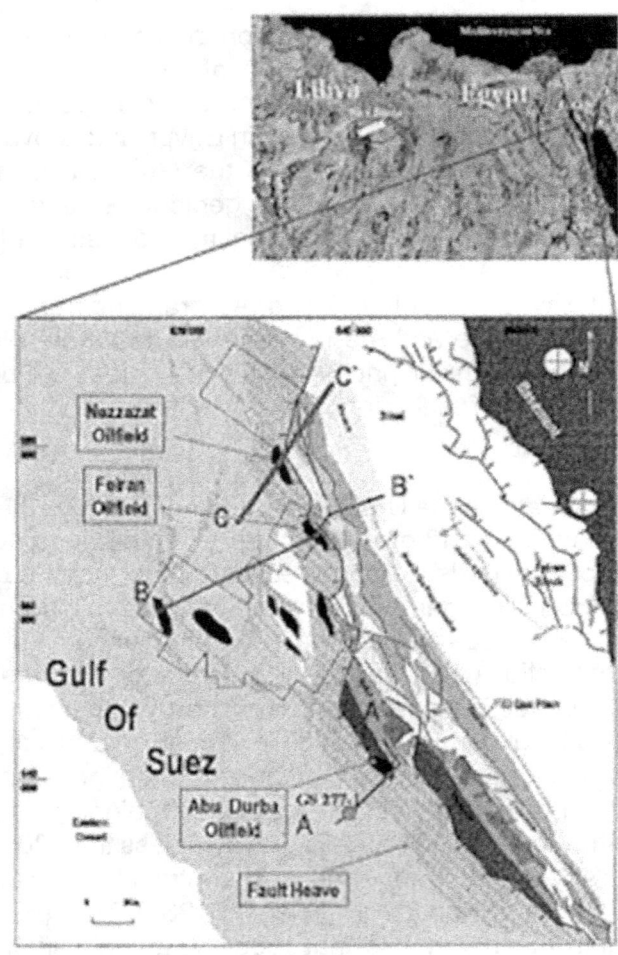

Fig 7B.1: location map of the study areas in both Egypt and Libya, showing the trends of the cross sections.

STRATIGRAPHY AND STRUCTURAL EVENTS

The stratigraphic section of the study areas ranges in age from pre-Cambrian to Recent. Different formation names are used according to the codes of stratigraphy in both Egypt and Libya **{Fig7B 2a&2b}**. We have adopted here the stratigraphic nomenclature of the Gulf of Suez according to National Stratigraphy Sub – Committee (1974) and the Sir! Basin according to Barr and Weegar (1972). Different formation names used by the operating oil companies in their log charts throughout the basin have been correlated referring to the lithologic description of above mentioned references.

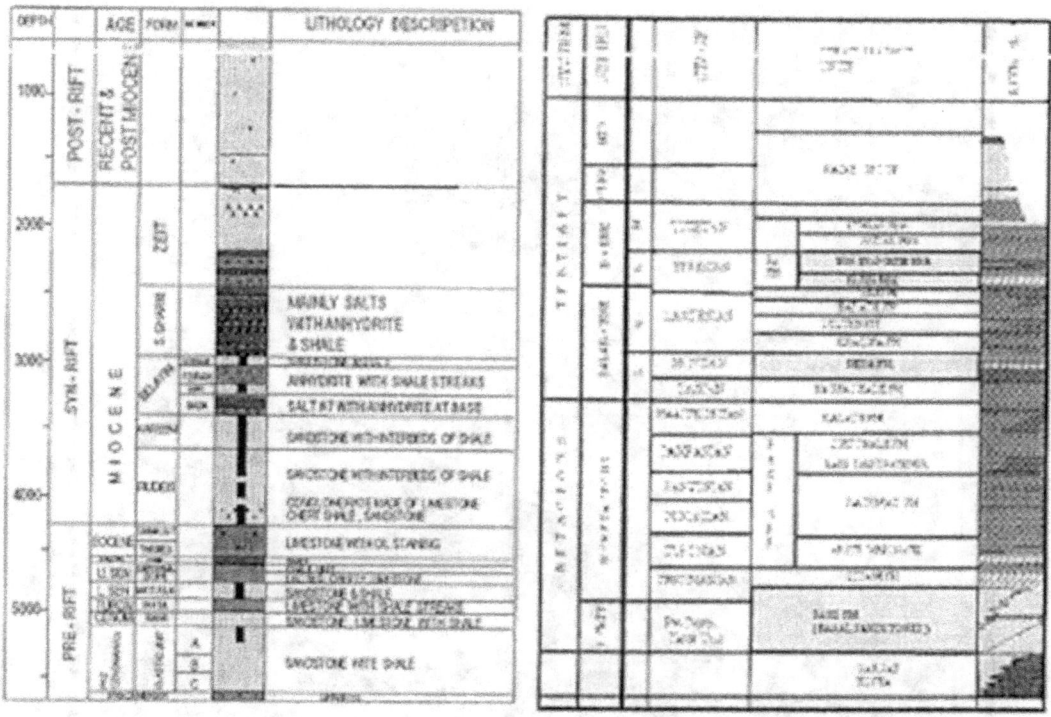

Fig7B.2: **Stratigraphic column of the study areas**

(a) Gulf of Suez, Egypt (b) Sirt Basin, Libya.

Rifting time and tectonic phases of the Gulf of Suez basin in Egypt differ than those of the Sir! basin in Libya. The thickness and type of deposits are mainly controlled by the structural events. This section in the Gulf of Suez is generally characterized by three mainrock sequences relative to the Miocene rifting. These are pre-rift (Early Paleozoic to Eocene), syn rift (Miocene), and post-rift (Pliocene-Recent) sequences. The three sequences are separated by two regional unconformities. The first and second sequences include important hydrocarbon source, reservoir and seal rocks. The Paleozoic-Lower Cretaceous sandstones of Nubia facies lie over Precambrian granitic rocks. The Upper Cretaceous marls and shales of the Raha, Wata and Matulla Formations are overlain by the Campanian- Maastrichtian Sudr Chalk which, in turn, is capped by both Paleocene Esna Shale and Lower-Middle Eocene limestones. The syn-rift (Miocene) succession is commonly subdivided into sandstones, marls and shales of the Rudeis and Kareem Formations evaporites with thin clastic interbeds of the Belayim, South Gharib and Zeit Formations. A blanket of coarse and fine clastics and, in some areas, oolitic limestone lies unconformably over the Miocene formations. This Pliocene-Recent blanket has widespread distribution and its thickness varies from a few meters (surface) to about 1500 m (Nabih and Abd-Allah, 1999).

(A) EXAMPLES FROM EGYPT

1- ABU DURBA OIL AREA GEOGRAPHIC SETTING

The Abu Durba area (about 1.9 km2) is located 8 km. to the south of the Belayim Bay on the Gulf of Suez (on shore) **(Figs. 1 and 3)**.

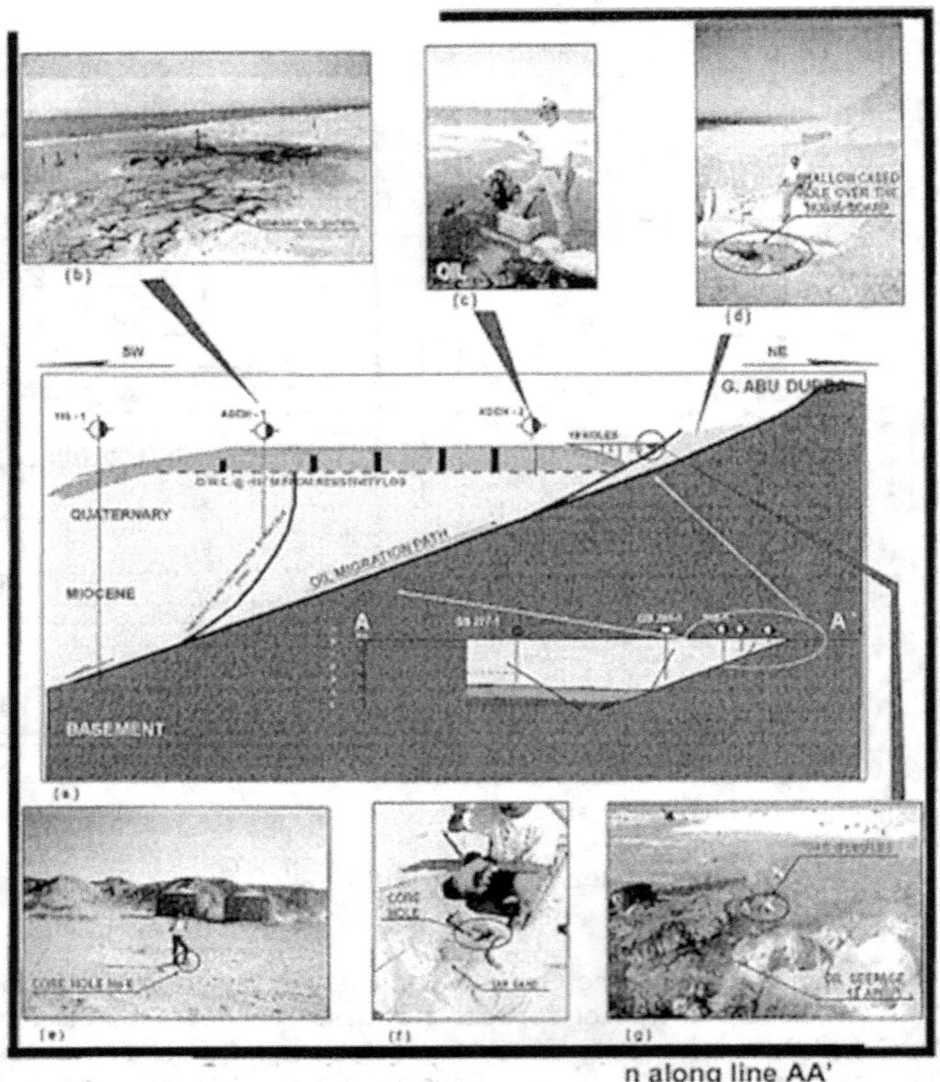

Fig7B.3: Strutural geologic cross section

    (a) Zooming of the eastern part of the cross section,
    (b) Abu Durba core hole No.1,
    (c) Abu Durba core hole No.2
    (d) Oil impregnated Nubia Sandstone,
    (e) Core hole No 6,
    (f) Zooming of core hole No 6
    (g) Abu Durba oil seepage.

7B.5 CONCLUSIONS AND RECOMMENDATIONS
The study of heavy oil reservoir in El Nezzzazat, Feiran and Abu Durba fields reveal the following:

1 - The estimation of the original oil in place concluded about 113 million stock tank barrel proven oil of API gravity ranges from 11.4o to 19° in pre-Miocene and post-Miocene sandstones.
2 - Similar potential could be repeated at the edges of the present Gulf of Suez.
3 - Due to poor seismic data near the shoreline Clysmic faults, the interpretation is mainly based on surface and subsurface data to construct the geological model.
4 - Heavy oil occurs in geologic settings which are unconventional by accepted standards, yet they are economically interesting project in the light of modern methods of potential production.
5 - The two main crude oils in Abu Durba Recent sediments and Kareem Formation sand of well GS 277-1 have different nickel / vanadium ratio, indicating derivation from two probable sources.

The study of heavy oil reservoir in a field in the western side of the Sirte basin, Libya reveal the following :

1 - The total recoverable heavy oil reserve is roughly estimated as 26 MMBBL (Hon Member in the western side of the Sirte basin example).
2 - The heavy oil reservoir in the example is laying in the downthrown block dip to SW and bounded to E by a major NW normal fault separating it from the footwall side.
3 - It is recommended to look for the sliver blocks along the major NW-SE fault separating.
4 - The small scale tectonic fractures density increase in the carbonate section near the intersection of the main NW faults with other fault trends specially toward the downthrown sides of the normal faults. Both tectonic fracture & dolomitization process are responsible for generating the secondary porosity & permeability.
5 - The early migration of oil from the source rocks to the carbonate
Open fractures **terminates** digenesis and prevents the damage of porosity.
6 - The Cainozoic **(Tertiary)** hydrocarbon systems are sourced mainly by the Rakb Shale to the structural and stratigraphic traps in the dolomite of Facha and Beda "C" Members. They are sealed by evaporite sections. While the heavy oil, which is trapped in the dolomite of Hon Member is sealed by evaporite interbeds and sourced from the carbonate evaporites.
7 - Subsuraface well data proves that heavy crude could be trapped in the carbonate and dolomite sediments.
8 - A second look at old unprofitable wells, using new technology to eke every last drop out of reservoirs due to the present high oil prices.

By
U.S.Geological Survey 2202-F
Director: Charles Groat

Cited References
PetroConultants (1996a) Petroleum Exploration and Prodution DataBase: Houston, Texas, PetroConsultants, Inc., {DataBase available from petroConsultants , Inc., P.O.Box 740619, Houston, TX.
---------------------- (1996b) PetroWorld 2: Houston, Texas, PetroConsultants, Inc. {DataBase available from PetroConsultants, Inc., P.O.Pox 740619, Houston, TX.772-0619}.
Larry W.Lake, Univ. Texas, Austin (1989)- Enhanced Oil Recovery, PreNice Hall, Englewood Cliffs, N..J. 07632

Elsevier (1972)- Oil and Carbonate rocks -Editors (George, V. Chilingar. Robert W.MANNON &Herman H. RickeIII

John M. Hunt (1996)- Geochemistry & Geology (2nd., Ed.) - Most important: Applications (part four) Glossary (pp633 706); References (pp642-707); Name Index (pp710-714);Subject Index(pp715-743).

L.P.Dake (1978) - Fundamentals of Reservoir engineering; Development in Petroleum Science Publishing CO.Inc.

OffshoreOil Development on the Gerges Bank (1976)-Continental OIICO;High Ridge Park, Stanfard Community; Com 06804.

----

sscanned cited References
pp 33a, 34b, & 35c

Ch7.C- The Sirte Basin Province of Libya-Sirte –Zelten –Total Petroleum System by Thomas S. Ahibrandt

U.S. Geological Survey Bulletin 2202-F
May, 8th., 2001

7.C.1 Foreword

The report was prepared as part of the World Energy Project of the U.S. Geological Survey. As previously mentioned in chapter (6), the world was divided into eight regions and 937 geologic provinces. Of these, parts of 128 geologic provinces were assessed for undiscovered petroleum resources.

The purpose of the World Energy Project is to assess the quantities of oil, gas, and natural gas liquids that have the potential to be added to reserves within the next 30 years. These volumes either reside in undiscovered fields whose sizes exceed the stated minimum-field-size cutoff value for the assessment unit (variable but, must be at least one million barrel of oil equivalent) or occur as reserve growth of fields already discovered. An assessment unit is a mappable part of a total petroleum system in which discovered and undiscovered fields constitute a single, relatively homogeneous population such that the chosen methodology of resource assessment based on estimation of the number and sizes of undiscovered fields is applicable. A total petroleum system may equate to a single assessment unit, or it may be subdivided into two or more assessment units if each unit is sufficiently homogeneous in terms of geology, exploration considerations, and risk to assess individually. A graphical depiction of the elements of a total petroleum system is provided in the form of an events chart that shows the times of: (I) deposition of essential rock units; (2) trap formation; (3) generation, migration, and accumulation of hydrocarbons; and, (4) preservation of hydrocarbons. Figure(s) in this report that show boundaries of the total petroleum system(s), assessment units, and pods of active source rocks were compiled using Geographic Information System (GIS) Software.

7.C.2 Abstract

The Sirte (or Sirt) Basin province ranks13th among the world's petroleum provinces, having known reserves of 43.1 billion barrels of oil equivalent (36.7 billion barrels of oil, 37.7 trillion cubic feet of gas, 0.1 billion barrels of natural gas liquids).It includes an area about the size of the Williston Basin of the northern United States and southern Canada **(490,000 square kilometers)**.

The province contains one dominant total petroleum system, the Sirte **zelten** based on geochemical data. The Upper Cretaceous Sirte Shale is the primary hydrocarbon source bed. Reservoirs range in rock type and age from fractured Precambrian basement, clastic reservoirs in the Cambrian Ordovician Gargaf sandstones, and Lower Cretaceous Nubian (Sarir)-Sandstone to Paleocene Zelten Formation and Eocene carbonates commonly in the form of Bioherms. More than 23 large oil fields (>100 million barrels of oil equivalent) and 16 giant oil fields (>500 rnilion barrels of oil equivalent) occur in the province.

Four assessment units arc defined in the Sirte Basin province, two reflecting established clastic and carbonate reservoir areas and two defmed as hypothetical units. Of the latter, one is offshore in water depths greater than 200 meters, and the other is onshore where clastic units mainly of Mesozoic age, may be reservoirs for laterally migrating hydrocarbons that were generated in the deep-graben areas.

The Sirte Basin reflects significant rifting in the Early Cretaceous and syn-rift sedimentary filling during Cretaceous through Eocene time, and post-rift deposition in the Oligocene and Miocene. Multiple reservoirs are charged largely by vertically migrating hydrocarbons along horst block faults from Upper Cretaceous source rocks that occupy structurally low positions in the grabens. Evaporites in the middle Eocene, mostly post-rift, provide an excellent seal for the Sirte-Zelten hydrocarbon system. The offshore part of the Sirte Basin is complete with subduction occurring to the northeast of the province boundary, which is drawn at the 2,000-meter isobath. Possible petroleum systems may be present in the deep offshore grabens on the Sirte Rise such as those involving Silurian and Eoeene rocks; however, potential of these systems remains speculative and was not assessed.

7.C.3 Introduction

The Sirte Basin province ranks 13th among the world's petroleum provinces, exclusive of the USA provinces, "With 43.1 billion barrels of oil equivalent: (BBOE) of known petroleum volume, and it ranks 15th, if U.S. petroleum provinces are included (Klett and others, 1997). The Sirte Basin province is considered a "priority" province by the World Energy Assessment Team as described in the Foreword. Sixteen giant (>500 million barrels of oil equivalent (MBOE) fields occur in the province; reservoirs range in age from Precambrian to Miocene. Exploration has focused on structural highs, principally the horst blocks such as the Waddan, Az Zahrah, and Zalten platforms (also variously known as Jebel Uddan, Beda Dahra, Al Hufra, and Zeltan platforms).

(**figs. 1,2**). These platforms are dominated by carbonate and bioherm Tertiary reservoirs of Tertiary age. In the eastem Sirte Basin, significant stratigraphic clastic traps superimposed on structural highs principally occur in Mesozoic clastic reservoirs such as at Sarir and Messla fields **(fig. 7. C1 or "fig. 1"**.

The Sirte Basin province is characterized by one dominant petroleum system, the Sirte Zeltan, which is subdivided into four assessment units. The Sirte Basin (also referred to as Sirt Basin) is a late Mesozoic and Tertiary continental rift, triple junction, in northern Africa that borders a relatively stable Paleozoic craton and cratonic sag basins along its southern margins **(Fig7.C2 or fig. 2)**. The province extends offshore into the Mediterranean Sea, which with the northern boundary drawn at the 2,000 meter (m) bathymetric contour. The thickness of sediments in the province increases from about I kilometer (km.) near the Nubian (known as Tibesti). Uplift on the south to as much as 7 km offshore in the northern Gulf of Sirte. The onshore area is relatively well explored for structures, which are dominated by regionally extensive horsts and grabens **(fig.2)**. Hydrocarbon resources are approximately equally divided between carbonate and clastic reservoirs (pre-Tertiary dominantly clastic, Tertiary dominantly carbonate reservoirs). The prospective area in the province covers about 230,000 km2 (Montgomery, 1994; Hallett and El Ghoul l996; MacGregor and Moody, 1998). The offshore portion **(figs.7.C3 or Fig 3)** is far less explored and its petroleum potential is largely unknown.

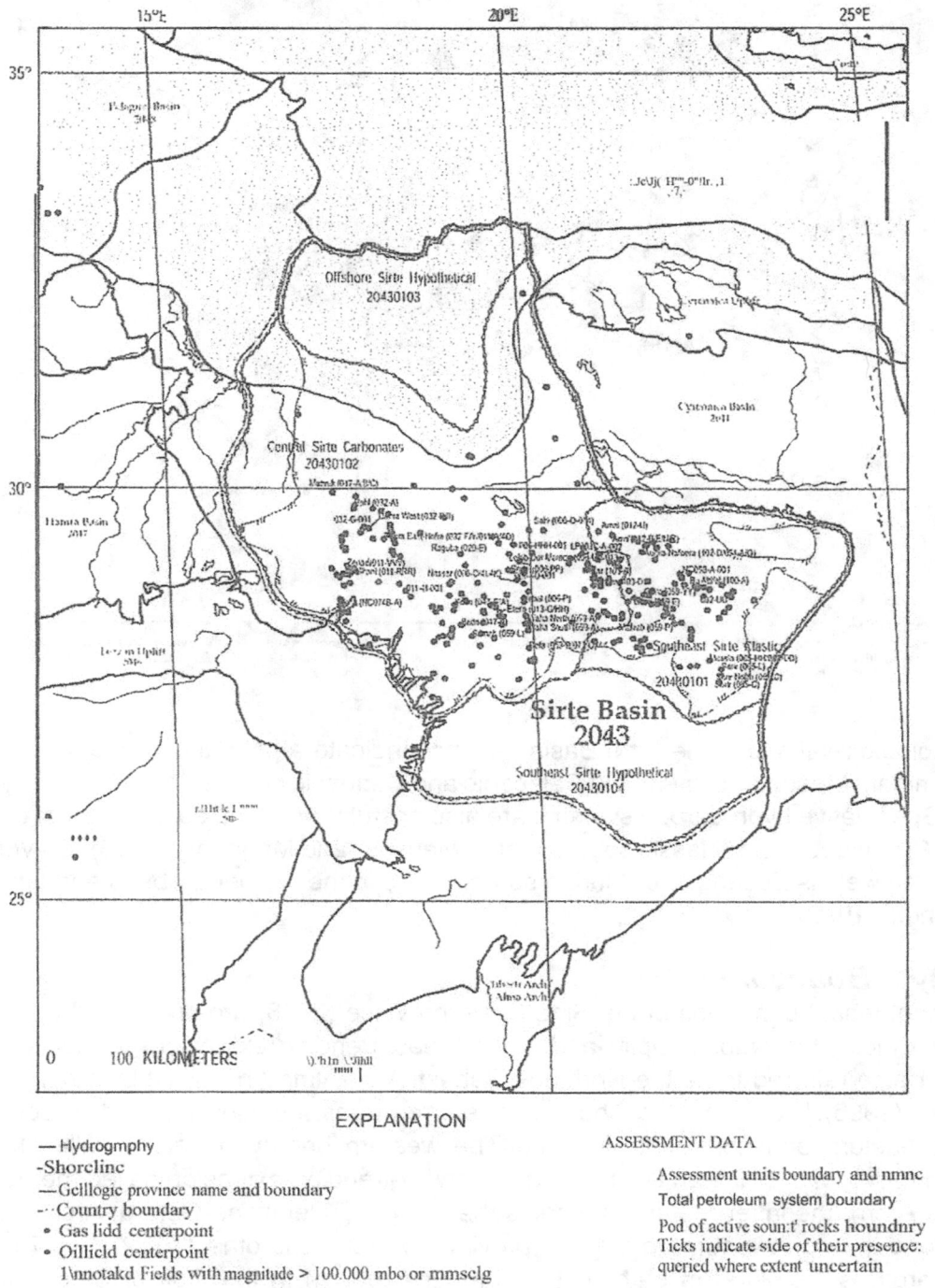

Figure 1. Sirte-Zelten Total Petroleum System showing boundaries of the total and minimum petroleum system, pods of active (thermally mature) source rock, four assessment units, and oil and gas field "centerpoints (named fields exceed 100 million barrels of oil equivalent). Note the 200m bathymetric isobath forms boundary between Offshore Sirte Hypothetical Assessment Unit (20430103) and Central Sirte Carbonates Assessment Unit (20430102). The 2.000 m bathymetric isobath forms northern boundary of province, total petroleum system, and Offshore Sirte Hypothetical Assessment Unit 20430103. Projection: Robinson. Central meridian: 0.

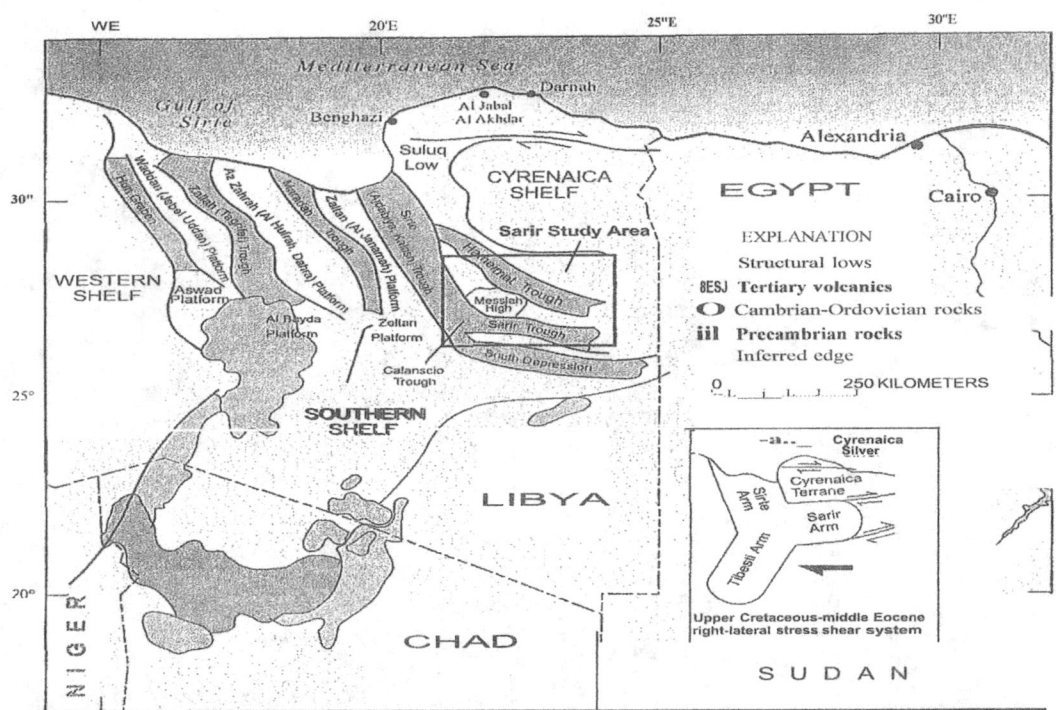

Figure 2. Structural elements of Sirte Basin. Troughs and grabens, platforms and horsts are synonymous terms. Individual horsts and grabens possess multiple names. For example, the Sirte (Sirt) Trough is also known as the Kalash or Ajdabiya Trough, as noted (modified from Ambrose, 2000). Barbs show direction of relative movement on faults.

Offshore, geologic relations in the Sirte Basin province indicate a potential for major reserves to be added by the dominant Mesozoic system, but Paleozoic and Cenozoic petroleum systems may be proven to exist as well. Speculative hydrocarbon systems are also postulated for the eastern part of the province, including Lower Cretaceous and Triassic source rocks (Mansour and Magairhy, (1996); Burwood (1997); Ambroze (2000), as well as Eocene and Silurian source rocks in the deeper grabens and offshore areas (Hallet and El Ghoul, 1996).

7.C.4 Geology & Boundary

The south and southeast boundaries of the Sirte Basin province (2043) are drawn at the Precambrian-Paleozoic contact along the Nubian Uplift and its northeast-trending extension, termed the Southern Shelf (figs.1,2) or also referred to as the Northeast Tebesti Arch/Alma Arch uplift by other authors (e.g. Futyan and Jawzi (1996).The Cyrenaica Shelf (also referred to as a platform, including both basin and uplift) forms the eastern and northeastern border.The western border, generally called the Western Shelf , is a combination of the Nubian Uplift and a northwest-trending extension called the Fezzan Uplift (Tripolias Sawda Arch); the latter feature intersects the Nafusa (Talemzane-Gefara) Arch, an east-west trending arch along the northwest margin of the province (Persits and others, 1997;(Fig 7.C3 or fig.3), The northern margin is the 2000 m bathymetric contour (isobath) in the Gulf of Sirte. Offshore, the province is separated from the Pelagian Basin petroleum province (2048) by the Medina Bank **(fig. 3)**.To the west is the Hamra Basin to the south of the Murzuk Basin, and the east the Cyrenaica Basin (2041). Alternative basin outlines have been drawn (for example, Montgomery, 1994; Futyan and Jawzi, 1996, Selley, 1997). However, the Sirte Basin boundary outline as just described was drawn upon surface geologic and subsurface geologic maps prepared at a scale of 1:5,000,000 by UNESCO and shown on supporting maps such as the African geologic map prepared for the World Energy Project (Persits and others, 1997).

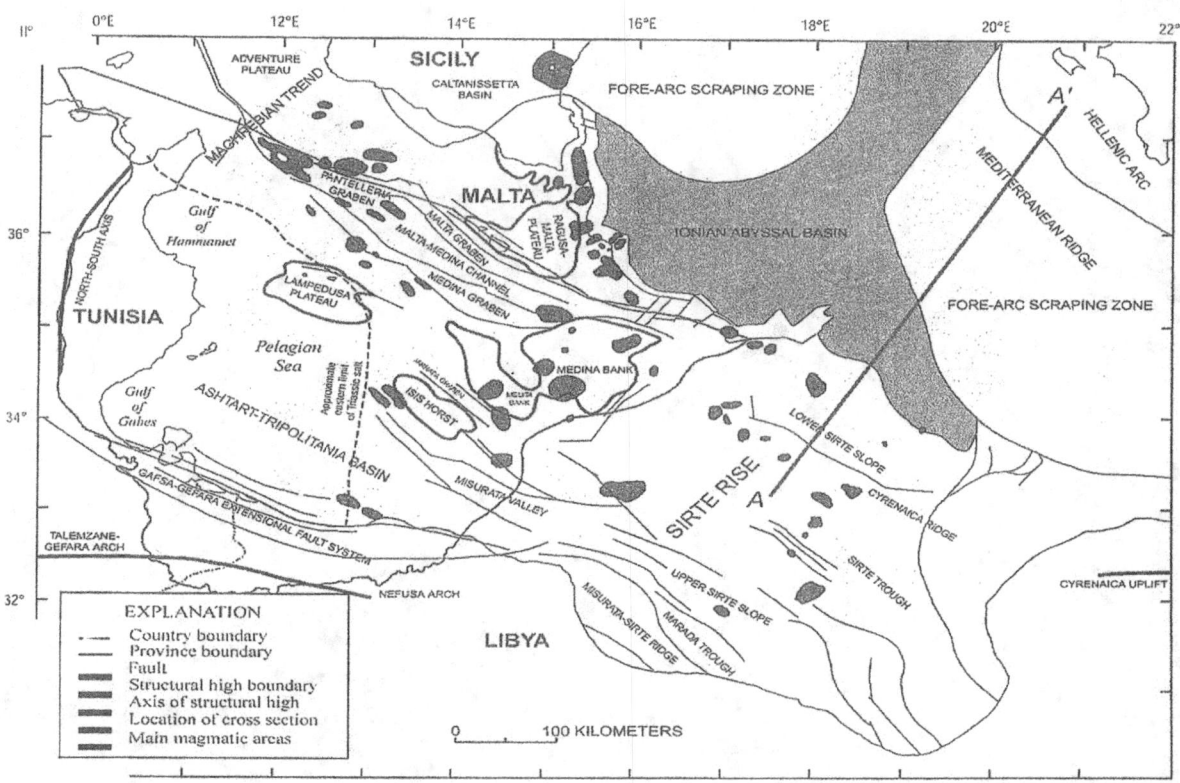

Figure 3. Central Mediterranean Sea, showing USGS-defined Pelagian province and offshore areas of Sirte Basin, the Sirte Rise. Structural high boundaries are based on seismic data and drawn on 1.0 and 1.8 second two-way travel time intervals (modified from Finetti, 1982; Bishop, 1988; Jongsma and others, 1985; and Klett, in press). Section A–A' is shown in figure 4.

7.C.5 Geographic Settings

The Sirte Basin is a triple-junction continental rift along the Northern margin of Africa (fig 2). It is bordered on the north by the Gulf of Sirte (Sidra) in the Mediterranean Sea. Although the Nubian Uplift rises to 3,000m south of the Sirte Basin, much of the land area in the bsin is characterized by desert steppes and includes eolean deposits of the Kalanshiyu and Rabyanah sand Seas of the Sahara Desert. In a relatively narrow, north coastal strip, some land areas are as much as 47m below sea level. The Sirte Basin is roughly the size of the Williston Basin In North America (~490,000 Km); Libya is the forth largest country in Afica, and the 15th largest country in the World. (http://geology.cr.usgs.gov/pub/bulletins/b2202-f).

BY

U.S. Department of Interior
Gale A. Norton, Secretary

U.S.Geological Survey
Charles G.Groat, Director

Version 1.0, 2001

PART IV

A Long Term Petroleum Project
"A Precise Research Study on a Virgin Location in NW Africa, correlated with Regional structures which led to the discovery of an
Oil-Ring within an important Gas Field"
-for the first time in the Middle East

Chapter 8
A Regional Petoleum Study of Algeria- A General Review

8.1-Historical Aspect

The National Algerian Petroleum Society (Sonatrach: Societé Nationale De Transport et de Hydrocarbures), increased its share with the French State "Sopefal" after its independence in 1962, whereas, Libya increased significantly its oil production after the military revolution in 1969; in the same period, Tunisia and Morocco, awarded large blocks off shore aereage to French companies; Russian and American oil companies discovered gas in the Nile Delta area and oil in the North western Coastal zone f Egypt. also, the National Society of Transport and Petroleum of Algeria "Sonatrach", started a precise survey of the existingmaterials for establishing prospects of petroleum hydrocarbons, and set up a work program for research in diverse regions (Technoexport 1971).

Alieve, M.M., et al (Technoexport, 1971- Sonatrach), the Head of the Soviet Delgate of geologists and engineers published two volumes on the geological structures, stratigraphy of the Saharan Atlas and the Saharan Platform; both volumes presented a tremendous effort that last for 15 years to systematize and generalize a great deal of geological and geophysical data over thousands of diagraphs, besides 750 reports on composite well logs, 300 geophysical reports and 250 Geology reports.

The Exploration activity in Algeria fluctuated through time due to three main events; a) The National Algerian struggle for independence from 1954 to 1962; b) The nationalization of oil industry in 1962, took a decade to gain the outcome and c) The political unrest and economic problems through 1980s, during which Sonatrach had been concentrating on the Triassic province in different areas throughout the Algerian Saharan Platform, urged the senior professionals to proceed research studies to provinces other than the Triassic reservoir.

Not all areas in Algeria are accessible for exploration as shifting sand of African Saharan deserts, presented technical difficulties in exploration, and hazards in production operations. Recent advances in gathering, processing, and reprocessing of seismic data, allowed exploration beneath sand-sea environ-ments such as the Algerian Grand Erg Oriental where the Ghadames (Berkine) and Illizi Basins are located.

New petroleum discoveries could be achieved without construction of major pipelines as the basic infrastructure had been already established (Sonatrach, 1972); Algeria has an extensive pipeline network that connects most of the major producing areas to port cities in Algeria and Tunisia, therefore facilitated transportation of oil, gas, and natural gasoline.

8.2 Location and Morphology

The Algerian territory covers an area between Latitudes: 20° 00'– 36° 00' N and Longitudes: 10° 00' E–08° 00' W, which is approximately equal to its neighbors combined. Algeria comprises the major portion of the Arabian Maghreb where it is bounded by Morocco and Mauritania from the west, Tunisia and Libya from the east, and the Mediterranean Sea that extends in the north for 1200 km of shoreline. The prime meridian (Greenwich) runs through the city of Mostaghanem. On the southern limit of the Algerian Sahara, the Hoggar range is 2000 km from the Mediterranean coastline, whereas, Ain Aminas in the East, is a long distance of ~ 1800 km, to Tindouf in the west.(**see Fig 6.3-Natural Boundaries of Algeria and Morphological Zones**). (http://www.mem-algeria.hydrocarbons. geology.htm).

Morphologically, the Algerian country is divided from north to south by 3- zones: (1) The Tellian Atlas (Tell), the northern most-steep relief, is flanked by rich coastal plains in the center, Chelif to the west and the Seybouse plains to the east; (2) The Epihercynian Platform includes the more stable northern High Plateaus of delineated features, and the southern active part of the Saharan Atlas; a mountain range of Alpine origin which is a chain of ENE/WSW oriented reliefs, extending from the Moroccan border to Tunisia; (3) The Algerian Sahara to the south, covered by large sand dunes, (east and west Erg; or "Oriental and occidental") and gravel plains (or rigs) with dispersed oases (e.g. El-Ouad, Ghardaia and Djanet). (**see Fig 6.5**- Regional Structure Map of the Algerian Sahara-Sonatrach, 1968).

8.3 Geological Framework of the Algerian Alpine

Technoexport (1971) discussed three tectonic zones of different structures from different ages and the evolution of the folded mountains: The Tellian region, the Epihercynian (mainly of the northern High Plateau and the southern Saharan Atlas), and the African Precambrian platforms.

The geology of the Algerian Atlas regions is mentioned in the publications of the "Service de la Carte geologique d' Algerie", 1972, and lately, in the Bulletin of the French Geological Society and in the Findings of the French Academy of Sciences together with the bibliography of works published before 1965, concerning various problems of the geology of North Algeria and the Sahara that were gathered in one volume by Merabet in 1969 (Sonatrach, Tecnoexport, 1971).

The northern Algeria, mainly affected by Alpine orogeny, is formed of young mountains during Tertiary times; the northern domain of the Algerian Alpine consists of a number of structural sedimentary units from north to south as follows:

8.3.1 The Northern Off-shore Area
The northern off-shore area is a reduced continental shelf of Tertiary and Quaternary sediments (1000-3500ms; of Mio-Pliocene objective) that overlies a metamorphic basement.

8.3.2 The Tellian Atlas
The Tellian Atlas is a northern complex area comprising a chain of nappes set in place during the Lower Miocene. The super imposition of the Nappes is observed in the areas of greatest development; their total thickness measures 1100m. Late Neogene's mountain basins (e.g. Chelif Basin), were deposited over these nappe structures (oil objectives are Middle Cretaceous and the allochthonous Miocene and Eocene formations).

Kieken (1962;Sonatrach, Tehnoexport, 1971) claims that the units of Nappe found in different regions under different names often have much in common in their stratigraphy and their tectonic settings. He has tabulated the information on the Nappe units and grouped them into six classes from top to bottom:

1) Numidian, 2) Sub-Numidian, 3) Epitellian, 4) south Tellian, 5) Infratellian, and 6) Flysch.

The Technoexport (1971) adds a separate classification of sedimentological clips, which are fragments of heterogeneous rock layers buried in autochthonal sediments. "Flysch Nappe" is shown as an independent entity, which is located between the Numidian and Sub-Numidian Nappes. The Technoexport explained the Numidian Nappes and Flysch, as been developed in the coastal zone at the northern slope of the Great Kabylie massif and on the southern side, probably originated from the Mediterranean "geosyncline" before the Miocene time.

Towards the beginning of the Alpine cycle, the essential geostructural elements of the Algerian Atlas primarily formed at the same time with the external and internal zones of the peri-Mediterranean mountainous orogenesis, e.g. The Kabyle geanticline and the Tellian Atlas trough at the border of the High plateau, in which the Mesozoic trough of the Saharan Atlas formed farther the south.

There are several stages of evolution corresponding to the sudden tectonic changes at the folded mountain domain of Algeria. These stages include the permanent periods of sinking or rising and were marked in time by the boundaries related to the folding movements or to orogenesis:

- The early geosyncline stage - Lower, middle and upper Jurassic
- Middle Geosyncline stage - Lower Cretaceous and Cenomanian
- Late Geosyncline stage – Turonian - Lutetian
- Early orogenic stage – Preabonian – Tortonian
- Late orogenic stage – upper Tortonian – Anthropogenic

8.3.3 The Hodna Basin
The Hodna Basin is a fore-deep basin of continental Eocene and Oligocene deposits at the base, overlain by Marine Miocene sediments (Eocene objective).

8.3.4 The High Plateax
The High Plateaus of delineated features in the center with large plains to the west (e.g. Oran Meseta) and highs to the east (e.g. Ain Reghada), are the foreland of the Alpine range bearing a thin sedimentary cover (Liassic objective).

8.3.5 The Saharan Atlas elongated trough
The Saharan Atlas pinched between the high plateaus and the Saharan Platform, is filled-in by thick Mesozoic sediments (7000-9000 ms). The Tertiary compressive tectonic stresses later modified the trough into a number of reverse faulted structures that led to the creation of the mountain range (Jurassic is the objective).

The folded range of the Saharan Atlas is differentiated from the adjacent High Plateaus area by several features **(Fig 8.1- A Tectonic map of the Saharan Basin in Northwest Africa, Sonatrach, 1968)**, the most important features of the Saharan Atlas structures are::

a) The Saharan Atlas-folded structures with an overall ENE/WSW trend, are characterized on its southern border, by a group of flexures and successive faults which define the Saharan flexure zone.
b) The Saharan Atlas consists of a mighty complex of Triassic, Jurassic and Lower Cretaceous rocks with great thicknesses (up to 12,000 meters) of sediments; in other words, the total thickness of the sedimentary formations of the Mesozoic regions of the Saharan Atlas in the northwest, undergone the maximum sinking that exceeds 10,000ms.
c) The increased thickness of Jurassic and Lower Cretaceous sections of the Saharan Atlas indicates a deep subsidence area on the southern unstable border of the Epihercynian platform.
d) At the end of the early Alpine cycle, thick sediments have been folded and faulted and narrow anticlines with dissected crests and vast, flat-bottomed synclines were formed (box-like folding).

J. Savornin (1931) who was the first to undertake the tectonic zoning of the Algerian Atlas Territory, established the following structural elements; tabular zones; alluvial plains; folded zones of the Saharan Atlas; outcrops of Pre-Cambrian basement; the undifferentiated Paleozoic (e.g. the Kabylies and others), eruptive and Plutonic massive rocks. On his map, he showed though schematically, the largest structural stages of the metamorphic basement and the un-differentiated sedimentary cover. The folded mountaineous Tellian region shows signs of the two last phases of evolution: the "geosynclinal" and the orogenic phases which were used as the basis for drawing up a tectonic map. Following up this metho-dology, the position of the old Paleozoic formation, the main anticlinoriums, and the synclinoriums could be easily indicated. The well known stages of the Algerian Atlas are: Early, Middle and Late Geosynclinal stages, also the early and late orogenic stages. The interval between the age of each of these stages and the corresponding structural units were determined by the essential phases of the geosynclinal evolution and its transformation into a folded terrain; these phases are: initial sinking during the course of the Jurassic; formational phase of the basic "geosynclinal troughs" of the Tellian region during the lower Cretaceous and Cenomanian; beginning of inversion in the geosynclinal troughs (pre-orogenic), corresponding to upper Cretaceous.-Eocene). During the orogenic phase, two stages can be distinguished in the folded Tellian area; early Orogenic (lower Oligocene-Miocene) and Late Orogenic (upper Miocene-Pliocene).

In addition to the geosynclinal and orogenic stages of a mountainous structural group in the small and Great Kabylies, there are the older Caledonian and Hercynian complexes, which make up the rocks of the folded domain.

The zone of the south Atlas faults, known as the Atlas Flexure, separates the Saharan Atlas from the Saharan platform; the age was set at tectonic phases between Lutetian and the Burdigalian. In addition to the large faults which extend hundreds or even a thousand kilometers, the sub Atlas system consists of numerous folds throughout north of Algeria: **"The folded Tellian region"**, where there are many outcrops of Triassic evaporites. The presence of the most important faults in the sub-Atlas system was indicated by seismology, loss of correlation and sometimes, shifting of horizons at the crossings of the faulted areas.

From a historical standpoint, the Saharan Atlas was considered by some geologists as a spread-out Mesozoic trough formed on the southern periphery of the Epihercynian platform during the end period of the Paleozoic geosyncline. The origins of the present structure, according to that view, are due to mountainous phases, of which the oldest is the Pyrenean.

C. Augier (1967; Technoexport, Sonatrach, 1971) considered the Saharan Atlas as a deep Mesozoic trough with neritic and continental sediments. However, he assumed that the present structure of the Atlas was formed by a series of tectonic movements during several successive stages. These uplifts, the most recent of which took place during the Cenomanian, caused marked the sedimentation gaps in the anticlinal areas and the first great folds in the Saharan Atlas trough. This hypothesis explains the Saharan Atlas rather by "geosynclines" with periods of uplift, inversion and by its own orogenesis, and related formations, magmatism and folding that may lead us to consider the Saharan Atlas an "intracratonic geosyncline".

The most characteristic peculiarity of the Saharan Atlas is of its complete tectonic origin that prevailed allover the North African Atlas and Tunisia. The tectonic setting is characterized by an alteration of anticlines which extends dozens of kilometers in length in the form of narrow crests and large flat ynclines. The summits are generally folded structures containing pockets of Triassic gypsum and anhydrites, e.g. the anticlines of Bou-Lerhfad and Dj. Mecharia to the southwest of the Saharan Atlas; Dj. Azreg and Dj. Ain Mahdi in the central region; and Dj. Zerga in the northeast.

The Mesozoic sediments of the Saharan Atlas trough, are quite different from the formational series of "typical geosynclines". Instead, they are similar to the structures of a sub-platform. The Triassic sediments are intercalated by basalts and rhyolites which are greatly developed in the Algerian and Moroccan Atlas. The Triassic has only been studied in the southwestern part of the Saharan Atlas. At the bottom of the section, terrigenous Triassic sediments are replaced by a rather thick salty formation.

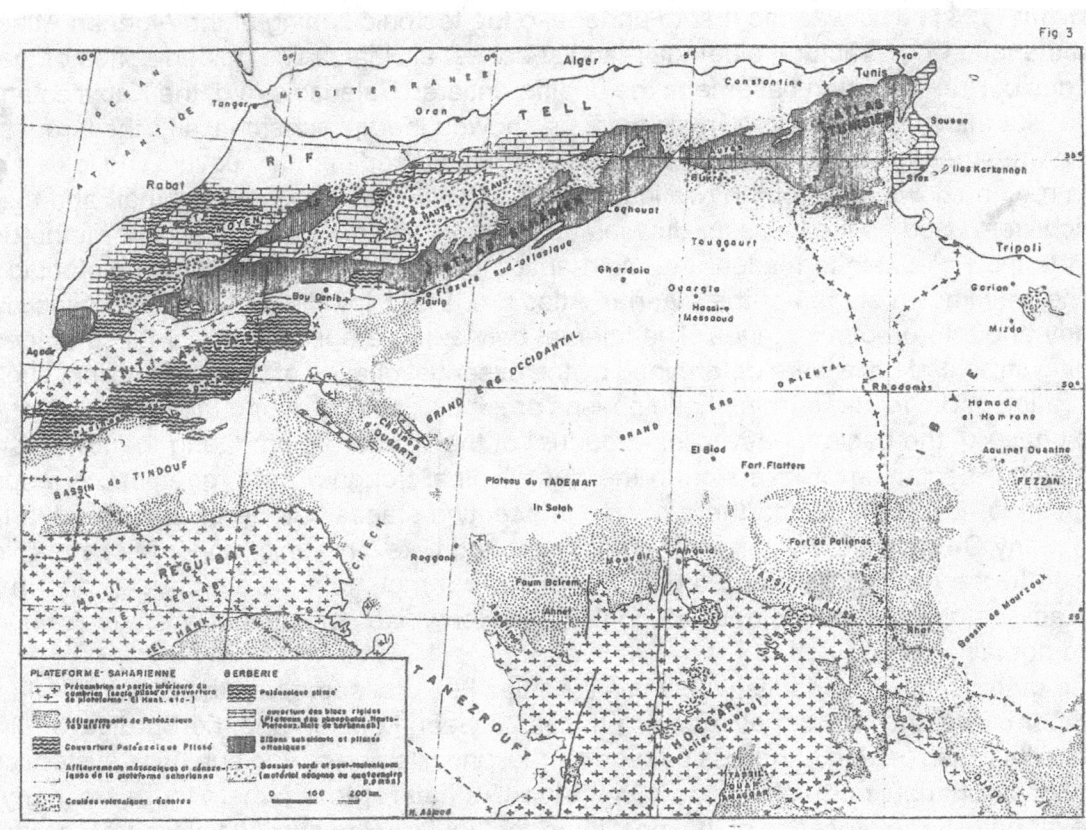

Fig 8_1a - A Tectonic Map of the Saharan Basin in Nortwest Africa, (Sonatrach, 1968)

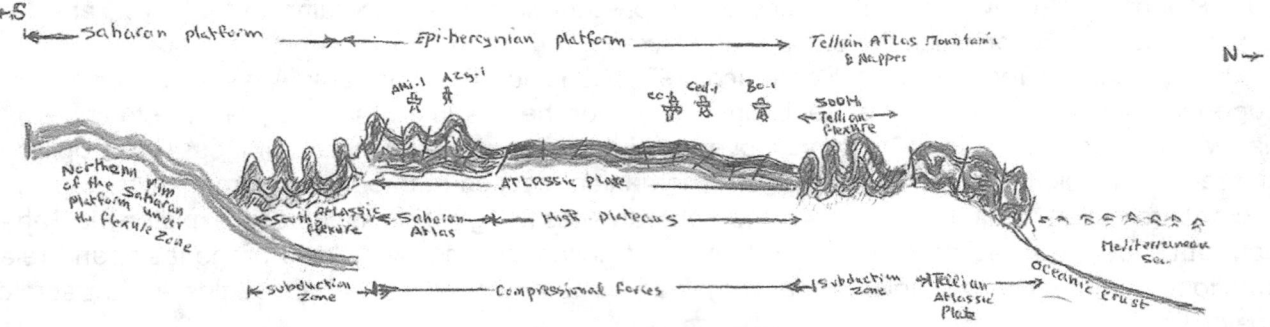

Fig 8_1b - A postulated sketch Diagram showing the Tectonic structures in Algeria

8.4 Geodynamic Evolution

Based on the global geodynamic process of plate tectonics, the sedimentary basin in Algeria can be divided into three main distinct domains, the Telliian Atlas, the Epihercyniain Platform (High Plateaus and the Saharan Atlas) and theSaharan Platform. The development of the Mediterranean Alpine Ranges of the Algerian territory is the result of the rotation of the African continent with respect of the Eurasian continental plate. Such rotation is a slow drifting of the continents. The tectonic environment due to the collision between Africa and Europe pruduced the Algerian Alpine that lay on the northern fringe of the African plate. The continental convergence w hich probaly started in Lower Jurassic times, can be clearly recognisable during the Upper Jurassic (150MY) when the first signs of drifting of Africa and Europe only became clear, even through the phenomenon can be probably intiated during early Liassic times (180 MY), being associated with the opening of the North Atlantic.

8.5 - Tectonic Outline

Rifting affected large parts of the area in the Cambro-Ordovician and is considered as the early Paleozoic *first extensional phase,* that had ended by the Hercynian event (Late Carboniferous) in the Atlas area due to a plate collision with Europe. Most of the Paleozoic basins in the interior of the neighboring countries are shown as broad shallow basins (Sag features), separated by many long structural highs (e.g. Tibesti-Sirte Ridge) over the earlier rifts (Boudjema, 1987). Inversions associated with the Late Hercynian compression event decrease in intensity from northwest across Algeria and Tunisia but do not seem to extend significantly into Libya. Broad Hercynian folding superimposed on early Paleozoic structures led to the development of arches on which Hassi Messaoud field is cited (Balbucchi and Pommier, et al- 1970); *a second phase of extension* commenced in the Early Mesozoic with the development of a series of rifts along the eastern margin of the Atlantic and the southern margin of the Tethys Ocean (Mengnoli Spinicci, 1985); rifting began on the Atlantic margin in the aTrissic, and extend to the Atlas area in the Liassic and into the eastern Mediterranean in the Middle Jurassic (Dixon and Robertson, 1984). Rifting continued to spread in the middle and late Jurassic in the Atlantic and western Mediterranean and in the Early Cretaceous, east of the Mediterranean. A trend of Mesozoic passive margin basins and rifts was thus created, extending from the Moroccan margin through the Atlas and Pelagian Basins (Tunisia and Libya) to the Abu-Gharadiq Basin in Egypt (Awad, 1984).

The opening of the eastern Mediterranean in the Early Cretaceous led to the formation of the multiphase rift structures of the Sirte Basin (Guiraud and Maurin, 1992). Movement on interior transform faults in the Aptian, led to localized structural inversions within the Ghadames/Triassic Basin (Budjema, 1987), while a change in the relative movements of Africa and Europe in the Late Cretaceous led to more regional-scale inversions, particularly in the Atlas and Western Desert areas of the Arabian Maghreb (Mengoli and Spinicci, 1985). All the Cretaceous tectonic events were significant in creating hydrocarbon traps in the encountered basins.

The effects of the Alpine orogeny were largely confined to the Atlas area and adjoining basins. The Alpine orogeny in the Atlas region has apparently led to at least partial destruction of pre-existing petroleum systems. This is clearly evidenced by the large numbers of seeps in the encountered area, particularly in the Saharan Atlas. In most regions of the world, there is a good relationship between seepage and reserves at a basin scale (Macgregor, 1993), but this is clearly not true further to the north; e.g. Thrusting and Nappe-formation occur in the northern Atlas (Tellian Basin), passing into milder inversions in the northern Pelagian Basin (Bishop, 1975) with additional minor effects in the Ghadames/Triassic area.

8.6 Updated Study on the Main Geological Regions of Algeria

8.6.1 Scope
A Geologic province is an area having characteristic dimensions of hundreds of Kilometers that encompasses a natural geologic entity, e.g. Sedimentary basin thust belt, or accreted terrain or some combination of contiguous geologic entities. Each geologic province is a spatial entity with common geologic attributes. Prnrovince boundaries were drawn as logically as possible along natural geologic boundaries,

although in some regions, they were located arbitrarily, e.g.along specific water-depth contours in the open oceans. **Fig 8.2- Major Petroleum Basins in N. Africa (Modified from Macregor, D.S. 1994, BP / UK).**

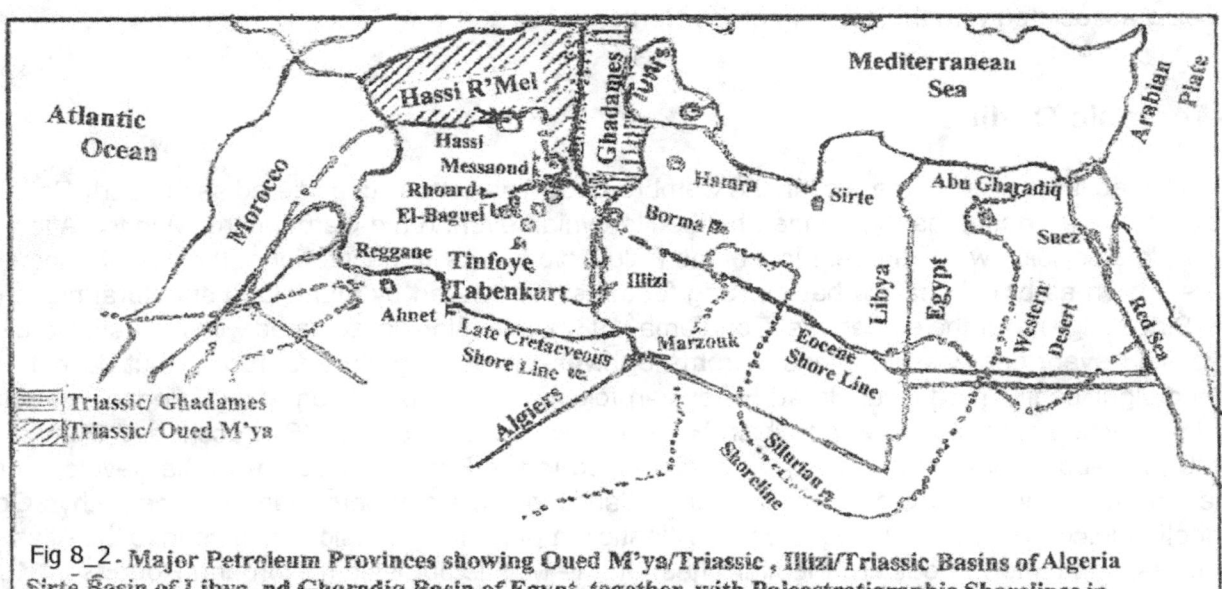

Fig 8_2- Major Petroleum Provinces showing Oued M'ya/Triassic , Illizi/Triassic Basins of Algeria Sirte Basin of Libya, nd Gharadiq Basin of Egypt, together with Paleostratigraphic Shorelines in North Africa (modified from Macregor, D.S., 1994 – BP. Explo. Uxbridge, UK.

The geological provinces might be defined in relation to the main region of the Algerian Saharan Platform: The Western , the Northeastern Trissic and the Eastern Regions.

8.6.2 The Western Saharan Region
The Western Saharan region mainly of the Grand Erg-Ahnet Province that includes Tindouf Reggae, and Bechar basins to the far western border of Algeria, whereas the Triassic basin encompasses the northeastern portion of the Grand Erg/Ahnet Province as well as the northwestern portion of the Trias-Ghadames Province.

8.6.2.1 The Tanezzuft-Timimoun Total Petroleum System (205801) of the Grand Erg/Ahnet Province- An example
   The **Grand Erg/Ahnet provi\nce-** is a geologic province where its northwest limit extends slightly into Morocco and encompasses approximately 700,000 Km.2. More than one total petroleum system may exist within each of the basins in the Grand Erg-Ahnet Province. The "composite" total petroleum systems are coincident with the basins in which they exist, e.g. Tnezzuft -Oued M'ya Total Petroleum system extends across both the Grand Erg /Ahnet and the neighboring Trias/Ghadames Province.
   Tnezzuft refers to the Tenezzuft Formation (Silurian) which is the oldest major source rock in the total petroleum system. Tanezzuft is then followed by the basin in which the total petroleum system exists **(See Table 6.1** – A numeric code to identify each region, province, total petroleum System, and assessment unit; the criteria are uniform throughout all other provinces(USGS, Klett, 2000).
   The Tanezzuft -Timimoun total petroleum System (205801) of the Grand-Erg /Ahnet province is given as an example; its prinicipal source rocks are the Upper Silurian (?) Tanezzuft Formation, and the Middle and Upper Devonian mudstone, which were deposited during a major flooding event and contains suprobelic and mixed Kerogen (Daniels and Emme, 1995, Makhos and others, 1997). The organic-rich graptolitic, marine mudstone of the (Lower?) Silurian Tanezzuft Formation **(or its lateral equivalents)**

overlies the Ordovician section. The present-day Total Organic Carbon (TOC) content ranges from about 2 to 4 percent across the Timimoun, Ahnet, and Mouydir basins (Makhous and others, 1997). The TOC content is presumably greatest at the base of the section as in the neighboring Ghadames (Berkine) and Illizi Basins (Daniels and Emme, 1995).

Petroleum hydrocarbon is presumed to have been generated during the Carboniferous, when the Paleozoic section was thickest, but was halted during uplift associated with Hercynian deformation (Makhous and others, 1997; Boote and others, 1998). Migration and charge are presumed to have occurred during the early stages of Hercynian Deformation, prior to major uplift and erosion. Petroleum most likely migrated vertically along faults or fractures (Boote and others, 1998). A later phase of dry gas generation is hypothetical occurred due to igneous activity in the Late Triassic, (about 200 M years ago), (Klett and others, 2000).

The Grand Erg/Ahnet Province contains more than 5,500 million barrels (MMB) of known "estimated total recoverable which is cumulative production plus remaining reserves" petroleum liquids (about 500 million barrels of oil "MMBO"; 5000 million barrels of natural gas liquids, "MMBNGL"); and approximately 114,000 billion cubic feet of known natural gas "114 x10^{12} CFG, or "114,000 BCFG", PetroConsultants, 1966a).

The Grand Erg /Ahnet province which encompasses the giant Hassi R'Mel field, is mainly of a Precambrian basement unconformably overlain by thick Paleozoic sediments, structurally defined by four basins (e.g.; sandstone reservoir rocks of Cambro-Ordovician, Silurian, Devonian, and Carboniferous age), ranging from Cambrian to Namurian (Upper Carboniferous), that are separated by high zones, followed by insignificant thickness of Mesozoic and Cenozoic sediments. **Fig 8.3** – USGS - Geologic province of Grand Erg/ Ahnet and the locations of two stratigraphic cross sections, Bulletin 2202-B **(cpg.cr.usgs.gov/pub/bulletin.html-49k). Figs 8.4.ab** – NE-SW and NNW-SSE stratigraphic cross sections passing by Grand Erg/Ahnet Province.

The major structural features of the Grand Erg/Ahnet Province constitute several structures including the Idjerane-M'Zab structural axis and the Tilrhemt Arch, and Maharez Dome, where Hassi R'Mel dome structure and most of the Oued M'ya Basins exist.

During Late Precambrian and Early Cambrian, erosion of the pre-existing craton occurred to the south due to an uplift, during the Pan African deformational event. Eroded sediments are deposited northward as alluvial and fluvial deposits of a thick Cambrian sandstone section. The sandstone is laterally equivalent to the Hassi Messaoud and Hassi Leila Formations, which are major oil and gas reservoirs in the neighboring Trias/ Ghadames and Illizi Provinces **(Pubs.usgs.gov/bul/b2202)**.

Applications of Petroleum Tools for Field Geologists

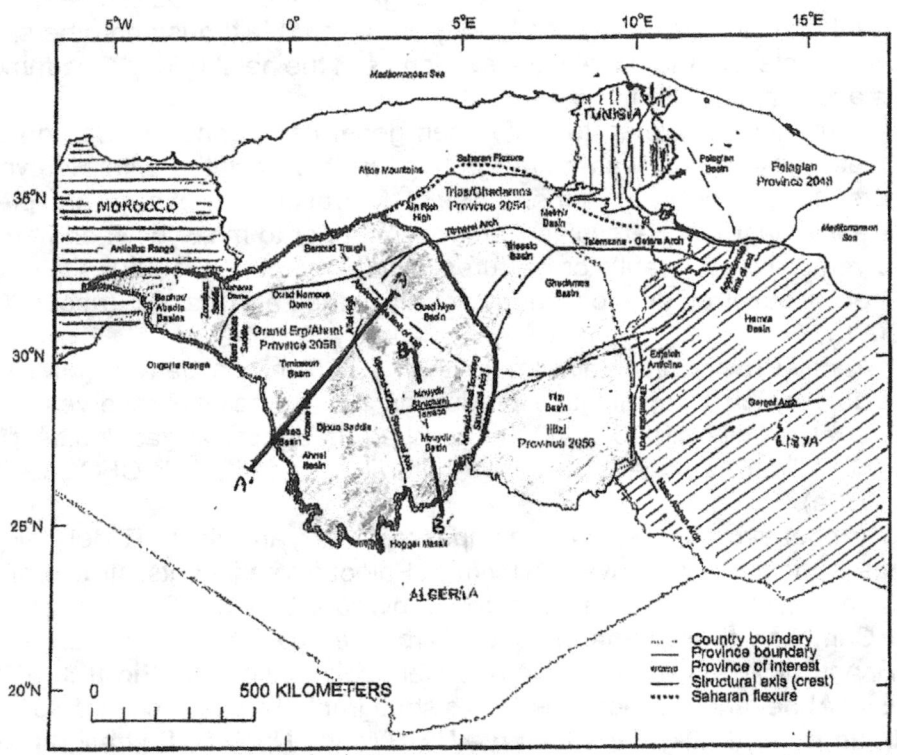

Fig 8.5 - North-central Africa, showing USGS-defined geologic provinces and major structures (modified from Aliev and others, 1971; Burollet and others, 1978; Montgomery, 1994; Petroconsultants, 1996b; Persits and others, 1997).

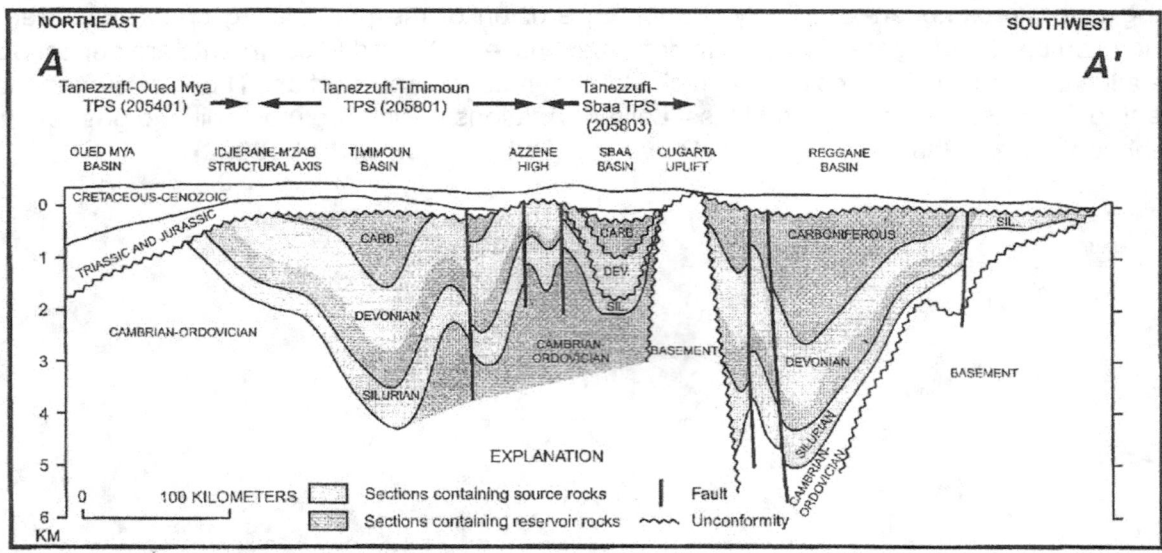

Fig 8.6 Stratigraphic cross sections through Grand Erg/Ahnet and neighboring provinces. A Northeast-Southwest stratigraphic cross section through the Timimoun, Sbaa, and Reggane Basins (modified from Aliev and others, 1971; Makhous and others, 1997).

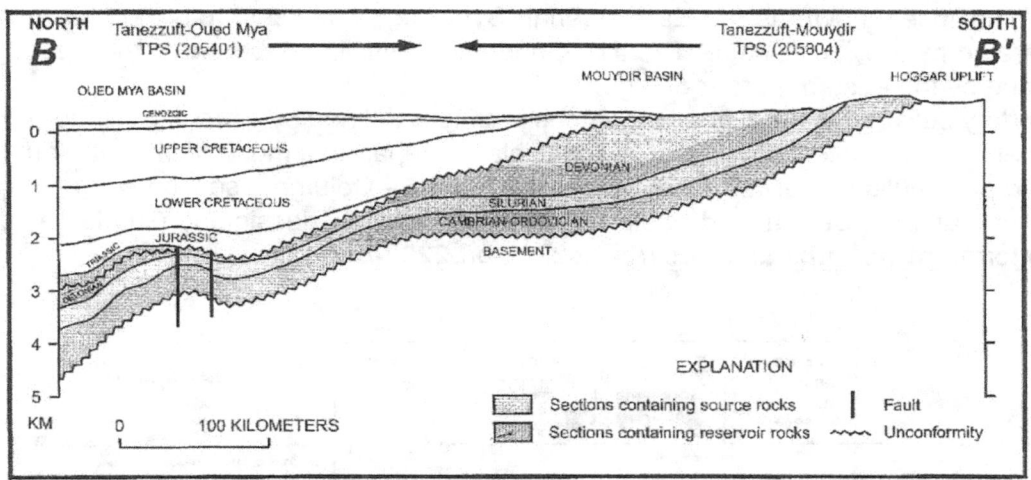

Fig 8.4.b -Stratigraphic cross sections. B, North-to-south stratigraphic cross section through the Oued Mya and Mouydir Basins (modified from Makhous and others, 1997).

8.6.3 The Northeast Central Triassic Region

The Triassic Region of Algeria divides the Northern Sahara into eastern and western basins where Paleozoic strata form a North/South trending ridges along the longitude of Algeria; the sediments transgressed the northeastern Sahara and northern Tripolitania from Permian up to Triassic.

The Triassic region, is a large, eastwest-oriented anticlinorium, located on the NE Central portion of the Saharan Platform, and comprises of four structural features:

a. The Tinrhemt Dome-Talemzane structure, maily contained in the Grand Erg/Ahnet Province that encompasses the Hassi R'Mel gas field and most parts of the Oued M'ya Basins.
b. The Djemaa-Touggourt structural system
c. The El-Algreb-Messaoud system where major oil fields are located in the Cambrian reservoirs of the Hassi Mssaoud field; it is mainly contained in the Trias/Ghadames province.
d. The Dahar Uplift.

The above structural features are separated by depressions, e.g. Oued M'ya where typical Triassic province formation is encountered and the Paleozoic deposits that are often eroded down to the Cambro-Ordovician level. The Oued M'ya basin which is infilled by thick Paleozoic and Mesozoic formations (~5000 ms), encompasses gas and oil in the Triassic reservoirs of Oued Noumer and Ait Kheir structures.

The Mesozoic rock section is thin across much of the Grand Erg/Ahnet province, but thickens to the northeast in the Oued M'ya and Benoud Basins. Triassic rocks include a lower clastic and upper evaporitic formation that grades into the lower Jurassic section. The approximate extent of Triassic Basin is defined by the extent of the Triassic-to Jurassic evaporites limits. The clastic formation which constitutes the main oil and gas reservoirs **(Table 8.1, Stoica and Assaad, 1981)**, is subdivided into three units: The lowermost continental (fluvial) shaly sandstone: "The Trias Argilo-Greseux Inferieur" that grades upward to; (2) Dolostone, dolomitic clays and anhydrite beds of: "The Trias Argilo-Carbonaté"; (3) The Upper alluvial clay, siltsone and fine to medium-grained sandstone of: "The Trias Argilo-Greseux Superieur".

The Uppermost Triassic-Lower Jurassic cyclic sequence, known as the saliferous units, mainly of interbedded salt, anhydrite, gypsum, dolostone, and shales, form a regional seal for many oil and gas reservoirs, and are thickest near the Saharan Flexure in the north, and thin southward.

8.6.4 The Eastern Geologic Provinces

The Eastern Province, known as the East Algerian Syncline, comprises both the Illizi and Ghadames basins, separated by the Ahara-ridge. The Illizi-Ghadames Syncline is bound to the west by Amguid-El Biod dorsal and by the eastern border of Algeria.

The boundary between the Illizi and Ghadames Basins is defined by a break or hinge line in the slope of the basement rocks. The hinge line was responsible for separating much of the petroleum generation, migration, and accumulation between the two basins. **Fig 8.5** – Columner section: A stratigraphic nomenclature and correlation of Central and Eastern Basins of the Illizi, Triassic and Ghadames basins of the Saharan Platform. (http://pubs.usgs.gov/bul/b2202-b/b2202-bso.pdf).

System	Stage	Illizi Basin (van de Weerd and Ware, 1994)	Triassic Basin (Boudjema, 1987)	Ghadames (Berkine) and Hamra Basins (Montgomery, 1994; Echikh, 1998)	Description (Boudjema, 1987)
					Hercynian Unconformity
Carboniferous	Stephanian		Tiguentourine	Dembaba	Mudstone, limestone, and gypsum
	Westphalian	F	El Adeb Larache		Limestone, gypsum, and mudstone
	Namurian	E	Oubarakat	Assed Jeffar	Limestone and sandstone
		D	Assekaifaf		Limestone and sandstone with concretions
	Visean	C		Mrar	Mudstone and sandstone
		B	Issendjel		
	Tournaisian	A	(Sbaa)		Limestone and mudstone
Devonian	Strunian	F2	Gara Mas Mulouki	Tahara (Shatti)	Sandstone
	Famen.-Frasnian		Tin Meras	Aouinet Ouenine	Mudstone — *Frasnian Unconformity*
	Givetian-Eifelian	F3			Sandstone
	Emsian	F4-5	Orsine	Ouan Kasa	Mudstone and limestone / Mudstone and sandstone
	Siegenian-Gedinnian	F6	Hassi Tabankort	Tadrart	Sandstone
					Late Silurian-Early Devonian Unconformity
Silurian			Zone de Passage	Acacus	Sandstone and mudstone
		"Argileux"	Oued Imirhou	Tanezzuft	Black mudstone with graptolites
			Gres de Remada		Sandstone
		Gara Louki	Argile Microgl.	Bir Tlacsin	Microconglomeratic mudstone — *Glacial Unconformity*
Ordovician	Cardocian		Gres d'Oued Saret	Memouniat	Limestone, sandstone, and mudstone
	Llandeilian-Llanvirnian	Edjeleh	Argiles d'Azzel	Melez Chograne	Silty black mudstone
	Arenigian	Hamra	Gres de Ouargla	Haouaz	Sandstone
			Quartzites De Hamra		Sandstone
	Tremadocian	In Kraf	Gres d'El Atchane	Achebyat	Sandstone and mudstone
			Argile d'El Gassi		Mudstone
Cambrian-Ordovician			Zone des Alternances		Sandstone and mudstone
Cambrian		Hassi Leila	Ro	Hassaouna and Mourizidie	Sandstone
			R2		
			R3		Sandstone and conglomerate
					Pan-African Unconformity
Infra-Cambrian			Socle	Infra Tassilian/ Mourizidie	Metamorphic and magmatic rocks

Fig 8.5 — Columnar section – A Stratigraphic Nomenclature and Correlation of Illizi, Triassic, and Ghadames Basins (Boudjema, 1987)

8.6.4.1 The Trias/ Ghadames Geologic Province

The Trias/Ghadames province, a geologic province delineated by the USGS; is located in east Algeria, southernTunisia, and westernmost part of Libya. It encompasses an area of approximately 390.000 kms^2, and coincides with the Triassic Basin; the province includes the Melrhir Basin, the Ghadames (Berkine) Basin and part of the Oued M'ya Basins; three composite total petroleum systems are known, each contains one or multiple total petroleum systems: a) The Tanezzuft -Oued M'ya extends into the neighboring Grand Erg/Ahnet province ; b) The Tanezzuft -Melrhir; and c) The Tanezzuft -Ghadames Total Petroleum System. The last two systems are located almost entirely within the Trias/Ghadames province.Each Total Petroleum system occurs in a separate Basin, and comprises a single cross section.

Fig 8.6 – Trias-Ghadames Province and the location of two Stratigraphic cross sections Figs 8.7AÁ, & 8.7 BB'

The Tanezzuft-Oued M'ya Total Petroleum System contains the giant Hassi Messaoud oil field discovered in 1956. The Ghadames basin was filled by a sedimentary section of Mesozoic formations that are characterized by a thick succession of upper salt and anhydrite deposits to the north and northeast parts; the Mesozoic formations of Triassic to Cretaceous and Mio-Pliocene clastics, are unconformably overlying the Paleozoic deposits.

The southern and southwestern boundaries of the Trias/Ghadames province represent the approximate extent of Triassic, and Jurassic evaporates that were deposited within the Triassic Basin **(pub.usgs-gov/bul/b2202-c-b2202, bso-pdf)**.

The main source rocks of the eastern province are the Silurian Tanezzuft Formation (or lateral equivalents) and Middle to Upper Devonian mudstone. Maturation history and the major migration pathways from source to reservoir are unique to each basin. The total petroleum systems were named after the oldest major source rock and the basin in which it resides. **Pub.usgs.gov/bul/b2202-b/b2202-bso.pdf**

The Trias/Ghadames province contains more than 15,000 million barrels of oil (MMBO), 1000 million barrels of natural gas liquids (MMBNGL), and approximately 25,000 billion cubic feet of natural gas (10^9 CFC or BCFG) (Petroconsultants, 1966a).

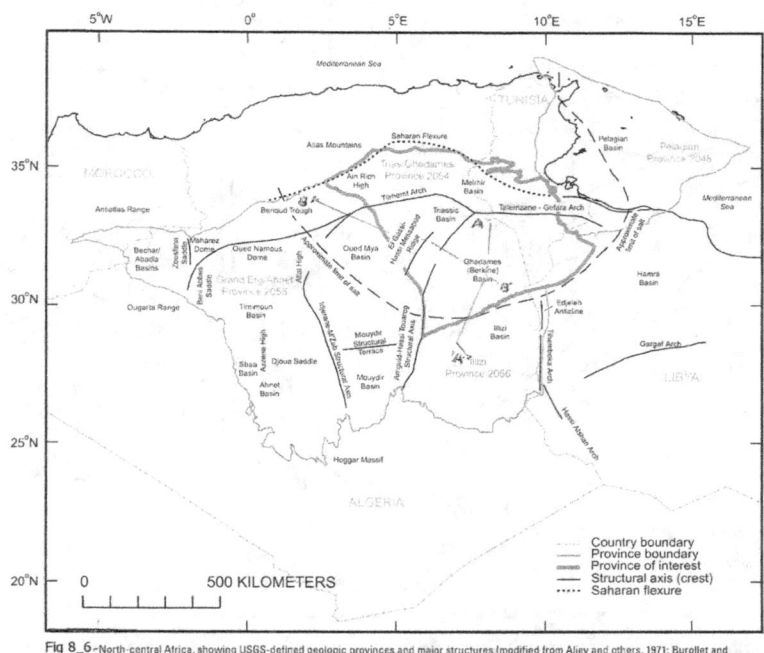

Fig 8_6 -North-central Africa, showing USGS-defined geologic provinces and major structures (modified from Aliev and others, 1971; Burollet and others, 1978; Montgomery, 1994; Petroconsultants, 1996b; Persits and others, 1997).

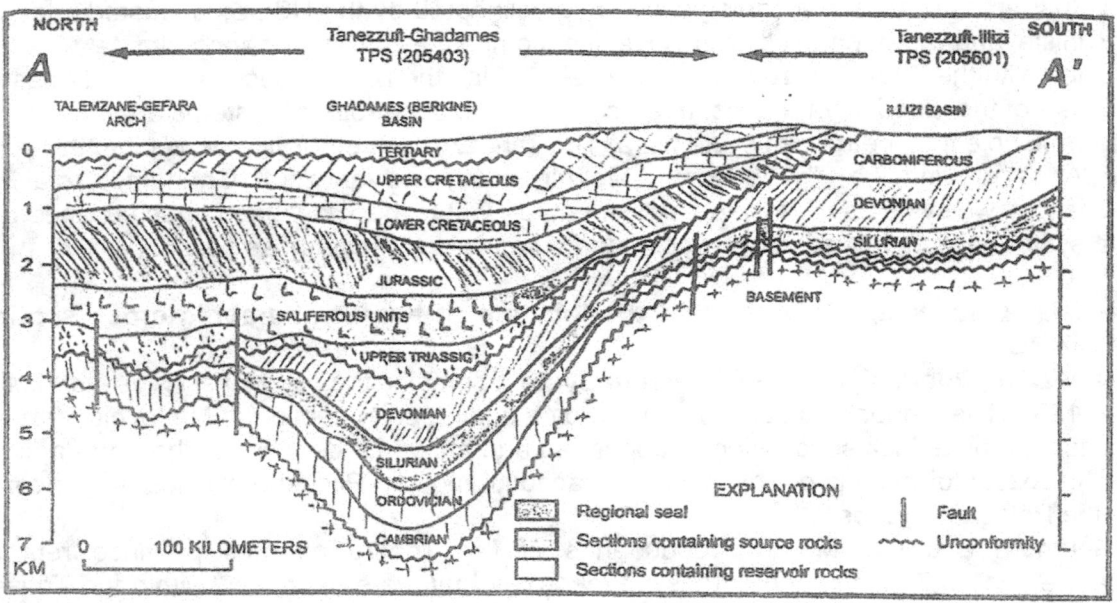

Fig 8.7a - Stratigraphic cross sections through Trias/Ghadames and Illizi Provinces. A, North-to-south stratigraphic cross section through the Ghadames (Berkine) and Illizi Basins (modified from van de Weerd and Ware, 1994, after Aliev and others, 1971).

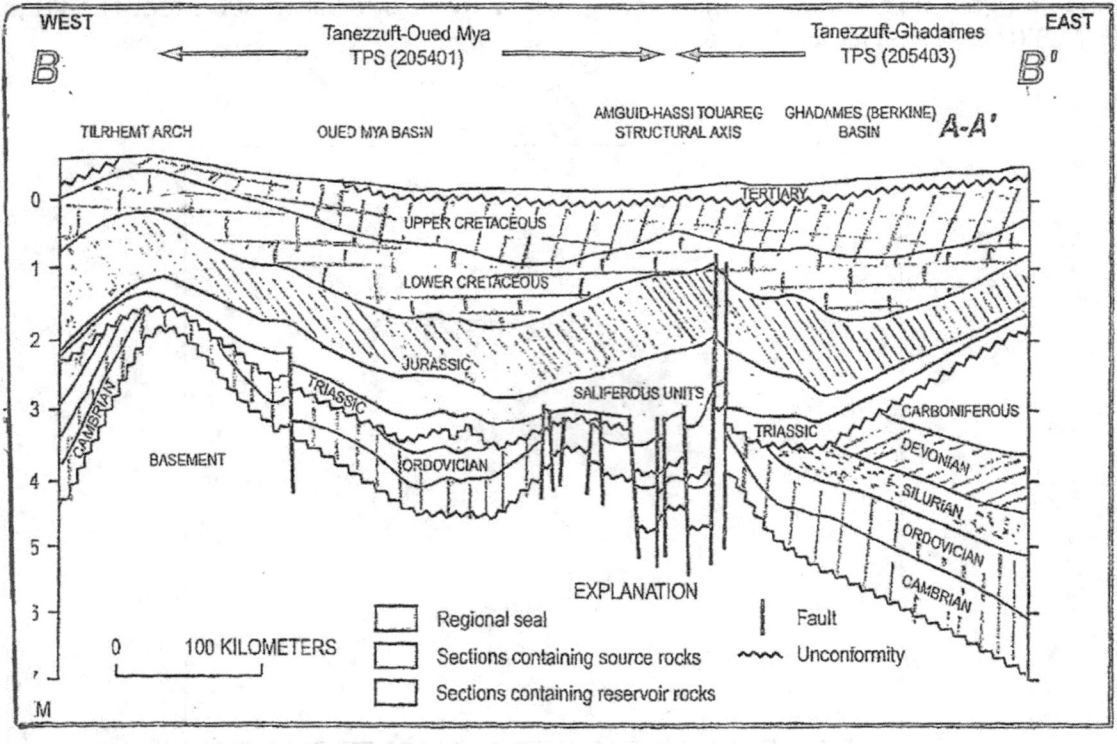

Fig 8.7b - Stratigraphic cross sections. B, West-to-east stratigraphic cross section through the Oued Mya and Ghadames (Berkine) Basins (modified from van de Weerd and Ware, 1994, after Aliev and others, 1971).

Minor amounts of petroleum have been generated in the southern portion of the Oued M'ya Basin during the Carboniferous when the Paleozoic section was thickest, but generation was halted during Hercynian deformation. The main phase of oil generation occurred after Hercynian deformation with the development of the new Triassic Basin depocenter, superimposed on the northern part of the Oued M'ya Basin. Oil generation and migration most likely started post Liassic orogenic movements and peaked in Late Cretaceous to early Tertiary. Oil generated in the Oued M'ya Total Petroleum System also charged the Hassi R'Mel area on the Tilrhemt Arch; the Hassi R'Mel area was either simultaneously or subsequently charged with gas from the north and west and is therefore included in another total petroleum system; pyrenean uplift and erosion terminated petroleum generation in the Oued M'ya Basin. Petroleum hydrocarbon most likely migrated laterally into adjacent or juxtaposition migration conduits and reservoirs. Some vertical migration may have occurred along faults or fractures in structurally deformed areas (USGS Bulletin, Klitt and others, 1997).

A regional seal of about 2,000 ms of Triassic to Jurassic evaporites, mudstone, and carbonate rocks provides a regional top seal for reservoirs in most Tanzzuft-Oued M'ya Total Petroleum System. The Triassic volcanic rocks provide the primary seal for some reservoirs, and intraformational Paleozoic marine mudstone provides secondary, lateral seals when in conjunction with the regional top seal (USGS Bulletin, Klett, 1997).

The Tanezzuft-Melrhir Total Petroleum System (205402) was not a major oil and gas producing unit and it was immature in terms of exploration. Oil may have generated as early as Carboniferous or Permian; however, the main phase of oil generation probably began in the Cretaceous and continued in Tertiary. About 4000 to more than 6000 ms of Mesozoic and Cenozoic overburden are estimated where the reservoir rocks are predominately fluvial to marine sandstone of the Ordovician Hamra formation and the fluvial sandstone of the Triassic "Kirchaou" Formation.

8.6.4.2 The Illizi Province

The Illizi Province is in eastern Algeria and a small portion of western Libya. The province and total petroleium system is described as the Tanezzuft-Illizi Total Petroleum System. In the Illizi basin, the total petroleum system coincides with the Illizi Basin. It encompasses approximately 200,000 km². The province is bounded on the north by the Triassic Ghadames (Berkine) Basin, on the east by the Tihemboka Arch (or uplift), on the south by the Hoggar Massif, and on the west by Grand Erg/Ahnet Basin.

More than one total petroleum system may exist within the Illizi Basin. One "composite" total petroleium system is described as the Tanezzuft-Illizi Total Petroleum System. In the Illizi basin, the Tanezzuft is sometimes referred to as Argillaceous "or "Argileux".

One assessment unit of the Tanezzuft-Illizi Total Petroleum System, is known as Tanezzuft-Illizi/Structural/Stratigraphic Assessment Unit. As in 1996, it contained 102 fields; of these fields, 51 are oil fields, 38 are gas fields, and 13 are unclassified fields having less than one million Barrels of oil equivalent (MMBOE), "based on USGS-oil and gas field definitions". The combined fields contain about approximately 3700 million barrels of oil (MMBO), and 900 million barrels of natural gas liquids (MMBNGL), and approximately 45,000 Billion cubic feet of natural gas (10^9 BCFG).

The Illizi province contains both the Tin Foyé-Tabankort and Zarzaitine oil fields. At the Illizi basin, the Paleozoic formation (about 3000 ms-thick) outcrops on the southern fringes of the basins at the Tassili location, whereas the Mesozoic formation outcrops to the center, and the Tertiary deposits outcrop on the NW flank of the basin **(Fig 8.8)**.

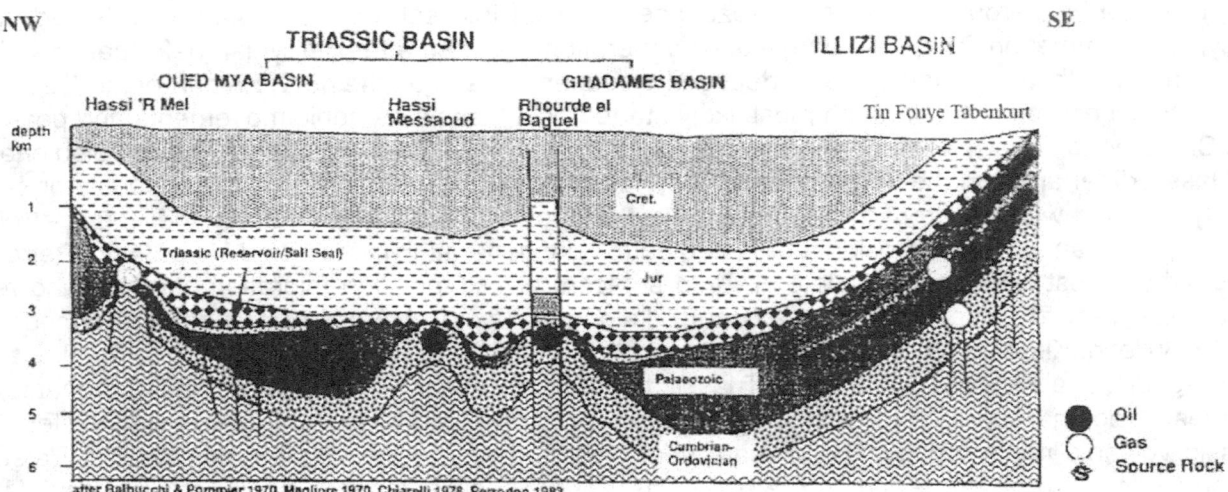

Fig 8_8 - Schematic cross-section of the Triassic/Illizi petroleum province. Note the significance of the Triassic evaporite seal and the Hercynian unconformity in controlling hydrocarbon habitat. Compiled from data and sections in Balbucchi and Pommier (1970), Magliore (1970), Perrodon (1983), Chiarelli (1978)

The Hercynian deformation surface started during Late Carboniferous and continued into the Early Permian. The lower Mesozoic transgression, which swept across the eroded Hercynian peneplain, from the northeast, did not reach the Illizi Province until very Late Triassic and Jurassic.

The Tanezzuft-Illizi Total Petroleum System (205601) is coincident with most of the Illizi Basin and the Illizi province. Paleozoic rocks, including the major source rocks, crop out along the Hoggar Maasif. An events chart summarizes the timing of sources, reservoirs, seals, trap development, and generation and migration of petroleum. Oil generation from both Silurian and Middle to Upper Devonian source rocks in the southern and western portion of the total petroleum system, probably started and reached peak generative phases in the Carboniferous, but generation was halted during the Hercynian deformational event. Erosion resulting from Hercynian deformation probably removed most or all of the petroleum generated prior to the Hercynian event, when the source rocks in this portion of the Total petroleum system were never again sufficiently buried to resume generation. Hercynian deformation was followed by Mesozoic subsidence and deposition to the north, which resulted in petroleum generation at the central, northern, and eastern portions of the total petroleum system.

In the Triassic/Illizi province, Petroleum most likely migrated laterally into adjacent or juxtaposed migration conduits and reservoirs. Some vertical migration occurred along faults or fractures in the structurally deformed areas (Boote and others (1998), therefore allowing Silurian oil to migrate upwards into either Devonian or Triassic sandstones, which then act as efficient lateral carrier beds and reservoirs. http://www.mem-algeria.org/hydrocarbons/geology.htm.

The primary reservoir rocks include Cambrian-Ordovician, Silurian, Devonian, and Carboniferous sandstone. Intraformational Paleozoic marine mudstone is the primary seal for reservoirs in the Tanezzuft-Illizi Total Petroleum System whereas, the Triassic to Jurassic evaporites, mudstone, and carbonate rocks that provide a regional top cap rock for reservoirs in the Ghadames (Berkine) Basin, do not extend into the Illizi Province.

Paleozoic rocks are thickest in the center of Illizi basin (approximately 2500ms), and the Mesozoic is thickest to the north and west (Boujema and others, 1987). The Triassic to Lower Cretaceous are represented by non-marine clastic rocks followed by marine carbonate rocks. The Cenozoic section is represented by thin, discontinuous Miocene-Pliocene non-marine rocks and Quaternary sediments.

Petroleum was generated within the Illizi Basin from the Middle to Late Jurassic to Early Tertiary. Secondary migration occurred, where many of the structural traps were formed by vertical movements of the basement during the Mesozoic and Tertiary deformational events.

Hydrocarbon accumulations within one of the largest oil fields of the Illizi Basin, Tinfoye-Tabankort, are presumed hydrodynamically trapped (Echeikh and others, 1989). Most of the hydrocarbon accumulations of the intraformational Paleozoic marine mudstone are within anticlines and Faulted anticlines.

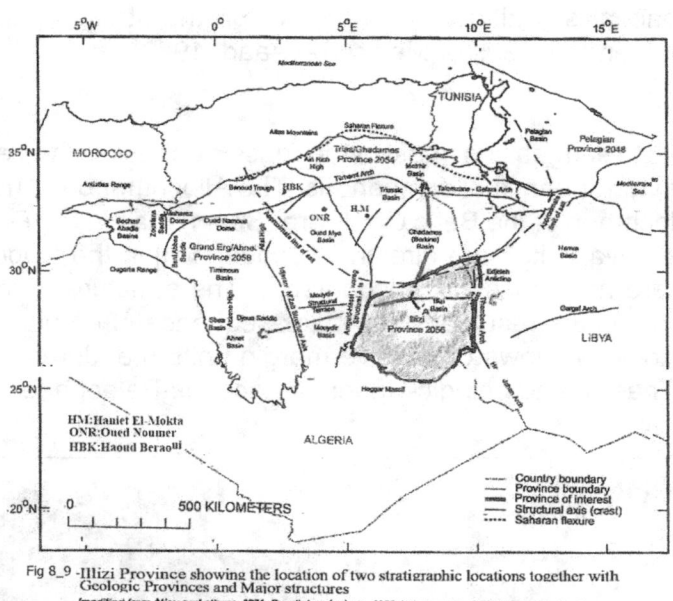

Fig 8.9 – The Illizi Basin with the geologic provinces and major structures of the Algerian Sahara (USGS 2000); AA'-CĈ are the location of two stratigraphic cross sections.

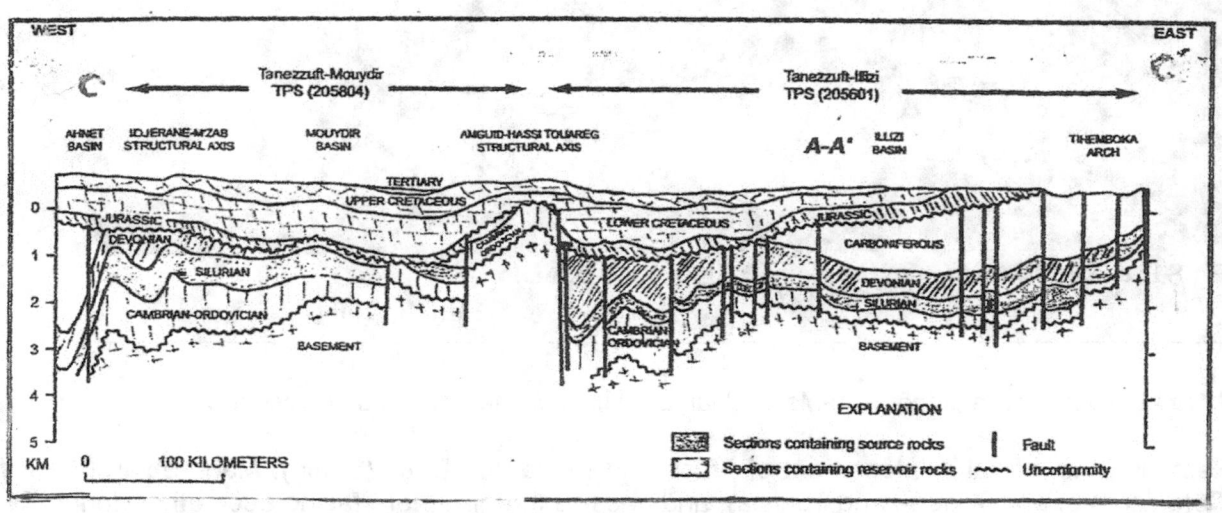

Figs 8.10 – A stratigraphic E/W cross section CC'-passing by Ahnet Province, Mouydir and Illizi Basin,

8.7 A Geological Aspect-The Allochthonous Triassic Evaporitic Region of the Betic-Maghrebian Domain

8.7.1 Introduction
Flinch (2004) submitted an exclusive article (#30025) on the Triassic Allochthonous Evaporitic Province within the Betic-Maghrebian Domain with a comparison of the present-day Gulf of Mexico. The study indirectly confirmed the concept submitted by the author on the stratigraphic relation of the Triassic-Liassic deposits in NW Africa and the Gulf of Mexico (Assaad, 1983).

8.7.2 Discussion
Field, seismic and well-log data suggest that most of the Triassic deposits of the external Betic-Maghrebian domain of the Western Mediterranean is allochthonous. The Allochthonous Triassic is much more abundant in the Guadalquivir Allochthon of the Betic Cordillera (Spain) than in the Perifaine Nappe of Morocco (Rif Cordillera). The Triassic evaporites are directly imbedded within the Upper Cretaceous-Paleogene deep-water sediments, where no Jurassic is encountered. The structure of the external domain of the Betic, Rif, and Tell Cordilleras is the result of piggy-back sequence of emplacement that resulted in the presence of numerous tectonic windows of passive-margin units (i.e. Jurassic and Lower Cretaceous) surrounded by intermixed Triassic and pelagic-Upper Cretaceous-Paleogene sediments (Flinch, 2004).

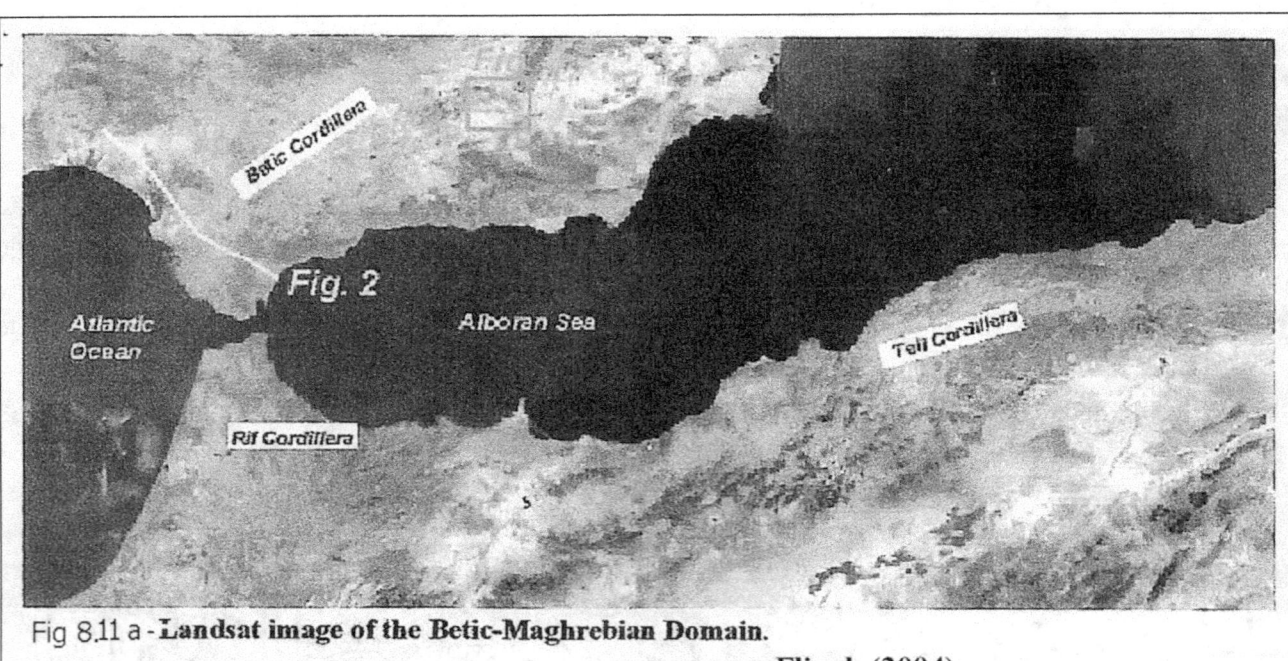

Fig 8.11a - Landsat image of the Betic-Maghrebian Domain; and the location of a cross section.

Restoring the lateral displacement of Africa and Iberia, the Betic (Spain), Maghrebian (the Rif and Tell Cordilleras of both Algeria and Tunisia), and passive margins were facing each other during Triassic and Jurassic times. During Neogene time, after the emplacement and collision of the Alboran block (Mediterranean), the allochthonous evaporitic province was dismembered and thrusted onto the passive margins of Iberia (Europe) and North Africa. Such observations suggest the South Iberian and Northwest African margins during the Cretaceous time, resembled the present day Gulf of Mexico.

8.7.3 Western Betic Cordillera (Spain)

A thick allochthonous Triassic evaporitic section is encountered within the Guadalquivir allochthon in the mid-west at Betica well 18-1, where it attained 1500 m in thickness. The well is also encountered sub-autochthonous Triassic salt, dolomite and anhydrite. Farther to the northeast, along the Gualdalquivir valley, where the Betica well 14-7 is located, the well attained several hundreds meters of salt imbedded within Miocene marls of the Gualdaquivir Allochthon. The Gualdaquivir Allochthon involves Upper Cretaceous deep-water pelagic deposits that are mixed with Triassic evaporites, mainly of gypsum and salts together with shales, siltstones and occasional sandstones.

Therefore, the Betic, Rif, and Tell Cordilleras could provide field analogs of exploration of the allocthonous evaporitic provinces:

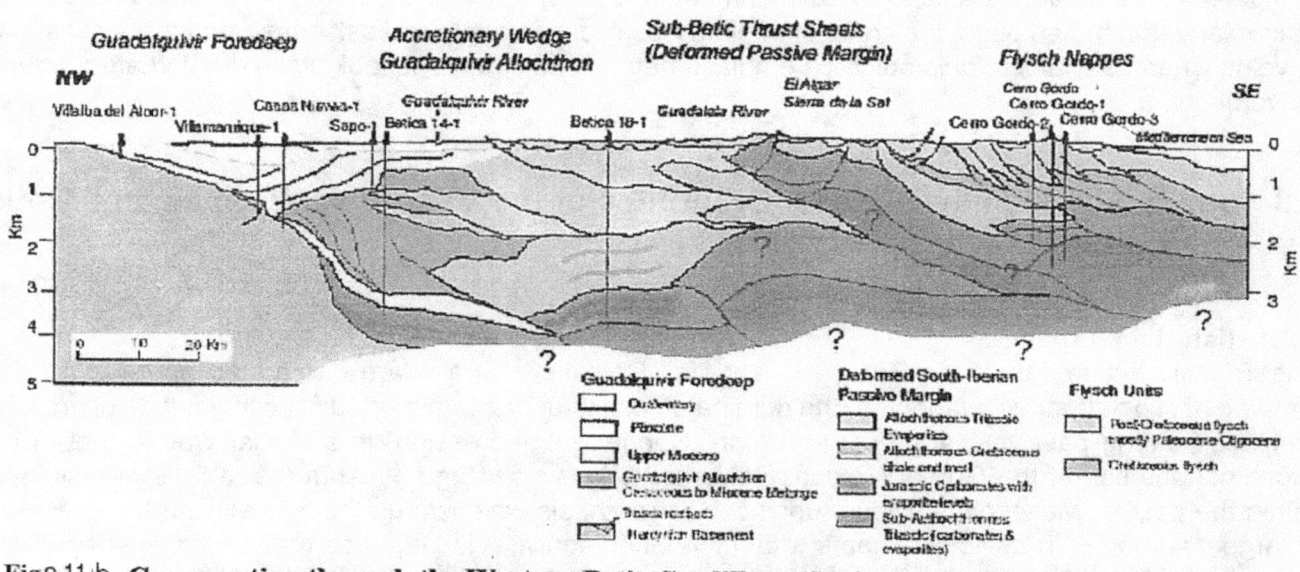

Fig 8.11 b - Cross-section through the Western Betic Cordillera (Spain), based on well-log data (Lanaja, 1987). (Flinch 2004)

-See Location of section in Fig 8.3a "Fig2"

Fig 8.11b – A cross section through the Western Betic Cordillera (Spain), based on well-log data (Lanaja, 1987/ Flinch 2004).

8.7.4 The Central Betic Cordillera (South Spain)

The Guadalquivir allochthon thins out towards the central Betic Cordillera, and mainly occupies the Sub-Betic thrust sheets, characterized by a thick Jurassic section. Triassic sediments constitute kilometric-scale thrust sheets to centimetric or decametric blocks imbedded within Cretaceous- Paleogene marls. The Triassic evaporites seemed to be emplaced during Upper Cretaceous time in a deep-water setting.

8.7.5 The Perifaine Nappe Rif Cordillera (Morocco)

The Prerifaine Nappe of the Rif Cordillera (Morocco) is equivalent of the Guadalquivir Allochthon and represents an Accretionary Wedge (Flinch 1996). The Triassic shales and evaporites, (mainly gypsum), are intermixed with deep-water; Upper Cretaceous marls, are referred locally as "marnes à gypsum" or marl with gypsum. Lower Eocene sediments unconformably overlie Triassic evaporites, suggesting that the emplacement had been already completed at that time.

In the Rif Cordillera, the Triassic Gypsum diapir, the so-called "marnes á gypse", that pierced the Cretaceous-Paleogene deep-water marls, suggest evaporitic re-sedimentation within the actual Perifaine Nappe of Morocco. Several Triassic "salt glaciers", emplaced farther to the east in Algeria and Tunisia, during Albine through Turonian time, have been described along the Tell Cordillera (Tellian Atlas of Algeria and Tunis).

8.7.6 "Zones des Domes" of Tell Cordillera (Algeria and Tunisia)
The Triassic salt glaciers of the "Zones des Domes" in Algeria and NW Tunisia, were emplaced during Albian to Turonian time, (slightly earlier than in the Betic Cordillera, "Spain". Even though, the allochthony of the Triassic salt glaciers have been subject of debate (Vila, 1995)". Actually, Assaad (1983) stated that during the Upper Cretaceous –Ypresian, salt domes were continued to develop due to halokinematics (within the framework of plate kinematics), and the absence of Cretaceous formation to the SW of Algeria, might be attributable to its development. In post Ypresian time, a time of collision between the African and the Iberian plate, the concept of plates is difficult to apply and salt dome structures ceased to develop whereas, Halotectonic outcrops emplacement of salt structures took place due to Late orogenic movements.

8.8 Comparison with the Triassic Evaporites of the Maghrebian Domain and Betic Cordillera (Spain)

8.8.1 Historical Aspect
The Triassic evaporites of the Betic, Rif, and Tell Cordillera "Spain" are interpreted as passive margin-type of allochthonous evaporites (the dominated constituents, transported from their place of growth), emplaced during passive margin stage (before Neogene compression) in a similar way to that of the allochthonous salt of the Gulf of Mexico. Reworked Triassic salt and gypsum (i.e. blocks and slumps) within the pelagic Mesozoic section, support the hypothesis which would help to explain the presence of large volumes of Triassic sediments within the Guadlaquivir Allochthon as well as the missing of the upper beds. The allochthonous Triassic detritals occupy the external zones of the Western and Central Betic Cordillera, extending for more than 200km. The allochthonous evaporites of the Gulf of Mexico consist mostly of Jurassic salt, whereas the evaporites of the Betic Cordillera are Triassic and partly of salt, with inclusions of gypsum, anhydrite and shale, (see Glossary)

Many observations suggest that during Cretaceous time, the south Iberian and North African regions resembled the present day Gulf of Mexico. Therefore, the Betic, Rif, and Tell Cordilleras could provide field analogs of exploration of allochthonous salt provinces **(review Chapter.8 /section 8.8).**

References
Assaad, F., (1983) An approach to "Halokinematics" and Interplate Tectonics (North- Central Algeria), Jr. Petroleum Geology, Vol.6, No. 1, July 1983.

Sonatrach (1992) – Exploration in Algeria: Algeria, Sur Presses Speciales U.A.F.A., 36p.- A Review of the East Algerian Sahara oil and gas (2000) - Historique **(http://www.mem- algeria.org.hydrocarbons/ geology.htm)**

Boudjema, A. (1987) Evolution Structurale du Bassin petrolièr Triassique du Sahara Nord-Oriental (Algerie), un-published Ph.D., Thesis, Universitè Paris-Sud, Centre d'Orsay, 290 p.

Flinch, J.F., (2004) A Cretaceous Allochthonous Evaporitic Province within the Betic-Maghrebian Domain: Comparison with the present-day Gulf of Mexico (Adapted from "extended abstract" for presentation at the AAPG International Conference, Barcelona, Spain, Sept. 21-24, 2003 **(http://www.Searchanddiscovery.net/sediments/2004/flinch/index.htm)**

------------ (1993) Tectonic evolution of Gibralter Arc: Unpublished Ph. D. Thesis dissertation, Rice University, Houston, Texas.

Flinch, J.F., Balley, A.W., and Wu, S. (1996) Emplacement of a passive-margin evaporitic Allochthon in the Betic Cordillera of Spain: Geology v.24, No. 1, pp67-70.

Alieve, M., Ait Loussine et al, (Sonatrach, 1971) - Geological Structures and estimation of oil and gas in the Sahara of Algeria: Spain, Altamira-Rotopress, S.A., 265p

Klett, T.R. (1997) Total Petroleum Systems of the Illizi Province, Algeria and Libya-Tanezzuft- Illizi, version 1.0., , USGS, Bulletin 2202-A.and 2202C **(http://pubs.usgs.gov/bul/b2202-a); http://greenwood.cr.usgs.gov/pub/bulletins/b2202-c/**

Vila, J.M. (1995) Premiére etude De surface d'un grand "glacier de sel" sous-marine: l'est de la structure Ouenza-Ladjebel-Meridef (confins Algéro-Tunisiens), Proposition d'un scénario de mise en place et comparisons: Bulletin de la Sociéte Geologique de France, v.166.

Marabet O. (1969) Bibliographie geologique annuelle de l'Algerie du Nord, du Sahara et des regions limitrophe, Bull. serv. Geol. Algerie, NLK ser., Bull. No. 39

Awad, G.M. (1984) Habitat of Oil in Abu –Gharandik and FyumBasins , Western Desert, Egypt., Am.Assoc, Petroleum Geol.Bull., 68, 546-573.

Mengnoli, S., and Spinicci, G. (1985) Tectonic evolution of North Africa (from North Sinai to Algeria).in: Proceedings of the Seminar on Source and Habitat of Petroleum in the Arab Countries, Kuwait, Oct. 1984, OAPEC, 119-174.

Dixon, J.E. and Robertson, A.H.F. (eds, 1984) The Geological evolution of the eastern Mediterranean Spec. Publ.Geol. Soc., London, No. 17

Macgregor, D.S. (1994) Relationships between seepage, tectonic, and subsurface petroleum reserves; in Petrol. Geol. 10, 606-619 (1994) The Hydrocarbon systems of North Africa; Conference, Malta, B.Exploration, 4/5 Long Walk, Stockley Park Industrial Estate, Uxbridge, UK. (Fig 1.2 and Fig 1.3).

Balbucchi, A. and Pommier, G. (1970) Cambrian oil field of Hass Messaoud, Algeria. In Geology of Giant Petroleum Field (ed. M.T Halbouty) Am. Assoc. Petrol. Geol. Mem. No. 14, 477-488; Figs 2&3 Marine Petroleum

Guiraud, R., and Maurin, J.C.(1992) Early Cretaceous Rifts of the western and eastern Africa, an Overview Technophysics 213, 153-168.

Shardanov, A. N. and Shumliova, M. B. (1983) Regional hydrocarbon migration as a factor in The formation of major petroleum accumulation zones. International Geol.Rev.25, 569-573.

Cited References

Assaad, F., (1981) A further Geologic Study on the Triassic Formations of North- Central Algeria with Special Emphasis on Halokinesis, Journal of Petroleum Geology, 4,2, pp. 163-176.

------------, (1972) Contribution to the study of the Triassic formations of the sersou-Megress Region (High Plateau) and the area of Daia M'Zab (Saharan Platform), No. 84 (B3), Eighth Arab Petroleum Congress, Algeria, 1972.

Guiraud, R., and Maurin, J.C.(1992) Early Cretaceous Rifts of the western and eastern Africa, an Overview Technophysics 213, 153-168.

Sonatrach (1993) Tectonic evolution of Gibralter Arc: Unpublished Ph. D. Thesis dissertation, Rice University, Houston, Texas.

Chapter 9

Discovery of a Triassic basin at a Virgin Area in northeast Algeria, Defined its trend and Boundary – Discussed the Lithology, Electric well Log, and Palynological correlation with that of the Triassic Province of the Saharan Platform, and determine its extention–Stratigraphy and Sedimentation, Stratighraphic Evolution

9.1 Introduction

Geological and Geophysical activities greatly increased in Northwest African countries, and in particular, Algeria, after the independence from the hardship era of the French colonialism; a precise interpretation and adequate correlation became possible after citing the proper stratigraphic level of different sedimentary formations, that helped determining the stratigraphic relation on a regional scale through the three main Algerian structural units of the High Plateaus, the Saharan Atlas, and the Saharan Platform, by reviewing hundreds of geological and geophysical documents, submitted by different French contractors, who had failed to occaisionally discuss their work as a team. **Fig 9.1-** A Geologic map of NE Algeria, and Location of cross sections.

The Triassic cycle began with continental deposits of fluvio-alluvial origin, whereas the detachment of Gondwanaland took place in a prevailing tropical climate, due to the successive southerly migration of the Equator that led to the formation of several lagoons during Late Triassic-Liassic periods. The Triassic/Liassic section could be correlated lithologically, palynologically and by electric log correlation between the two widely separated tectonic units, the High Plateaus and the Saharan Platform, particularly among the available wells drilled at the High Plateaus, e.g. wells of Chott Chergui (CC-1), Doghman (DOG-1), Cedraia (CED-1), together withthose of the Saharan Platform at Oued Nomer (ONR-5), Oued N'SA (ONS-1), and Megadine (MGD-1); taking into account, the Late tectonic events that prevailed in the Saharan Atlas, in much more recent time, possibly during the Early Alpine orogenic period (Post Triassic-Liassic period).

It should be mentioned that the pronounced disharmony between the subdued and flat Triassic pre-salt rocks and the strongly disturbed post salt rocks was due exclusively to the mobility of the Triassic-Liassic salt that played the most impact rôle of the post-Triassic salt structural development of the Saharan Atlas trough and caused the outcrop of saliferous shales which concentrated along the major faults on top parts of the faulted anticlines.

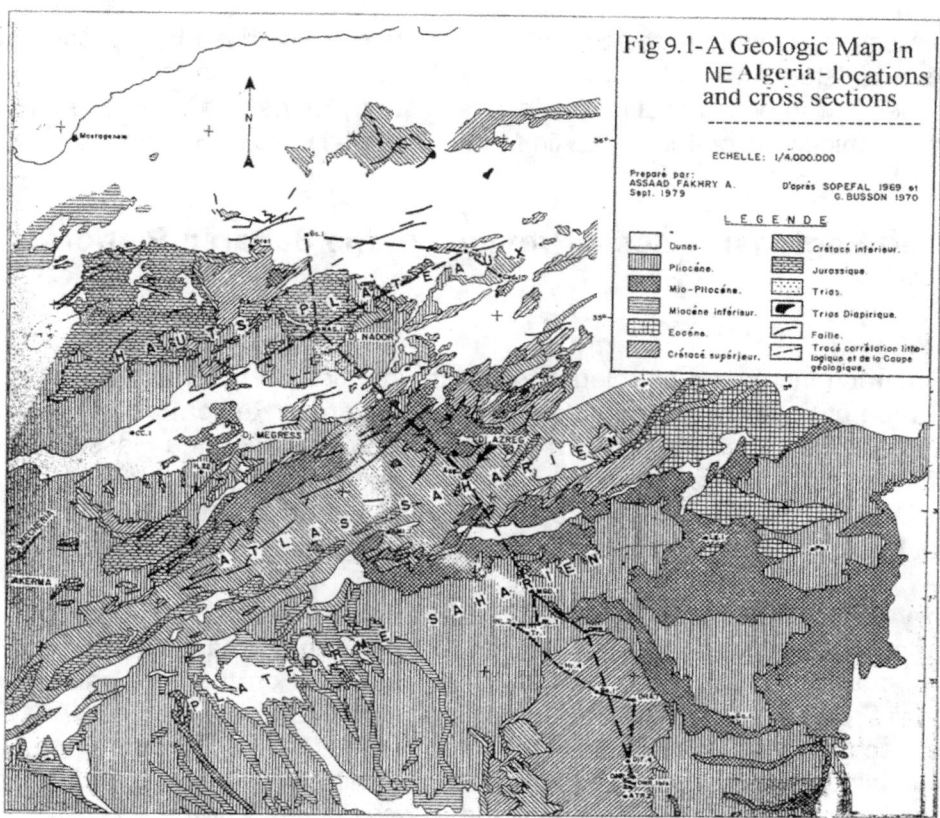

9.2 Triassic Deposits and Triassic Outcrops

The Triassic detritus is mainly of shaly sandstones, whereas, the Liassic sediments consist of shaly evaporites and carbonates. At the end of the Lower Liassic period, the deposition of evaporites ceased in northeast of the High Plateaus and the Saharan Platform; marine transgression inaugurated, when carbonate facies deposited later; the Triassic formations, which unconformably overlie different Paleozoic units, due to variable degrees of Hercynian erosion, were transgressively overlain by carbonates of the Lower Jurassic.

The Triassic/Liassic deposits in North Algeria, are generally recognized by salty clay outcrops, along Great flows or in the reaches of faulted anticlines. According to experimental and hydrogeological wells, one can verify the autochthonous nature of the Triassic deposits in certain sections on the southern slopes of the Kabely- ranges of the High Plateaus (Technoexport, 1971).

In the northwestern part of the High Plateaus, the Triassic deposits outcrop in the horsts of Ghar Rouban and in Tiffrit, as well as in the neighboring regions of Morocco, where volcanic formations are represented by reddish brown conglomerates of eruptive rock pebbles (calcites, granodiorites and rhyolites), green and red clays, and basalts with rare dolomites and marl.

A French Petroleum Company, Sopefal (1970), noticed that the detrital Triassic were pinching out to the SW, according to the few available drilling data from the old wells. To the east, in Hodna Mountains (BouThaleb, Dj. Tafourer, Dj. Dabba), the lower Liassic, of 150-200m.thick, consists of dolomites and dolomitic limestones.

In the Saharan Atlas, the Lower Jurassic outcrops in the region of Mecheria and in the vault of the anticline of Dj. Bou Lerhfad of the Ksour Mountains. The Lower Jurassic is penetrated by both the exploratory wells of Ain Mahdi (AMI-1) and Azreg (AZG-1) in the southeast periphery of the central part of the Saharan Atlas.

The upper Jurassic sediments of the Saharan Atlas are characterised by a great variety in lithology; terrigenous rocks, sandstones and clayey sandstone that predominate in the southwest regions of Bou Lerhfad, and Mecheria, with occasional intercalations of sandy marls and limestone; The Jurassic sandy sediments attained a thickness of 4965m at Ain Mahdi w (AMI-1) and of 3342m at Azreg well. (AZG-1).

9.3 An extended Regional Triassic Province of the Saharan Platform

9.3.1 Historical Aspect

Since 1980s, Sonatrach increased petroleum research activities on the Triassic reservoir potentialities at the eastern portion of the High Plateaus (ex-Sersou – Megress Region) and at the Tilrhemt Dome – Talemzane structure of the Saharan Platform (ex-Daia M'Zab includes the Hassi R'Mel High structure and Oued M'ya basin); both covered an area between Latitude 31° 30´ and 36° 00´ N, and Longitude 00° 00´ to 6° 00´ E). Different discoveries were discussed and led to an extended Triassic Province of the Algerian Saharan Platform.

9.3.2-A Discovery of a new Triassic Basin, NE of the High Plateaux

The author cited a key well location (NAS-1) at Nador Sud virgin area, Northeast of the High Plateaus of Algeria, carried out Palynological, and palaentological studies on the core samples and cuttings of the Triassic deposits recovered from drilling operations; a composite well log of NAS-1 was completed together with the electric well log correlation of the nearby wells. A Montageprogram report was prepared according to the available seismic section (the Isochrone map); knowing that the precipitation of anhydrites on top of salt deposits denotes the last stage of development of the Triassic basin in NE of the High Plateaus.

The NAS-1 well is located in the center of a local Triassic basin northeast of the High plateaus, **Fig 9.2-** is a stratigraphic correlation cross section among nearby wells of the proposed exploratory well (NAS-1); **Fig 9.3a** – A seismic cross section wrongly considered as a "noisy area:" and had been also *mis-anticipated* by a Yougoslave senior geophycisist as a faulted area **(Fig 9.3b--** An isochrone map on Top of Triassic deposit at Nador Sud area, was mis-interpreted as a faulting.area,), it was also misinterpreted by a senior geologist (A.H.), as a multi-reversed faults, though, the drilling of the NAS-1-well, revealed the presence of unexpected thickness of salt deposits, in comparison with the nearby wells; e.g. CC-1 well, in the southwest; CED-1 & Dog-1 wells to the northeast and north respectively (Assaad, 2018). The tectonic history of the complexity of the Saharan Atlas shows that it is elongated trough pinched between both the northern High Plateaus and the southern Saharan Platform; the trough of the Saharan Atlas is filled by thick sediments of much younger periods of the Middle Mesozoic (Dogger); it has been developed at the early or0genic period. **Fig 9.4** – The lithology well log of NAS-1. **Fig 9.5** – A Geologic cross section of nearby wells of NAS-1 well.

Discovery of a Triassic basin at a Virgin Area in northeast Algeria, Defined its trend and Boundary...

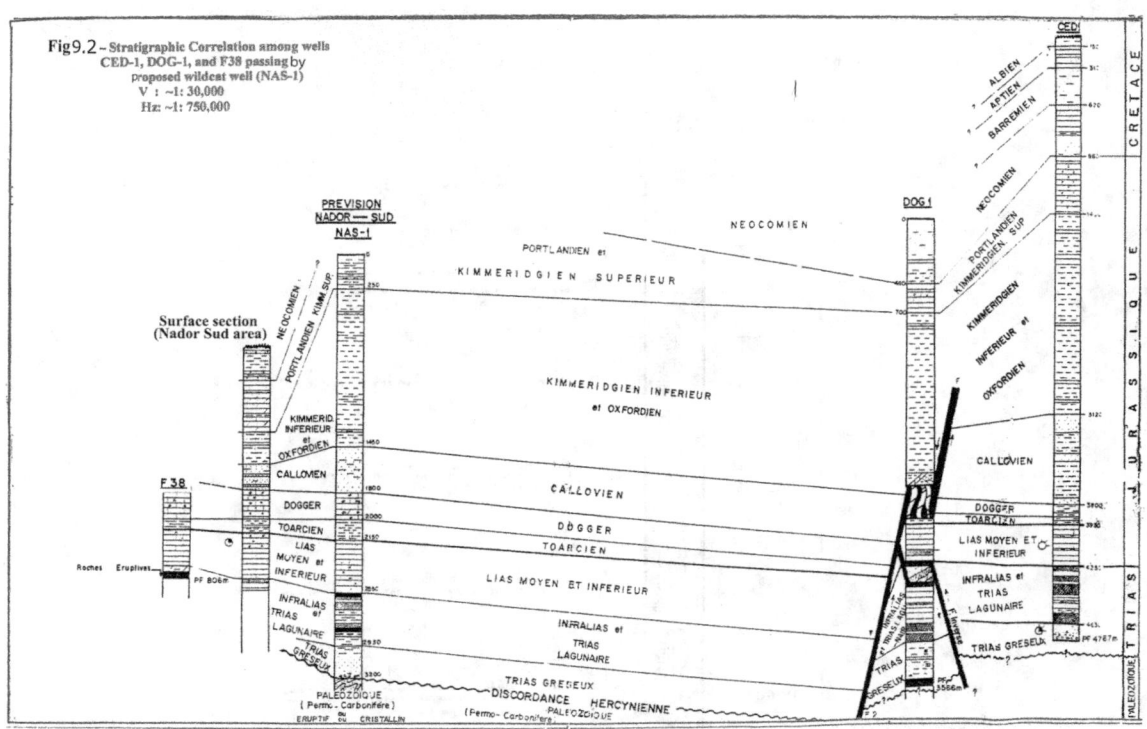

Fig 9.2 – Stratigraphic Correlation among wells CED-1, DOG-1, and F38 passing by proposed wildcat well (NAS-1)

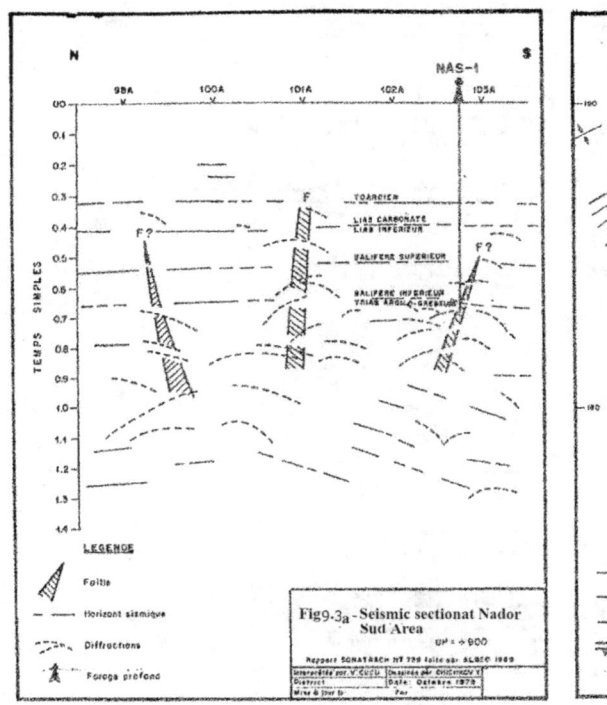

Fig 9.3a – Seismic section at Nador Sud Area

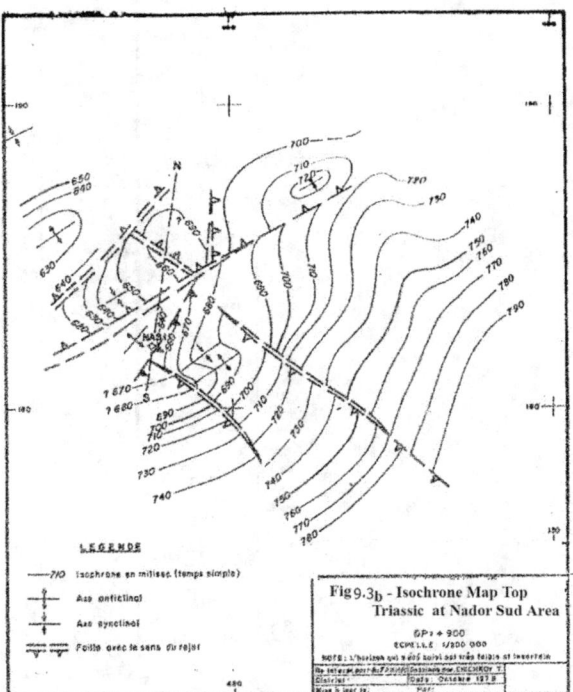

Fig 9.3b – Isochrone Map Top Triassic at Nador Sud Area

Applications of Petroleum Tools for Field Geologists

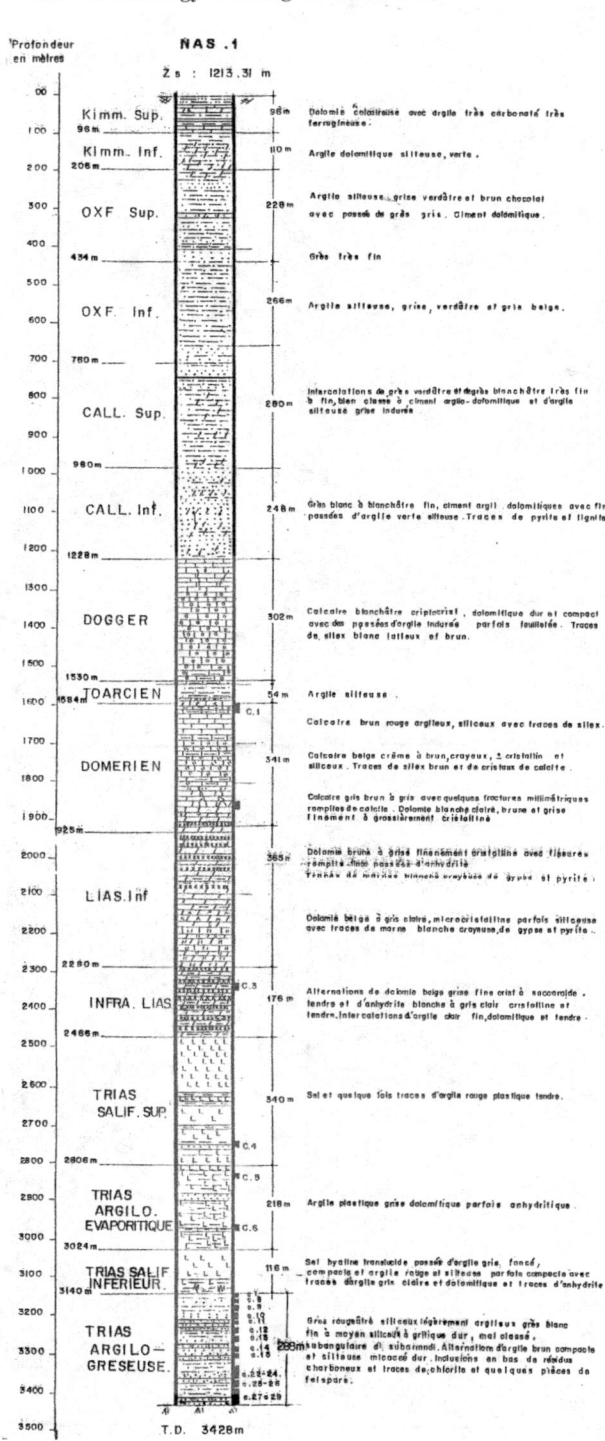

Fig 9.4 - A Lithology Well Log of NADOR SUD.1

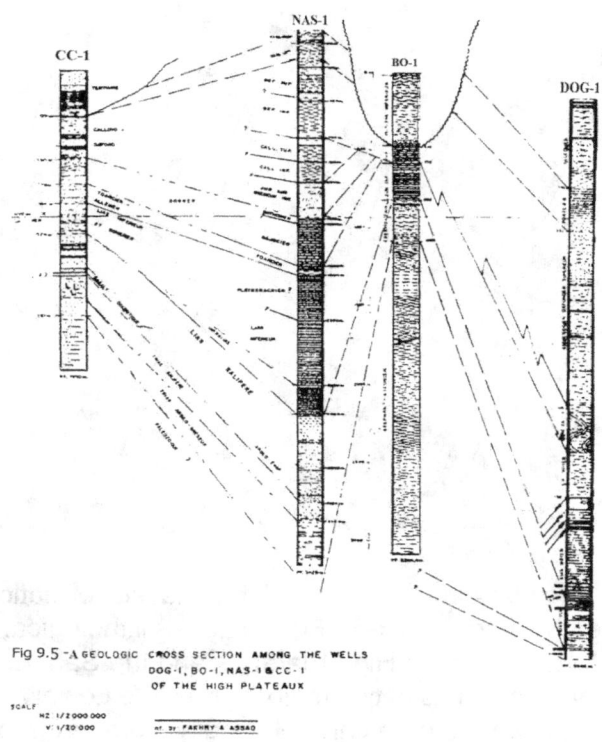

Fig 9.5 – A GEOLOGIC CROSS SECTION AMONG THE WELLS DOG-1, BO-1, NAS-1 & CC-1 OF THE HIGH PLATEAUX

To the North of the High Plateaus, at Bourlier well (BO-1), the so called Permo-Triassic had been defined as the Triassic Shaly sandstone that lies with an angular unconformity over the Permo-Carboniferous (Stephano-Autunian by Ashab, 1971).

9.3.3 The Triassic section to the NE of the High Plateau is a mirror image of that of the Saharan Platform

It is worthy mentoning that theTechnoexport (Sonatrach, 1971) characterized the lithology of certain parts of the sedimentary section by using terms such as "complex", "series", "formation", "member", to designate the whole sediments that can be distinguishable by composition and thickness.

Surprisingly, the Triassic section at NAS-1 well, was a mirror image of ONR-5 well of the southern Saharan Platform and could be lithologically and by electric well logging correlatable with the Oued M'ya basin (MGD-1 and ONR-1-wells), regardless of the complicated structure of the Saharan Atlas in between, at wells AZG-1 (Azreg area) and AMI-1 (Ain Mahdi area), where the Jurassic reached at several thousands meters. **Fig 9.6** – A Geologic cross section of wells of the High Plateaus, Saharan Atlas and the Saharan Platform.

To the east of the high Plateaus, the subsurface section of the Triassic formations varies in thicknes and ranges from 250m Bourlier (BO-1) to 1028m at Nador Sud area (at interval 2400m-3428m); at CED-1, the thickness is of 410m, and at DOG-1, it is of 341m in thickness.

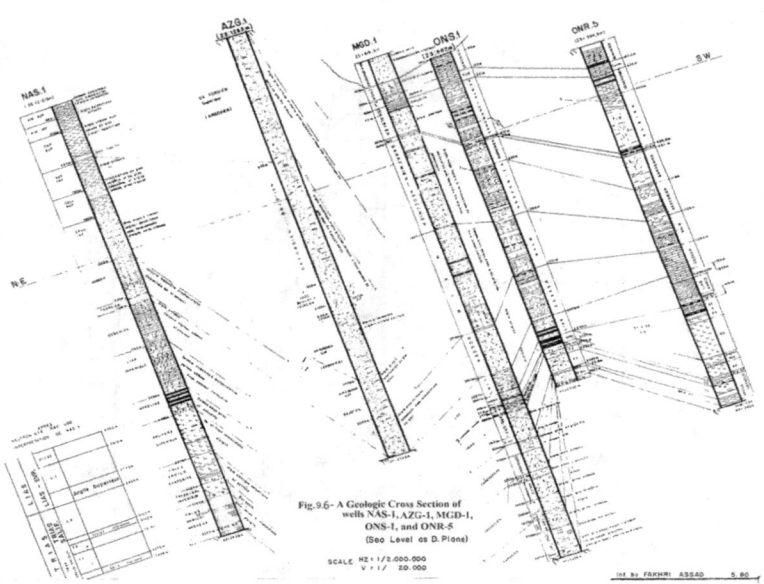

Fig.9.6- A Geologic Cross Section of wells NAS-1, AZG-1, MGD-1, ONS-1, and ONR-5 (Sea Level as D. Plane)

In the Saharan Platform, the subsurface section of the Triassic formation ranges from 283m at Oued N'SA (ONS-1) to 780m at Oued Noumer (ONR-5). **Fig 9.7ab** – A llithological aithostrarigraphic correlation of Trias/Lias beds among wells drilled in the High Plateaus and the Saharan Platform.

The Triassic/Liassic Formation consists of two major series; the Lower Argilo-sandstone detrital series and the upper evaporitic deposits, and can be chronologically stated from bottom to top:

1-The shaly (Argilo-) sandstone constitutes of the Lower series (Sl-series or "serie inferieur") at the base, followed upward by the detrital series "T-1 and T-2 series", with frequent local unconformities. The Triassic S4- salt deposits, are overlain by the Upper Shales of the Liassic deposits; followed upward by the S3-shaly salt deposits, and on top, are the S1 + S2-salty shale deposits (updated); **Table 9.1** - Old Classifications of the Triassic among several wells in NE Central High Plateaus **(Assaad, 1981).**

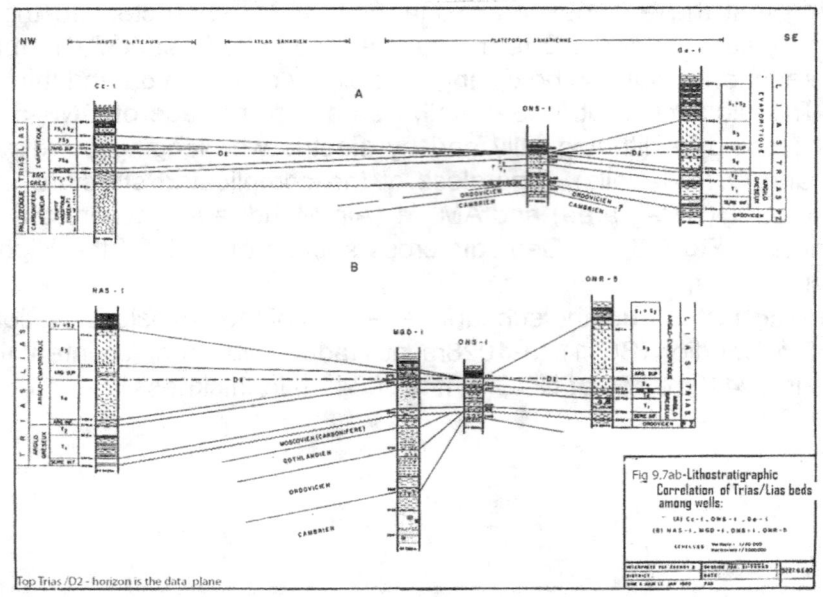

Fig 9.7ab-Lithostratigraphic Correlation of Trias/Lias beds among wells

Table 9.1 – FORMER CLASSIFICATIONS OF THE TRIAS PENETRATED BY SOME EXPLORATION AND HYDROLOGIC WELLS OF THE EASTERN PART OF THE HIGH PLATEAUX CONFRONTED WITH THE PRESENT WORK

Assaad, 1972

Wells	CARATINI 1970 Thick. m	CARATINI 1970 Formation	TECHNOEXPORT T.1 1970 Thick. m	TECHNOEXPORT T.1 1970 Formation	PRESENT WORK Thick. m	PRESENT WORK Series	Remarks
Ced-1	34	Rhetian	107	Lias-Keuper	112	Infra-Lias	The Trias section (510m) is not complete and the total depth of the well is 4789 m
	171	Keuper	98	Keuper	270	Anhydrite Trias	
	41	Muschelkalk	175	Muschelkalk			
	135	Upper Werfenian					
	139	Lower detrital formation	139	Buntsandstein	77	T1 + T2	
					60	Lower Series	
Dog-1			62	Lias-Keuper	69	Infra-Lias	The Trias section is complete. The Lower Carboniferous (Viséen) is reached in this well & the T.D. is 3566 m
			333	Keuper-Muschelkalk-Buntsandstein	51	S1 + S2	
					286	T1 + T2	
					4	Lower Series	
BO-1			251	Lias-Keuper-Muschelkalk-Buntsandst.	260	Trias argilo-sandstone	The Trias sect. is complete & overlies unconformably the Upper Carbon. TD = 3094 m
CC-1			52	Lias-Keuper	160	S1 + S2	The Trias section is complete and overlies unconformably the Lower Carboniferous (Viséen) – T.D. 1950 m
	150	Upper saliferous bed	168	Keuper	60	S3	
	31	basaltic bed	31		31/25	Upper shaly series	
	120	lower saliferous bed	165	Muschelkalk	110	S4	
					30	Lower shaly series	
			90	Buntsandstein	101	T1 + T2	

9.3.4 A Lithostratigraphic Correlation of the northern Triassic basin with that of the Triassic Province of the Saharan Platform

The lithostratigraphy of the Triassic reservoirs were carried out among wells located northeast of the High Plateaus at Sersou-Megress area and that of the Algerian Sahara at Oued M'ya-M'Zab area **(Tables, 9.2 & 9.3, Assaad, 1972)**. The boundaries between different Paleozoic and Triassic beds are well marked on the electric and lithology well logs, but those between the top of the Triassic and the base of the Liassic had been a subject of a long debate. **Fig 9.8** - Electric Log Correlation of the Triassic Formation at NAS-1 well of the High Plateaux and those of the Saharan platform.

Table 9.2 - CLASSIFICATION OF TRIAS IN THE PRESENT WORK VERSUS THAT OF G. BUSSON, 1970 (modified, 1971)

CLASSIFICATION OF TRIAS IN THE PRESENT WORK (1970)									CLASSIFICATION OF TRIAS IN THE DAIA-MZAB REGION AFTER G.BUSSON, 1970			
RESULTS OF LITHOSTRA-TIGRAPHIC CORRELATs	THICKNESS OF TRIAS SERIES								Base of Horizon B			
	WELLS ON THE NE OF HIGH PLATEAUX					WELLS ON DAIA-MZAB (SAH.PLATF.						
SUCCESSIVE SERIES OF TRIAS	NAS1	BO-1	Ced1	Dog1	CC-1	ONR5	ONS1	Mgd1				
Infra Lias	64	-	112	69	-	190	65	-	a_2B interval : salt + anhydrite		Principal saliferous Trias	TRIAS - INFRA LIAS
S1 + S2	115	-	270	51	160	139	66	53				
S3	218	-	-	-	60	279	-	32	base of shaly zone a_2			
Upper shales	106	-	-	-	56	75	-	29	d_2-a_2 interval : massive salt ; base of hor.d_2			
S4	263	-	-	-	110	74	-	8	infra D_2 evaporites - mainly massive salt - base of evap. infra D_2			
Lower shales	38	-	-	-	30	18	39	51	shaly bed		Upper sandstone series of Trias	SANDSTONE TRIAS
T2	75	251	77	286	101	48	47	164	T2 3d group	sandst. A		
T1	158					92	112		T1 2d group	shales between A & B		
									1st group	sandstones B & shales betw.B & C		
Lower series	55	-	60	4	-	55	19	47	Lower ser.	Lower series		Lower S/
TOTAL THICKNESS OF :												
Upper evaporites	740	-	270	51	416	585	105	173				
Lower detrital	288	251	137	290	101	195	178	211				
T.T. of TRIAS	1028	251	407	341	517	780	283	384				

Table 9.3 - COMPARISON BETWEEN OLD AND RECENT CLASSIFICATIONS OF THE TRIASSIC FORMATIONS OF THE STUDIED WELLS

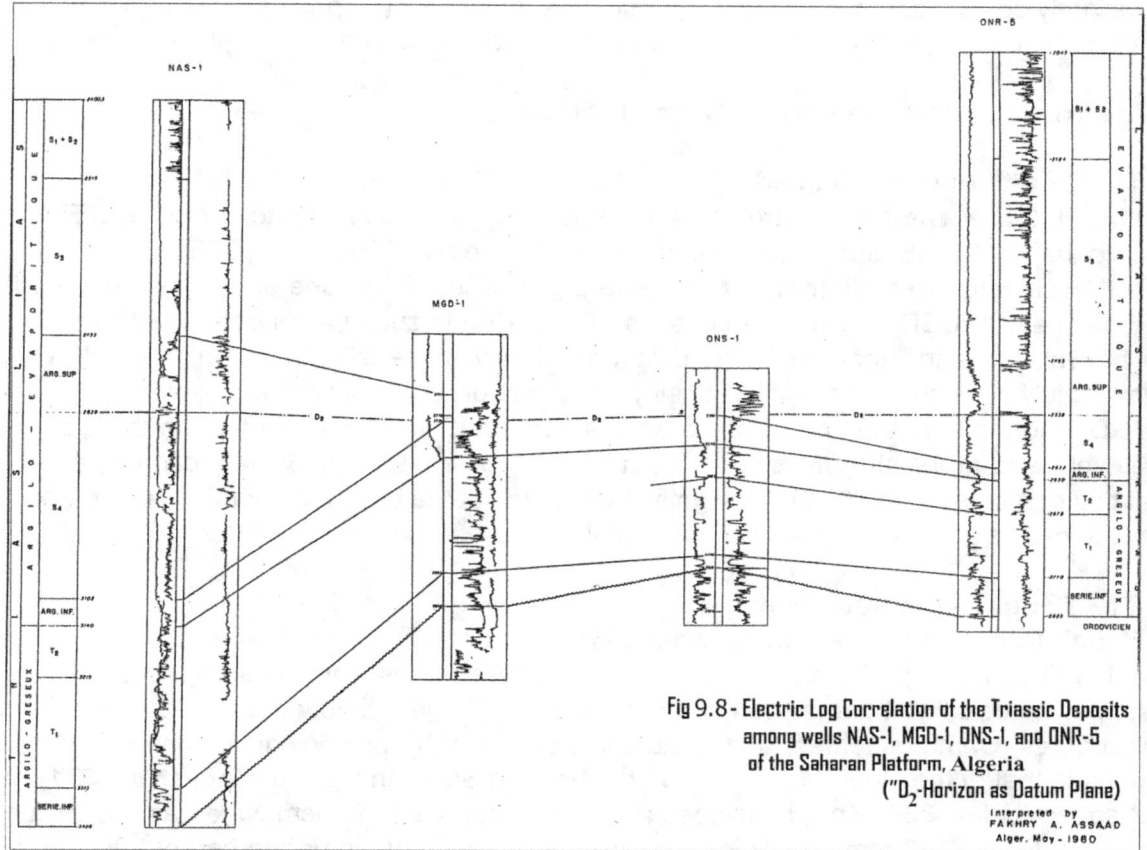

Fig 9.8 - Electric Log Correlation of the Triassic Deposits among wells NAS-1, MGD-1, ONS-1, and ONR-5 of the Saharan Platform, Algeria ("D_2-Horizon as Datum Plane)
Interpreted by FAKHRY A. ASSAAD
Alger. May - 1980

The sequence of the Triassic sediments can be described chronologically, where the SI-series, reworked from local sources of fluvio-alluvial origin, is overlain by upper sediments of fluvio-marine deposits with local unconformities that occur intermittently over the detrital sediments, followed upward by vast lagoons or shaly evaporitic deposits that accumulated in the center of the Triassic basin, at Nador Sud area (NAS-1), predominantly under conditions of increasing subsidence.

9.3.4.1 The Lower Series (Serie Inferieur: "SI")

In the east of the High Plateaux, at Nador Sud NAS-1-well, the lowermost series (SI-series), of 55ms thick, is characterized by two conglomerate beds, one very thin at the bottom and of 30cm -1 on its top; another one of one meter thick, showing local unconformity; in general, the lower series is mainly formed of alternations of micaceous green and red silty shales, with inclusions of reddish brown, fine to coarse, subangular to subrounded quartzite, sometimes with feldspars; traces of pyrite and chlorite are also encountered, together with inclusions of gray shale and siltstone on the upper beds.

At BO-1 to the north of Nador Sud area, the lower series (SI) is represented of conglomerates with rhyolitic fragments; to the northeast at "DOG-1 and CED-1; it is minly of conglomerates of 4m and 60m thick, respectively, denoting local unconformity; to the southwest, at Chott Chergui CC-1 well; it was replaced by eruptive "andesites", previously attributed to the Lower Carboniferous (Visean?), though, more likely to the lower detrital Triassic series (SI).

In the studied area of the Saharan Platform, at ONR-5-well, the Lower series is of 55m thick, (same thickness of NAS-1 well), and is mainly of pale green silty-shales, and red plastic micaceous shales intercalated with greenish white, subrounded sandstones, micro-conglomerates and breccias. To the northeast at ONS-1 (Oued N'SA), the lower series is 19ms thick and comprises of chocolate brown argillites, slightly dolomitic, silty and few conglomerates; at MGD-1 (Megadine), it is 47m in thickness, red and green in color, with chocolate silty shales, slightly micaceous, with dark gray, very fine sandstone.

9.3.4.2 The Triassic Argilo-sandstone Series (T1+T2)

(1) The T1-T2 unit –Undifferentiated
At the High Plateaus, the T1-T2- unit of the Triassic basin is un-differentiated at BO-1, CED-1, DOG-1, and CC-1 to the N, NE, NE and SW, and attained thicknesses of: 250m, 77m, 286m, and 101m, respectively; at BO-1, it comprises alternations of friable argillaceous sandstone and green shales with green dolomitic nodules; at CED-1, it mainly consists of coarse quartzitic sandstone; at DOG-1, it is fine to medium brown hard sandstone, with intercalations of silty shale, slightly dolomitic; and at CC-1, the T1-T2 unit mainly consists of brownish red, silty shale and greenish gray shaly siltstone.

At MGD-1-To the northeast of the Saharan Platform, the T1- T2 unit is of 164m thick, and is represented by yellow to greenish gray, shaly sandstone, very fine, subangular, with carbonate cement and intercalations of brown hard silty shale, slightly micaceous; at the base, conglomerates are encountered, together with quartz granules, black shale and coal beds.

(2) The T1 +T2 units - differentiated:
a) The T1-unit (C-series is overlain by B-series)
- At NAS-1 well of the High Plateaus- The T-1 unit (B+C series) is well developed and attains 158m in thickness; palynologically, it contains in its upper part of C-series (below depth of 3331m), an association of Disporites, Camerosporites, and Enzonalosporites of the palynological zone of "P1+P2" which corresponds to that of the Oued M'ya basin of the Algerian Saharan Platform (Achab in 1971); the lower part of C-series (below 3361.4m), it includes a conglomerate bed of 40cm, whereas, on the upper part of B-series (at depth 3282.5 m), there is a dark red and green conglomerate bed of 30cm in thickness, showing two local unconformities.

At Nador Sud well (NAS-1), the lower T1-C series is quartzitic sandstone, turns to medium and coarse grained, rounded to subrounded sandstone, with dolomitic cement and traces of feldspar granules and abundant ferruginous minerals; the upper T1-B series manily consists of quartzitic red brown, very fine to medium, well sorted, subangular to subrounded hard sandstone, with traces of magnetite, mica and pyrite.

-**At ONR-5 well, of the Saharan Platfotrm, the T1-unit (series B+C)** has an average thickness of 92m; The lower part of T1 unit (C-series), is coarse and micro-conglomeratic, whereas, at the base, it turns to gray; very fine sandstone, shale, and chocolate brown siltstone; the upper part **of T-1 unit (B-series)** mainly consists of grayish white to dark gray, subangular to subrounded fine sandstone, well sorted, friable with argillaceous and carbonate cement.

- **At ONS-1 well, to the northwest,** the **T1-unit attains a thickness of 112m,** mainly composed of chocolate brown sandstone, slightly dolomitic with inclusions of chlorite and microcrystalline dolomite; conglomerates and micro-conglomerates are encountered at the base.

b) The T2-unit:
- **The NAS-1 well of the High plateaus-** the T-2 unit (series A) is of 75m thick, represented by brown hard silty shale, very fine to medium quartzitic sandstone, micaceous, subangular to subrounded, with intercalation of chocolate red to brown micaceous compact siltstone at the base.

- **At ONR-5**, of the Saharan Platform, the T2 -unit of 48ms thick, manly consists of shaly siltstone and very fine to fine, gray argillaceous and siliceous, subrounded sandstone.

- **At ONS-1,** the T2-unit of 47m in thickness, is represented by gray shale, slightly dolomitic, and red plastic shales with intercalation of dark gray fine siliceous sandstone

9.3.4.3 The Lower Shales

In the northeastern part of the High Plateaus, at NAS-1, the lower shales attain a thickness of 44m and consist of alternations of pale reddish brown shales, sometimes plastic, slightly salty with traces of hard compact siltstone. The Lower shales are not represented at BO-1, CED-1, and DOG-1, whereas, at CC-1, they attain 30ms in thickness and consist of red brown shale.

In the Saharan Platform, the lower shales have a thickness of 18m, 39m, and 51m, at ONR-5, ONS-1, and MGD-1 respectively. **At ONR-5,** the shales are red brown, saliferous shales; **at ONS-1**, they are represented by white salty and multicolored saliferous shale, and at the base, by plastic red to dark gray hard shales; and **at MGD-1**, the lower shales comprise of dark gray and red brown hard shales, sometimes slightly calcareous.

9.3.4.4 The Evaporite Deposits

At Nador Sud area (NAS-1), NE of the High Plateaus, the Triassic/Liassic deposits attained unexpected thickness of anhydrite and salt deposits that overlay the Triassic detritus and could be correlatable with that of the Algerian Saharan Platform. A "faulting or noisy area", **was wrongly interpreted** on the isochrone map due to the abnormal thickness of the evaporites.

The evaporite deposits are also described in a chronological sequence:

(1) The Triassic S4-bed (TS4):

-**At the northeast of the High Plateaus,** it is not represented at BO-1, CED-1, and DOG-1, whereas at NAS-1, it attains a thickness of 263m, characterized by hyaline salt and pale green dolomitic shales, slightly plastic, with some inclusions of salt. The appearance of debris associated with "*disaccites air bags*" or "*sacs aeriférs de Dissacites*" in the cuttings, at depth of **2**967m, indicates the palynological zone "P3", correlatable with the S4 bed of the Algerian Saharan platform. At CC-1, the S4-series attains 110m thick, mainly of massive salt.

- **At the Saharan Platform,** the S4-salt bed of ONR-5 well, attains a thickness of 74m and consists of white rosy hyaline salt, with intercalations of red brown shales, slightly plastic, and traces of pale gray shale. At MGD-1, the salt bed is of 8m thick, of white limestone, slightly chalky, and at ONS-1, it is not represented.

Generally, the S4-bed is directly overlain by a good marker of thin dolomite bed "Horizon D2", with slightly shaly white limestone.

(2) The Liassic Formation

a) **The Upper Shales of the High Plateaus-** are not represented at BO-1, CED-1 or DOG-1, whereas at NAS-1 and CC-1, they have a thickness of 204 m and 56m respectively. At NAS-1, they are represented by alternations of salt, red shale and fine sandstone with intercalations of hard compact limestone and dolomite. At CC-1, they are characterized by red brown shales, overlain by a sill of greenish doloritic basalt; **denoting their submarine effusion**.

At the Saharan platform, the upper shales are of 75m thick at ONR-5, and comprises of red brown hard shales; slightly sandy and saliferous. At MGD-1, they are of 29 m thick and consist of gray and brown microcrystalline dolomites with predominant red brown dolomitic shales. The upper shale bed is not represented at ONS-1.

b) **The shaly dolomitic salt "S3" series:** At NAS-1 well, the series is of 218m in thickness, mainly of massive salt with some traces of red shale; at CC-1, it attains a thickness of 60m and consists of rosy, massive salt and red brown sandy shales. It is not represented at BO-1, CED-1, and DOG-1.

In the Saharan Platform, at ONR-5, the series is of 279m thick, and consists of massive salt and gray shale to greenish gray saliferous shale and shaly dolomite. It is missing at both wells of ONS-1 and MGD-1.

c) **The S1- S2 - series: At NAS-1,** the series are of 115ms in thickness, and consists of pale gray to brown dolomites, fine crystalline, white anhydrite and pale gray shale with salt intercalation downwards. At the interval of 2460-2860m, the palynological studies show **the presence of a microflora sp.,** essentially formed of Classopolis of the Palynological zone Q, and corresponds to S1- S2+ Upper Triassic saliferous shales of the Saharan platform. The series were updated as a Lower Liassic Age (Achab, Sonatrach, 1971).

At CED-1, DOG-1 and CC-1, the S1-S2 bed has a thickness of 270m, 51M and 160m respectively. At BO-1, the S1-S2 bed is missing;

At CED-1, the S1-S2 -It is characterized by predominant shaly anhydrites with a fauna of Myophoris sp. af. elongata Oueb"; it is given a Muschelkalk Age by Technoexport (1971), whereas at Caratini, C. (1970) considered the lower 135m of the S1-S2 bed at CED-1, as Upper Werfenian and the overlying 35m (old Infra Lias ?) had been given a Muschelkalk Age due to the presence of Ceratites Binodosus *(see Table 9.1, modified by Assaad, 2012).*

At CC-1-series, it is characterized by dark gray anhydrite and multicolored shales with inter-calation of siltstone and dolomites and is given a Lias-Keuper age by Technoexport (1971), whereas, C. Caratini (1970) gave the term "Upper Salifirous Shales" as S1+S2+S3 series.

- **At the Saharan Platform,** the S1-S2 bed, a thickness is of 139m **at ONR-5**, mainly consists of alternations of white salt, dark crystalline anhydrite, reddish gray saliferous shales, pale gray dolomites and hard pale gray limestone. **At MGD-1**, the "S1-S2-S3 bed" of 84m thick, is characterized by white massive anhydrite sometimes brown to gray dolomite at the upper 53m (S1 - S2 bed); in the lower 31m (S3-series), it is mainly white massive pale gray salt that turns to hard, multicolored shale; **at ONS-1**, the S1-S2 bed is of 66m in thickness, mainly of massive black anhydrite with intercalations of gray shales.

d) **"The Infra Lias bed" – Mushelkalk bed –** At the NE Central High Plateaus, it is underlain by S1-S2; the Infra Lias bed is of thickness: 64m, 112m, and 69m at NAS-1, CED-1 and DOG-1, respectively, it consists of alternations of micro-crystalline dolomites and veinules of secondary white dolomites, to dark green shales and anhydrites.

At CED-1, of the High Plateaus, the bed is considered as Lias-Keuper by Technoexport (1971), and as Rhetian by Caratini, C. (1964), because of the presence of rich Lamellibranchs.

9.4 Stratigraphy and Sedimentology

9.4.1 Scope

Over virtually the whole territory of North Algeria, the Triassic period commenced with a phase of continental deposition. The Kabylian massifs were thought to have been the major source of terrigenous rocks. Both on the High Plateaus and on the central Moroccan Hercynian basement massif, there were probably uplifted portions of the folded Hercynian basement and massifs of magmatic rocks that supplied detrital materials.

In the North of the High Plateaus region, the Paleozoic substratum remained high and level at the beginning of the Triassic period; coarse detrital rocks accumulated in shallow depressions, while on the southern flanks of the High Plateaus, as well as in the Saharan Atlas, red shales and sandy beds were deposited, probably originating from the Paleozoic highs, southwest of Morocco.

Caratini, C. (SPHP, 1964), considered the Lower Triassic (SI) shows a littoral or very often fluvial, while in the studied area of the Algerian Sahara, it is reworked from local fluvial or fluvio deltaic deposits. The lower Triassic detrital series (SI) of the NE High Plateaus, were deposited transgressively over different Paleozoic units due to variable degrees of Hercynian erosion period.

9.4.2 Sedimentology (A Summary)

The deposition of the main shaly sandstone sequence (T1+T2), was followed upward by the accumulation of (S3+S4) salt series in vast lagoons, predominantly under conditions of increasing subsidence; whereas the last stages of deposition are represented by (S1+S2) series of shales and anhydrites.

From the examination of the Triassic detrital beds of the High Plateaus, one can determine a vast structural basin, where thick salt deposits are encountered at NAS-1 well, representing its center, followed upward by shaly anhydritic deposits, representing the last series of development of the basin; both wells of CED-1 & DOG-1 are cited at its NE rim of the basin, whereas its southwest at CC-1, well, thinner massive salt (TS4), deposited, while the basement remained relatively high at the end of the Triassic period.

It is worthy mentioning that the drilling of the wildcat well of "NAS-1" NE of the High Plateaus, revealed the problem of **"the missing Triassic cycle"**. **Fig 9.9a-** A Facies Map of the Triassic basin in NE Central High Plateaus of Algeria.

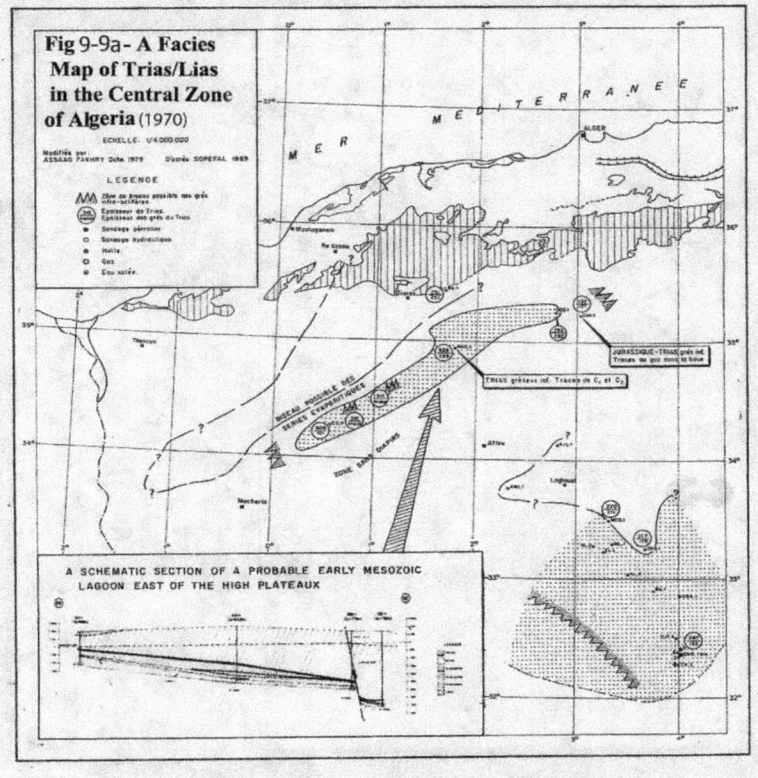

At the Saharan Platform, the T1-Triassic detrital bed, disconformably overlies the lower series (SI) and is overlain by the T2-unit; at MGD-1, the Upper Argilo-evaporitic sequence is only represented by hard brown shales, slightly calcareous in the lower part, followed upwards by shaly limestone, as a lateral facies variation of the S4-series, overlain by brown microcrystalline dolomites and red brown dolomitic shales, defining the northern rim of the vast Triassic basin of the Saharan platform, where open sea sedimentation prevailed. At ONS-1 well, the north-northeast rim of the basin, is mainly of massive and black anhydrites with intercalation of gray shales.

At the end of the Triassic/Liassic period, the salt deposition ceased, the open Tethys Sea sedimentation partly prevailed and the proportion of the carbonate rocks increased gradually.

Fig 9.9b– A postulated regional facies map of the extended Tariassic Province that covers both the High Plateaus and the Saharan Platform.in Algeria (Assaad, 2012).

Most of the geological syntheses of the Algerian Sahara fixed the top of the Triassic, at the base of Horizon "B", as a stratigraphic equivalent level of the *"Avicula Contorta"* beds of Libya, characterizing the Rhetian stage. Late palynological analysis of core samples taken from drilled wells of the Algerian Sahara, Ms Achab (1971), found the Triassic Saliferous Series of Alsace, and concluded that only the basal part of the S4-salt series, should be considered as Triassic "palynological zone of Disaccites", whereas the Upper evaporitic series together with "Horizon B", were stated as Liassic *"Palynological zone Classopolis"*.

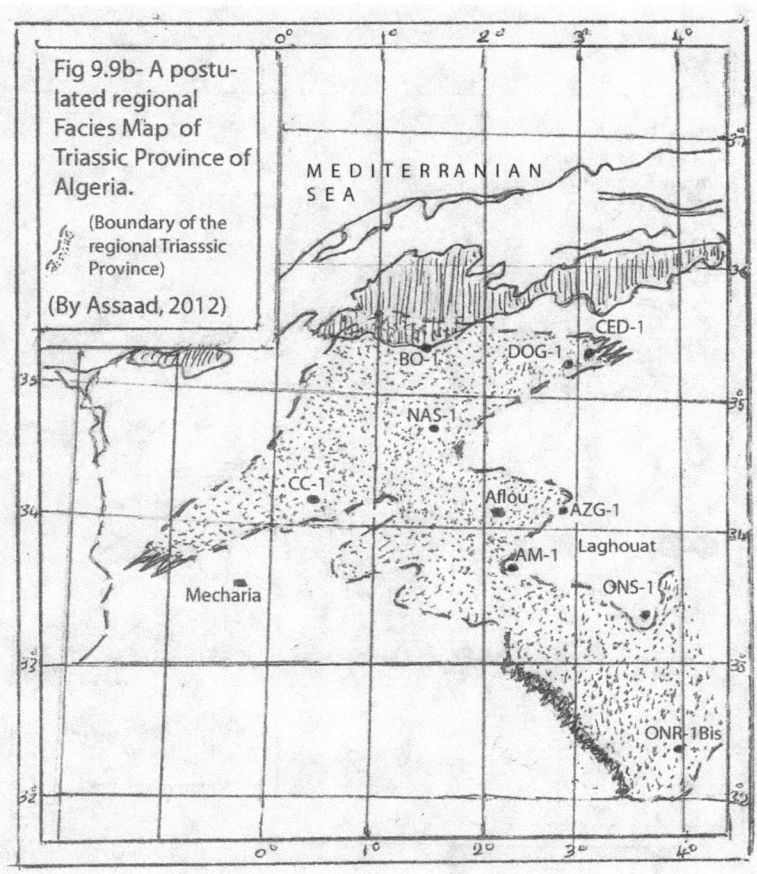

Stoica (1972) who later became the Petroleum deputy Minister of Romania, applied the data provided for his correlation of the Triassic beds for the different wells drilled in the Algerian Saharan Platform and defined the Trias/Lias boundary at the base of D2-dolomite bed on top of S4-series **(see Fig 9.7b)**; the D2 horizon is a marker bed, of more or less vugular dolomites, and is determinable in most of the studied wells of the Triassic Province.

The Jurassic/Cretaceous boundary was conventionally taken above the Kimmeridgean carbonate bed that is situated at top of lithologic unit "IV" (Butaz, personal communication).

The subsurface Triassic section of the High Plateaus, varies in thickness and ranges from 251m at Bourlier (BO-1) to 600m at Nador Sud (NAS-1), with an average of about 425m, whereas, in the studied area of the Algerian Sahara, it ranges from 217m at Oued N'sa (ONS-1), to 287m at Oued Noumer (ONR-1), with an average of about 250m.

From the lithostratigraphic examination of the Triassic formations of the studied areas, we may state th following:

(a) Local unconformities occur in the Lower series as well as in the Median Triassic detrital series.

(b) The Median Triassic series contain an association of Dissacites Camerosporites and Enzonasporites of the palynological zone P1-P2, which corresponds to that of the Saharan wells (Achab, 1971).

(c) From palynological studies, the so-called Permo-Trias at Bourlier (BO-1), consists of the lower Triassic sandstones that overly the Stephano-Autunian with an angular unconformity.

(d) The appearance of sacs aeriféres de Dissacites", in the cuttings of the drilled Nador Sud well (NAS-1) at a depth of 2,967m., indicates the palynological zone P3, which corresponds to the Lower Saliferous series (TS4).

(e) At (NAS-1) well, of at depth intrval of 2,460-2,860m, the cuttings are essentially formed of Classopolis of Palynological zone Q, corresponding to the Upper Saliferous series and lately, given a Lower Liassic Age.

(f) In the southwest, at Chott Chergui (CC-1) well, and also to the northeast, at Doghman (DOG-1) well, there are eruptive andesites attributable either to the Lower Carboniferous (Visian?), or more probably to the Lower Triassic detrital series.

(g) In the North at (BO-1), no Triassic evaporites have been recorded, whereas, to the northeast, at Cedraia (CED-1), and Doghman (DOG-1), the principal saliferous series of the Triassi/Liassic section are represented only by anhydrites and shales, representing the NE border of the Triassic Basin; in the southwest, at Chott-Chrgui (CC-01), thinner massive salts were deposited, as the basement remained relatively elevated locally until the end of the Liassic. Accordingly, the Triassic basin is believed to have a northeast/southwest trend; the (NAS-1) well, is located in the midst of the basin, where thick salt accumulations followed by shales and anhydrites **(see Fig 9.9a)**.

(h) In the studied area of the Algerian Saharan Platform, the Principal Saliferous series of the Triasic -Liassic formations, are well developed in the Oued Noumer well (ONR-5), but poorly developed at the Oued N'sa (ONS-1), and represented by shales and anhydrites of 100-200m in thicknerss.

In the NNW, at MGD-1, theTriassic-Liassic formations are represented by massive anhydrites and dolomitic limestones with intercalations of salt. The marker bed of Horizon "D2" at interval of 2,650-2,657m., is represented at the top by greenish-gray shales, while at the base, it is represented by white, chalky limestones; the salt layer" TS4" is absent , as it was close to the the southern boundary of the lagoonal basin.

9.4.3 Stratigraphic Evolution

The Algerian Saharan Platform belongs to the African Platform, which remained relatively stable during the geological time. Numerous basement uplifts accompanied by faults, have affected the Paleozoic sediments, but the tectonic phase had never reached the violence of the Alpine folds which disturbed the northern part of the shield, producing the folding sheets in the Saharan Atlas mountain and disengaging the Mesozoic sediments from the plastic saliferous sequence of the Triassic formation.

The primary sedimentation has been disturbed by the deep-seated movements and the Saharan basin had been consequently separated into several, more or less well individualized basins, each had its own specific geologic history that varied slightly from other basins, and consequently formed different problem for oil exploration.

Pamerol (1975) in his "Stratigraphy and Paleogeography of the Mesozoic Era" stated that the break-up of Gondwana Land took place within the Upper Triassic and continued up to the present due to sea-floor spreading.

During the Mesogene, the climate of the North Africa was tropical with warm seawater that led to the deposition of carbonates. The successive southward migrations of the Equator were accompanied by the deposition of evaporitic sediments, and many of the neritic basins became lagoonal with strong variations in salinity. The Alpine orogeny had its early phases in Jurassic and Creataceous times and reached its maxinmum in the Cenozoic.

The history of the Stratigraphic/tectonic evolution of the encountered area may be summerizd as follows (Tecnoexport, Sonatrach, 1971):

1) **Towards the end of the Pre-Cambrian period, and at the beginning of the Paleozoic period,** the Epihercynian platform was assumed to subside, in relation of the Saharan platform, therefore contributing to the formation of the south Atlassic flexures of the Lutetian-Portgalian Age through the beds of the transitional zone between areas.

   A system of NE/SW faulted structures or flexures was formed at the southern edge of the Saharan Atlassic Mountains", where terriginous deposits of the vast Paleozoic "Geosyncline" accumulated over a considerable thickness showing a gradual southward pinching out.

2) **Late Paleozoic -** To the far north of the High Plateaus, at the Bourlier Graben (BO-1, old Sersou Permit), thick terrigenous subcontinental sediments of flysch facies (black shales with dolerite veins) of Late Paleozoic age (Stephano-Autunian), associated with faulting at the end of the Caledonian phase during the Hercynian orogeny, and became filled with thick Paleozoic deposits of the Carboniferous, and attained a thickness of 2304 meters (interval 1080-3384). The total Triassic deposits at BO-1 well, are represented only by the lower detritus series (SI- series) that does not exceed 250 meters and consists of conglomerates with Rhyolitic fragments. The Paleozoic section is uncoformably overlain by a virtually horizontal bedded continental Triassic section of sandstone and sandy eruptive rocks.

   At the end of the "Visean period", the lower Triassic detritus series (SI-series) was marked by basalt, andesite and rhyolite extrusions, e.g. at Chott Chergui area (CC-1) to the southwest and Doghman area (DOG-1) to the northeast of the Triassic basin.

3) **During the Hercynian movements**, the Paleozoic section of the High Plateaus, was folded and affected most intensively the Central portion of Algeria, though, their effects were widespread all over; it was then lifted above sea level at the end of the Paleozoic period; when the axis of maximum subsidence became displaced towards the rim trough of the Saharan Atlas (Technoexport,1971). Simultaneously, with the uplift that was achieved by the Hercynian movements, the rejuvenation of the existing faults, resulted in the fracturing of the pre-Alpine basement in North Algeria. There is no doubt of the ancient origin of such faults as they preceeded both the formation of the "Paleozoic geosycline" of North Algeria and that of the Mesozoic troughs.

4) **The Paleozoic cycle was then followed by the early Alpine cycle** that formed the shape of the Mesozoic Trough of the Saharan Atlas within the High Plateaus, up to the northern limits of the Saharan Platform (or the Saharan Flexure of the southern part of the Epihercynian Platfiorm).

   At the Saharan Atlas, tremendous thicknesses of the terrigenous Mesozoic sediments, accumulated due to the strong prevailing epeirogenic vertical movements that extended to the Tunisian Saharan Atlas, and support its stratigraphic separation from the relatively thin sediments of the southern Saharan plate, e.g. the section of 5020ms and 3342ms of Jurassic formations were deposited at Ain Mahdi area (AMI-1) and Azreg area (AZR-1), respectively.In the earliest"Geosynclinal Phase", the Alpine Cycle passed through the Lower, Middle, and Upper Jurassic periods.

   The beginning of the Jurassic Period is marked in the Saharan Atlas trough by a transgression of the Pliensbachian sandstone. A thick carbonate formation accumulated, followed in the upper Liassic period by important arrivals of terriogenous material, forming a marl-carbonate series (Liassic sediments, reached 800m in thickness approximately). At the Saharan Atlas, vertical movements were strongly prevailing during the Liassic periods, and terrigenous deposits accumulated over considerable thicknesses (1200m at Ain Mahdi, 1740m at Dj.Azreg).

Fig 9.10- A NE/SW-Schematic Structural cross section showing the evolution of Sedimentary Formation among some wells of the Epihercynian platform and the Saharan Platform **(See Key Map - Fig 9.1; Assaad, 1981)**.

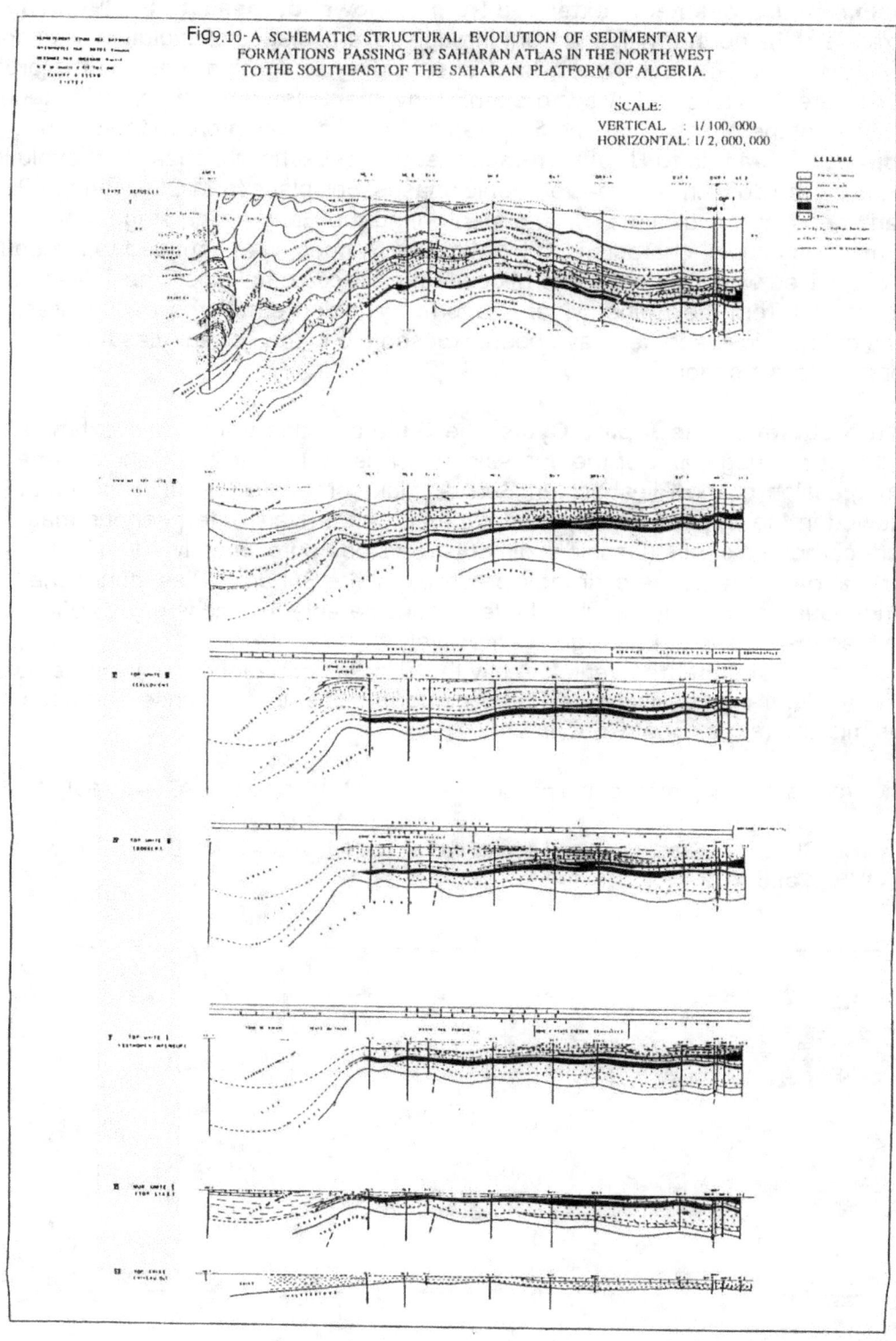

Fig.9.10 - A SCHEMATIC STRUCTURAL EVOLUTION OF SEDIMENTARY FORMATIONS PASSING BY SAHARAN ATLAS IN THE NORTH WEST TO THE SOUTHEAST OF THE SAHARAN PLATFORM OF ALGERIA.

In the High Plateaus area, the Liassic sediments mostly overlie the Triassic formations directly, but in some horsts, they overlie transgressively the Paleozoic formations. In such cases, the base of the Liassic is absent from the section e.g.Ghar Reuban)

5) **The main Alpine movements extended from the lower Jurassic to the Miocene** - The Alpine folded range of the northern Algeria went through several stages of evolution, with abrupt tectonic changes, that covered lenghly periods of of subsiding or uplifting, besides the orogenic movements. and are then followed by two orogenic movements from the Upper Paleogene to Miocene. At the outset of the Alpine cycle, the Saharan Atlas trough was more to the south.

From Late Jurassic to Upper Cretaceous, the "geosynclinal phases" of evolution Prevailed during the Tellian domain, whereas orogenic phases took place from Oligocene to Pliocene times. The early geosynclinal troughs were formed during the Early Alpine stage, and later, the folded Tellian mountain chain, characterized by Miocene Nappes, was bounded to the south by a marginal trough filled with Oligocene and Miocene molasses complexes. The Tellian mountain chain was separated from the Epihercynian platform by folded structures (or flexures) where many outcrops of Triassic evaporites have been exposed along the Tellian Atlas faults.

Upper Jurassic periods:

a) **At the outset of the Alpine Cycle,** the Saharan Atlas trough had extended more to the south, at the beginning of the Jurassic, which is marked in the Saharan Atlas trough by a transgression of the Pliesbachian Sea. A thick carbonate formation accumulated and was followed in the Upper Liassic period by important arrivals of terrigenous material, forming a marl- carbonate series (Liassic sediments reached approximately 800m in thickness).

b) **Vertical movements** were strongly prevailing at the Saharan Atlas, during the Liassic, where terrigenous deposits accumulated over a considerable thickness e.g. 1200m, and 1740m at Ain Mahdi area and at Dj. zerga , respectively.

c) **The Liassic sediments** overlie directly the Triassic formations, but in some horst structures, they overlie transgressively the Paleozoic formations; in such case, the base of the Liassic was missing (e.g.at Ghar Rouban area).

Fig 9.11a – Lithostratigraphic Diagrams for the encountered wells: NAS-1, AMI-1, AZG-1, MGD-1, and ONR-5

Fig 9.11b–a NW/SE Structural Cross section among wells: NAS-1, AZG-1, MGD-1, NL-1, **NL-2, HR-4, BE-1, DRA-1, DJF-1, ONR-1, and AT-2.**

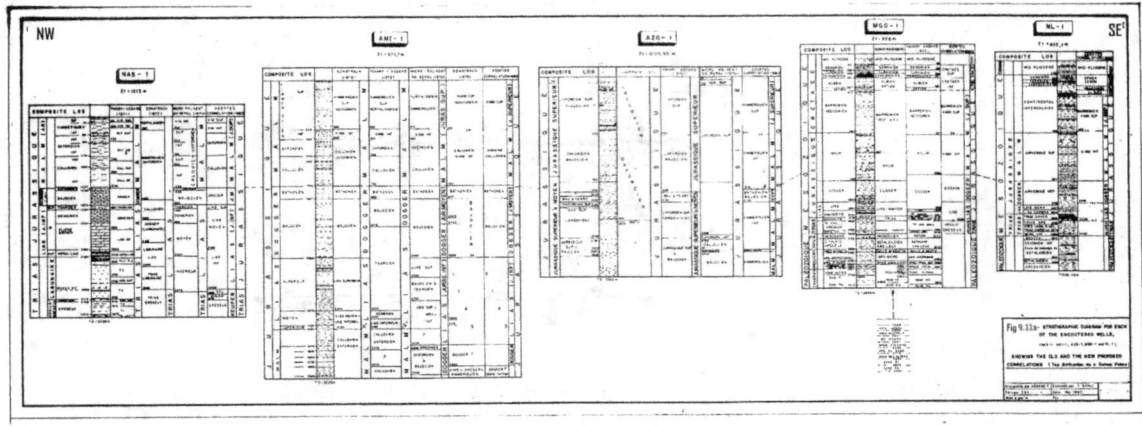

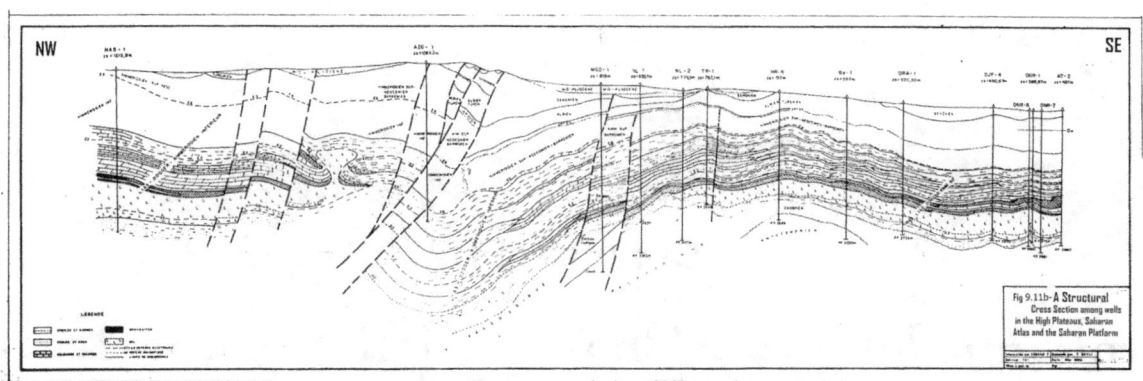

Fig 9.11b- A Structural Cross Section among wells in the High Plateaux, Saharan Atlas and the Saharan Platform

References

Assaad, F., (1972) Contribution to the study of the Triassic formations of the sersou-Megress Region (High Plateau) and the area of Daia M'Zab (Saharan Platform), No. 84 (B3), Eighth Arab Petroleum Congress, Algeria, 1972.

Aliev, M.M. et al, Technoexport, Sonatrach (1971) – Geological Structures and Estimation of Oil and Gas in the Algerian Atlas and the Algerian Sahara, Algiers.

Caratini, C., SHIP (1964) Etude Geologique des Monts Du Chellala Reibell, SPHP.

Busson, G., (1970) Le Mesozoique Saharien. Essai de Synthése des données des sondages Algero-Tunisians

SN RÉPAL (1961) Les séries Permo-Triassiques dans le Nord Sahara. Études Pétrographique du cycle Détritique. Études du Cycle Salifére (texte et planches).

Pomerol, Ch., (1975) Manuel de Stratigraphique et Paleogeographie;:Era' Mesozoique.

SNPA (1964) Étude Microstratigraphique et Sédimentologique des séries du Trias á Tertiaire recontrées par les Sondages de l'Hydraulique des Chotts Rharbi et Chergui.

------- (1963) Essai de Correlations et étude des variations de facies du Trias au Tertiere, entre le sondage de Chott-Chergui et lesforages H.18, H.34 & H.41.

Achab, A., 1971 Le Permo-Trias Sahariene, sonatrach, Algiers

Stoica, I. an Assaad, F.A. (1972) Geological studies on the Triassic reservoirs of Gas and Condosate at Hassi Er R'Mel Field (Internal Report, Sonarrach).

Selected References

Michel R (1970) Petroleum Developments in North Africa in 1969, The American Association of Petroleum Geologists Bulletin, v. 54, No. 8, Fig.1/p. 1458, Geneva, Switzerland.

Labouyris, M.& Tortochaux, F.(1963) Etude lithostratigraphic des Formations Triassiques et Jurasiques dela bordure des Hauts Plateaux et de L'Atlas Saharienn (algerie Ouest) (Janvier.-Juillet 1963).

Chapter (10)

Petroleum Reservoir Characteristics of the Triassic Province of the Algerian Saharan Platform- Emphasizing the Dome-like structure and the Discovery of an Oil Ring structure at the base of Hassi R'Mel Gas field – Application of the Gussow Theory- Tectonic settings of the related structures- The Evolution of Hydrocarbon Migration- Methodology

10.1 The Triassic Province and Oil and Gas Fields of the Algerian Saharan Platform

The Triassic province of the Saharan Platform comprises mainly of large structural elements such as the Oued M'ya depression, the Ghadames (or Rhadames) depression, Hassi Messaoud horst-anticline and El-Biod-Hassi Tourag Horst-anticline as well as the Djamaa-Touggourt structures, and the Tinrhemt high structure that includes the Hassi er-R'Mel field (Technoexport, 1971).

The Triassic detritus constitutes the main productive reservoirs of the M'Zab-Oued M'ya region of the Algerian Saharan Platform that encompasses an area of approximate 70,000 sq. km., and covers Hassi R'Mel, Oued Noumer fields and Haoud Berkaoui field.

There are three large oil and gas bearing systems within the sedimentary Triassic section of the M'Zab-Oued M'ya: a) the Hassi R'Mel oil and gas dome –like structure; b) The Oued Noumer area (Oued Noumer and Ait Kheir oil and gasoline fields), and c) The Haoud Berkaoui oil structure further to the southeast. **(See Fig 6.6** – Geologic Provinces and major structures of the Algerian Saharan Platform).

The source rocks constitute of highly bituminous shale and carbonate rocks of disseminated organic and inorganic matter of soluble liquid petroleum hydrocarbons, algae, diatom, soluble asphalt and insoluble kerogen that produced petroleum through diagenesis by bacteria under temperature and more effectively under pressure gradients. The Ordovician, upper Devonian, Gothlandian, and Carboniferous formations are the main source rocks (or non-reservoir rocks) of the Triassic reservoirs of Oued M'ya basin and possibly of the Cambrian reservoir of Hassi Messaoud field during the Pre-Early Alpine (Pre-epiorogenic movements in which the Paleozoic formations at Ghadames basin are probably the source rocks for oil fields farther to the east at Rhoud Nouss, Rhoud Yakoub and Borma fields (Techno-export, 1971).

Productive fields are isolated petroleum hydrocarbon accumulations that occurred in systematic determined groups of independent petrophysical zones with major structural features of defined regularities. Oil and gas pools are essentially controlled by large positive tectonic features that were characterized by a long period of evolution and large concentration of hydrocarbon reserves, e.g. Tinrhemt arch, Tinfoye' semi-arch and the Northern portion of Amguid –Hassi Messaoud Horst-anticline. Large zones of oil and gas pools are located in high trends of less sharply defined regional negative areas such as Illizi semi-depression (Alieve et al, Technoexport, Sonatrach 1971).

The analysis of the Triassic detritals has been proved to be azoic except some rare fossil-bearing beds, at HR-1, HR-4 and OA-1. The Triassic series of the M'Zab-Oued M'ya area is hardly favorable to palynological studies because of the following two facts: (1) As early as the Paleozoic time, the pollen material shows advanced degree of carbonization more than that of Erg-El-Hassi Messaoud farther to the east; and (2) The presence of anhydrite, that is hardly attackable by acids, might be a factor for the azoicity of the Triassic formations.

The eruptive flows at the lower most series of the Triassic detritus have a considerable significance, by the intensive filling and perhaps even by the reliefs they have produced as they controlled the later deposition of the post-andesite flows of the lower series of Hassi R'Mel, which were deposited at a later time than the Berkaoui lower series sandstones, and influenced by the eruptive substratum; in other words, the lower series should be regarded as a piedmont series, formed by reworking of the local substratum and by filling with materials from the dismantling of the nearby reliefs. The lower series first deposited at Hassi R'Mel and then later at Haoud Berkaoui fields, with the setting of important eruptive flows (Tecnoexport, 1971).

The overlying evaporites and the eruptive magmatic rocks constitute the most efficient seal for the reservoir rocks by the molding effect over the deformations that might occur within the Triassic detritus, forming a proper cap rock for the underlying reservoirs. Salt movements affected greatly the Triassic Structures; the pronounced disharmony between the mature configuration of the subdued and flat pre-salt Triassic rocks and the strongly disturbed post-salt rocks were exclusively due to the mobility of salt rocks during the post Liassic early Alpine orogeny (according to the latest classification of the Triassic deposits, Assaad, 1983).

10.2 Petroleum Reservoir Rocks of the Algerian Saharan Platform

10.2.1 Introduction

The Triassic Province comprises a vast region of the Algerian Saharan platform, where the Triassic shaly-evaporitic (Argillo-evaporitic) deposits constitute the main sealing cap rock of oil and gas reservoirs. To the west, the northern extremity of the Idjerane –M'Zab Dorsal is part of the Triassic Province, and covers the Oued M'ya Depression to the southeast, the northern parts of the Amguid-El-Biod –Hassi Messaoud Dorsal complex and the eastern Algerian syncline; there are evidences to suppose that the Meso-Cenozoic movements have led to renew the leveled Paleozoic tectonic "plane" that determined both the size of the above reservoirs, and the concentration of oil and gas deposits above the Hercynian discordance. The size of oil and gas fields reflects the hydrodynamic relationship between the hydraulic complex of the Cambro-Ordovician-Silurian and that of the lower series (Sl), (Technoexport, Sonatrach, 1971).

The oil and gas reservoirs at the Amguid-El-Biod-Hassi Messaoud dorsal complex are controlled by structural traps with consideable closures mostly affected by important faults. The Ordovician reservoir rocks are normally saturated with gas; whereas, in certain cases, they present a unique natural reservoir together with the Lower series of the Triassic reservoirs, so peviously called "Permo-Trias", in the central part of the Dorsal.

10.2.2 Petroleum Characteristics

The petroleum characteristics of Paleozoic and Mesozoic reservoirs of the Triassic province can be summarized (Technoexport, 1971)

10.2.2.1 The Paleozoic Reservoirs

The Paleozoic reservoirs consist of several beds of practical interest for oil and gas reservoirs; though, its petrophysical characteristics are low (the porosity is from 5.7-12%, and the permeability is 2.5md).

a) The complex of the Cambrian reservoirs - often shows intense fissures, both horizontally and vertically. Large oil fields such as Rhourde –El Baguel are linked to the fissured zones, which almost cover the Cambrian reservoir.

The Technoexport (1971) concluded, from the petrophysical and lithostratigraphic composition, that the oil fields of the Paleozoic reservoirs (at the Hassi Messaoud High, El-Gassi-El Agreb and Hassi Chergui- Rhourde El-Baguel ridges), are of secondary origin, either granular or fissured, and are linked to the granular reservoirs of Cambrian formation of which the predominant rôle belongs to the epigenetic transformations of the argillaceous cement, whereas from the tectonic point of view, the fissured reservoirs of Cambrian and Ordovician are mainly linked to the most active zones at the eastern part of the Amguid –El Biod –Hassi Messaoud dorsal and the Mega-Ridge of Erg Djouad-Haoud Berkaoui.

b) The Dorsal complex of the Amguid El-Biod- Hassi Messaoud, of which the Amguid Hassi Messaoud anticline structure is the principal element, comprises the main Cambrian reserves of oil in the Algerian Sahara; its area covers 1600km^2, of which 1150 km^2, are properly petroliferous. The Hassi Messaoud "dome-like structure", and El Agreb El –Gassi Ridge in the north, control El Agreb – Hassi Messoaud zone of oil and gas accumulation, as well as its four oil fields: Hassi Messaoud, El Gasi, Zotti, and El Agreb.

Both reservoirs of the Cambrian: "Ri and Ra", are the most active petroleum productive beds; the porosity of sandstones and fissured quartzites are fairly good from 5 to 10 %; the permeability of 0-100 md; sometimes reaches more than 1000md; the average saturation in oil was 84%; and the average net pay thickness of both "Ri and Ra" are of 85m.

In general, the oil/water interface of the reservoir is at 3380m subsea, and the whole yield was in average of 300-350 m^3/d, and in certain wells, it reached 1500m^3/d.

At the southern high zone of El-Agreb-Hassi Messaoud, are the oil fields of El-Agreb, Zotti and El-Gassi. The Rhourde El-Baguel-Toual zone of oil and gas accumulation lies to the northeast of the Amguid –El-Biod –Hassi Messaoud dorsal and is controlled by Rhourde –El Baguel -Hassi Chergui ridge and by the structural axis of Marfague –Gassi El-Adam.

The petrophysical properties of the Cambrian are not constant either vertically or laterally; they are mainly controlled, in different degrees, by the epigenetic process within the reservoir rocks; the regressive epigenesis, shown by the change of the mineralogical composition of the cement of the petroliferous sandstones at Hassi Messaoud and the local tectonic fissuration, played an important rôle in the formation of the granular and fissured "secondary" reservoirs.

It is noticed that the processes of regressive epigenesis only occurred at the Amguid-El-Biod-Hassi Messaoud Dorsal; towards the north, the Cambrian reservoirs, are mainly oil productive in the Algerian Sahara; to the west, the Cambrian sandstones of the Oued M'ya basin are illitic with little kaolitic cement; whereas, further to the northwest at Hassi R'Mel, the Cambrian rocks are of argillaceous chloritic cement.

10.2.2.2 The Mesozoic Reservoirs

The Triassic sandstones are the only Mesozoic productive oil and gas reservoirs of the Algerian Sahara. Lithologically, the Triassic reservoirs comprise of an alternation of sandstone and argillaceous sandstone rocks; the percent of sandstone increases towards the base where gravels and conglomerates are often found; the total thickness of the sandstone rocks is of 17-75m; it reaches 100-110m. in the Rhourde Nouss region and to the west at El-Borma structure; the total thickness diminishes towards the southwest of El-Borma region, and the Ghadames depression, and to the north and northeast towards the axial line of the Dahar high. The Mesozoic reservoirs can be grouped into two categories (Sonatrach, 1971):

(1) The Complex of "Permo-Triassic" reservoirs (compared to the Lower series "SI")– An old term that comprises three reservoirs, developed in different regions of the Triassic Province, and are distinguished from each other by their lithology, petrographic composition of rocks and in certain cases by their age:

a) The Nezla sandstones of terrigenous rocks, are generally defined as the "lower argillaceous sandstone member "SI-series".
b) The Haoud Berkaoui sandstones, developed in the big structural highs of El Agreb- Hassi Messaoud and Dahar, lies in discordance over different horizons of the Paleozoic formation and is sealed on top by effusive overflowing rocks. The Haoud Berkaoui sandstone reservoir, attained a good permeability and medium porosity (15-20%); whereas the lower sandstone series of the basin produce a small commercial oil field;
c) The Guerrara sandstones are developed on the southeast border of the northern part of the Idjerane M'Zab Dorsal and of the Oued M'ya basin. The sandstone rocks are of 10-51m. thick, mainly represented by varieties of fine and very fine quartzites whereas gravels appear at the base of the section.

(2) The Terrigenous Complex of the Triassic reservoirs- originally classified in two main horizons in chronoligical sequence **(see Fig 6.4- modified)**:

a) **The T1-unit**, developed over most of the north of the Triassic complex, at the Tinrhemt High; it mainly comprises of two members of sandstones **("B"+ "C"- series)**; the B-series varies from 34m at the Hassi R'Mel field to 5-6m in the Guerrara and Ouargla regions. The underlying C-series of the unit, is well developed and increases to the northeast and reaches 60-75m; it is only missing in the highest part of the Hassi R'Mel "and Hassi Messaoud structures";
b) **The T-2 unit-on top (or "A"-series); mainly of sandstones (3a reservoir, overlain by 3b clay bed)**, is extensively developed along the northwest border of the Triassic complex and on the northern dip of the Idjerane –M'Zab High. The unit is overlain by a clay bed (3b), intercalated by one or two shale beds

The reservoir characteristics diminished regularly towards the northwest of the line of extension of the Hassi R'Mel sandstones, whereas, there is a sudden drop of porosity and permeability from the east to the west of the complex; the highest porosity was recorded of 22-25% on the southern part of the Hassi R'Mel structure; it diminishes for about 15-20 % to the north at Haoud Berkaoui structure. The T2-unit "A-series" developed to the southeast of the Tirhemt high and in the northwest part of Hassi R'Mel, with a permeability of more than 1000md.

The yield of Hassi R'Mel gas wells was considerable, reaching 2.7 M m^3/ day in the beginning of 1970's and the depth of the productive horizons is 2100-2400m; The possibility of discovering gasoline and oil should not be execluded in the so-called the "Permo-Triassic sandstones" of the eastern and the northeastern borders of the Tinrhemt High.

The Upper Argillaceous Sandstone bed (3a+3b) extends to the southwestern part of the Amguid -El Biod -Hassi Messaoud Dorsal; it is of 160m thick, well developed farther to the south with a predominance of hard detritus sandstone rocks in its lower part (3a), of high porosity of (15-25%) and permeability of (100-1000md); Petrographically, the sandstones are quartzose, and granulometrically fine and sometimes with a predominance of medium sandstones, but of a tendency to diminish from bottom upward; in the upper part, the (3b-series) is usually of argillaceous- carbonate of (20-35%), with an argillaceous material of (5-10%).

10.3 The Main Productive Structures

The Triassic Province of the Algerian Saharan Platfom, comprises of several high structures: Tinrhemt, Allal High, Djamas, Touggourt system of structures, besides the Oued Mya Depression, Amguid El-Biod –Hassi Messaoud Dorsal complex, and Rhadames (or Ghadames) Depression:

10.3.1 The Tinrhemt High
The Tinrhent High lies in the northern part of the Idjerane –M'Zab Dorsal where the vast Tinrhemt zone of oil and gas accumulation exists with a good commercial value. It includes the following two petroleum fields:

A. The Hassi R'Mel gas field – was updated by the author as *a dome-like stucture, which* covers a surface area of 55 x75km^2, and of 140m in closure; it lies in angular discordance over different eroded beds of the Cambro-Ordovicean; the thickness of the main reservoir ("A" or (3a), is of 25m in average, and is the most important Triassic gas reservoir of Hassi R'Mel field; the reservoirs "B" and "C", are of areas 994km^2 and 1252 km^2 respectively. The gas/water interface was at a depth of 1505 ms subsea, determined by three productive wells, The average saturation of gas is 80%; the porosity for 'A' is 14%, for "B" 13%, and 17% for "C"- series. The argillo-saliferous complex of the upper Triassic/Liassic deposits is of 400m thick.

Oil Ring Discovery at the Base of the Hassi R'Mel Gas Field - The Hassi R'Mel gas field was previously interpreted by the a French contractor "SN-Repal, 1970" as an Anticlinal structure, yet the author primarily considered it as a dome-like structure, due to assunption by the author that the high topography of an area, might reflect a high subsurface structure as well.

While replacing a wellsite Bitish geologist on the Drilling Rig; for drilling one well, cited to the far east of Algeria nearby the border of Tunis, the author realized the high location of Hassi R'Mel Gas Field at the Tinrhemt Structure attained an extreme high altitude (of 700ms).

An exploration well, south of Hassi R'Mel ("Hassi R'Mel Sud": HRS-1), had been later proposed and a Montage report was prepared; oil shows were luckily recovered from the Triassic reservoir. Few years later, by checking the palenological results of one well at the Lab of S.N, Repal (at Gharmoul, Alger), two black crude oil samples were found on an old shelf, marked as Hr-4, and, Hr-8; located east and west of the Hassi R'Mel structure respectively; The isobaths maps on top of the reservoir showed that both samples were on the same subsea elevation, which were then updated as a dome-like structure; knowingly from the old drilling reports, Hr. 8-well, showed a feeble belt of oil near the gas/water interface (Technoexport, 1971)", with a yield of oil 0.34 m^3/day, and a density of 0.824 at 15ºC.

B. The Bordj Nili Oil and gas Field comprises of Moscovian sandstones of the Carboniferous formation, located on the northern flank of the Tinrhent High. The field is penetrated by two wells; NL-2 and NL-3, with depths of 2525 -2555m; NL-2 well is oil productive with a yield of 192m^3/d, whereas, oil water interface is at depth of about 1910m subsea; and the reservoir pressure is 322 kg/cm^3. At NL-3 well, it gives a yield of 1000 m3/day of gas and a weak yield of oil, 0.25m^3/day.

10.3.2 The Oued M'ya Depression
The Oued M'ya Depression, productive of oil and gasoline, comprises of "two subzones": The Erg Djouad -Haoud Berkaoui and Oued Noumer oil fields:

The Erg-Djouad – Haoud Berkaoui "Subzone"- of oil and gas accumulation belongs to a system of two ridges that complicate the eastern border of the Oued M'ya depression. The southern border of the subzone coincides with that of the reservoirs of the "Permo-Triassic" (Lower series) and the Triassic Detritus (A, B, and C-series).

Four oil fields were discovered in the sub-zone of Haoud Berkaoui, Ben Kahla, Oulougga, and Guellala, where numerous indications of oil and gasoline were found. **Fig 10.1**- A location Map shows different Permits and petroleum Fields of the Algerian Saharan Platform.

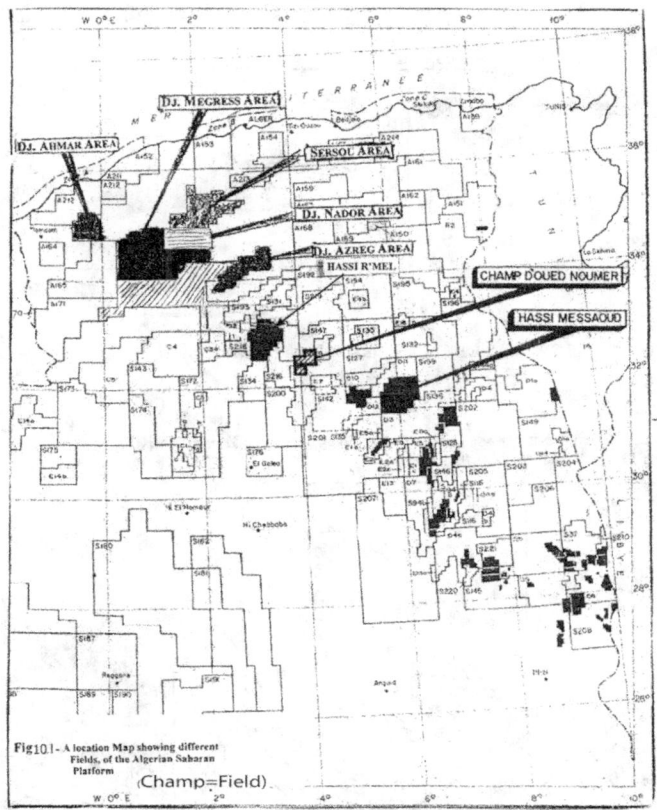

Fig 10.1 - A location Map showing different Fields, of the Algerian Saharan Platform (Champ=Field)

Table 10.1 – General well Data of the Triassic reservoir rocks in different oil and gas fields

Well Name	Zt/ms	Elevation reservoir ms	Thickness of overburden ms
Hassi R'Mel Field	700	-1500	2200
Oued Noumer and Ait Kheir Fields of Middle Triassic reservoir	400	-2300	2700
	400	-2325	2725
Haoud Berkawi Field (lower Triassic series)	200	-2500	2700

Table 10.1- General well data of the Triassic reservoir rocks at Hassi R'Mel, Oued Noumer and Haoud Berkaoui fields;

Fig 10.2 – NE/SW structural cross section passing by the Haoud Berkaoui oil field;

Fig 10.3a – An isobaths map on top liassic at Haoud Berkaoui and Ben Kahla oil fields and the location of three sections: I-I II-II and III-III (Technoexport, 1971);

Fig 10.3b- Three structural cross sections at both Haoud Berkaoui/Guellala and Ben Kahla oil fields, (Sonatrach, Technoexport, 1971).

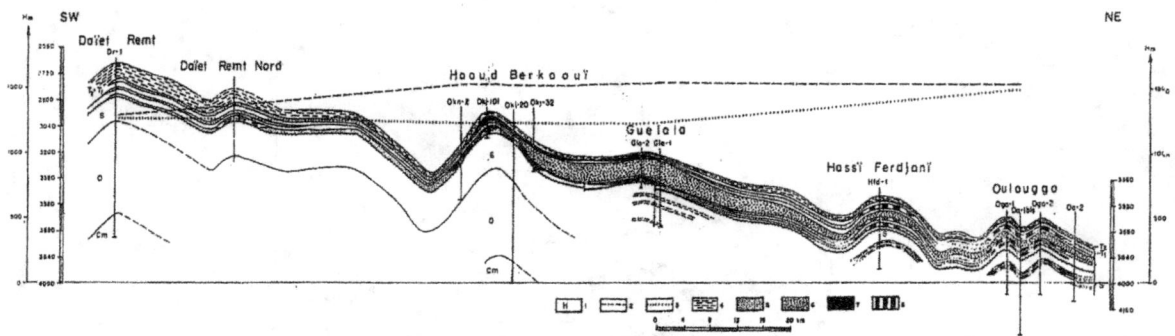

Fig10.2 - NE/SW Structural Cross Section in Haoud Berkaoui oil Field

Fig10.3a – Isobaths Maps: (A) Top Liasssic "bASE Hz B"; (B) Top Shaly Sandstone Triassic reservoir; (C) Isobaths Top Lower Sandstone series (S1).

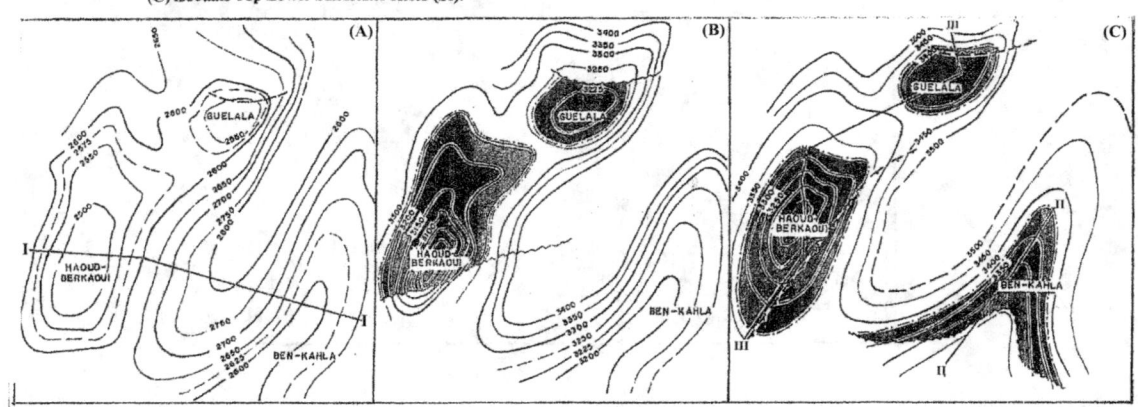

Fig10.3b – Geologic Cross Sections:
-- E/W Structural Cross Section I-I – Fig10.3a/Plate A; passing by Haoud Berkaoui and Ben Kahla Ridges;
-- NNE/SSW stratigraphic cross section II-II of Top Shaly Sandstone unit (A) - Fig. 10.3a/ Plate C; passing by Ben Kahla Ridge;
-- NNE/SSW Structural Cross Section III-III showing the Triassic Shaly Sandstone Reservoir- Fig10.3a/ plaste B; passing by Haoud Berkaoui and Guelala Ridges

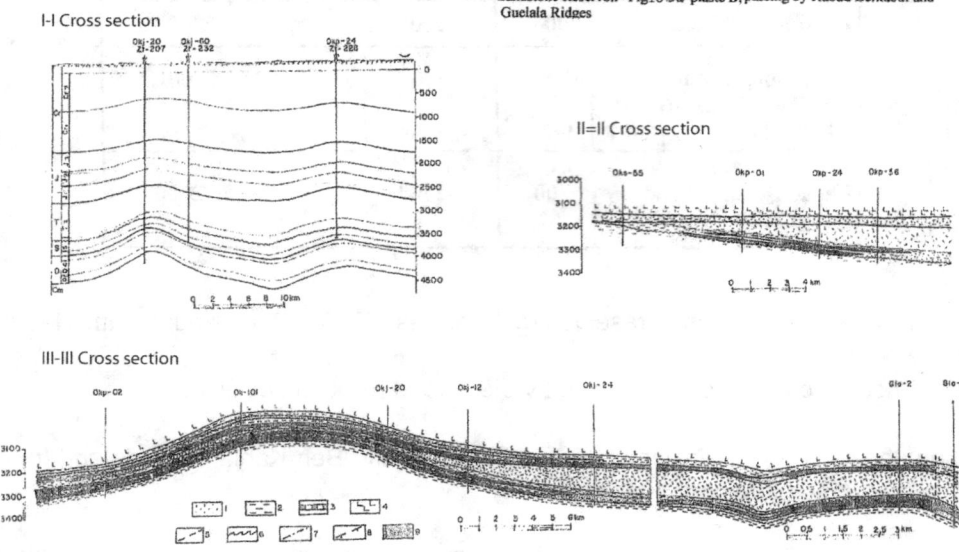

The Ben –Kahla, Haoud Berkaoui and Guellala oil fields – discovered on the anticlinal folds, and were well represented on the isochrone and isobaths maps, and were established at both markers of the base of Horizon "B" and top Hercynian discordance. The dimensions of the structures at the level of the Permo-Trias (lower series) were of: 0.5x19.0Km; and a closure, of 250m in thickness.

The discovered productive units were linked to two reservoirs of the Permo-Trias (Lower series) and the median Triassic reservoir (T1-unit); oil was found at different levels of oil/water interface.

The Lower Series- The refilling coefficient of the trap of the Permo-Trias (SI) reached to 0.80. The oil/water interface of the Permo-Trias reservoir (Lower series) lies at 3302m subsea in the northern part of the structure, at 3320ms subsea, in the south-eastern part of the structure, a possible due to the existence of a fault. The thickness of the productive lower sandstone series reached an average of 45-50 ms., and the net pay sandstone was equal to 20-35m. The yield of oil in the wells of the Permo-Trias reservoirs (SI) was in average of 350-380 m^3/day.

The T1-resrvoir- The refilling coefficient of the trap of "T1-unit" was equal to 1.0, the Triassic sandstone has an average thickness of saturated oil from 6 to 8m, whereas, the net pay sandstone varied from 2.6 to 13m. The yield of oil in the wells of the productive zone of the Trias (T1), reached 60-75 m^3/day.

In the summit of the Ben Kahla structure, a yield of oil and gas has been obtained "oil yield was calculated at 43m^3/day" from the fissured quartzite of the Ordovician (Saret unit). The northern part of Haoud Berkaoui field was almost unexplored before late 1990's, the position of the oil/water interface of the Triassic layer was not then determined and the northern closure was not yet drilled either (Sonatrach 1983); the available seismic data supposed that the Guelala structure was probably complicated as that of the Haoud Berkaoui and could attain the same reservoir characteristics.

The Guelala Structure- Previous studies concluded that the structural complication of Guelala and the Hercynian discordance presented a small fold of a dome-like structure with dimensions of 6.5 x11 km; with a closure of 40ms. The Trias "T1 +T2 –units" of the Guelala reservoir, yield 3 m^3/hr. of oil and a gas of 200m^3/hr; the T1- reservoir was tested in another well and found to be saturated with oil. The Permo-Trias reservoir (Lower series), was oil productive from 1.8 to25.14 m3/hr and of gas in the order of 1359.8 m3/hr.

At GLA-1-well located on the northern border of the structure, the analysis of one core sample showed oil indications from the lower sandstone series at a depth of 3588ms (or 3390.8m subsea), whereas the Chromatographic analysis of the gas in the mud, showed a certain percent of heavy hydrocarbons in the lower sandstones. By considering the previous facts, the position of the oil-water interface of the lower sandstones series was taken at 3385m subsea.

B. The Oued Noumer "Subzone"- of oil accumulation lies on the western border of the Oued M'ya depression, and is separated from the Tinrhemt high by flexures with a gentle slope.

According to preliminary seismic data, the discovered reservoirs are contained in an elongated horst-like structure with dimensions of 5 x10km at the Hercynian discordance; and of closure less than 75m in thickness.

The subsaliferous Triassic sandstones are saturated with oil and gasoline; the yield of gasoline was tested in the order of 1.8 M m^3/d and that of oil equals to 19.3 m^3/hr; whereas, the gasoline/oil interface was at 2303m subsea and the productive reservoir at depths between 2600 and 2700m. In addition, the seismic survey evidenced about 10 local prospects.

10.3.3 The Rhadames Depression (or Ghadames as "pronounced by the French Geologists")

It comprises a considerable part of the Triassic Province. However, the available data were very poor till 1990's. Different high structures were included in the depression; e.g. El-Borma oil field

The El Borma oil field had been explored in the Dahar High on a large sub-meridianal brachy-anticline; the greater part located within the Libyan Territory whereas, only the western part lies in Algeria; its closure reaches about 100m; the northwest of the structure is limited by a normal fault with a downthrown in the order of 200m.

The oil sandstone bed lies at the base of the Permo-Trias (the lower series), and includes five productive zones, intercalated by thin beds of siltstones and clays. The net pay in the oil zone is nearly 80m, the average net pay saturated in oil is 37.8m, the oil/water interface lies at 2200m subsea; the porosity increases, vertically, from bottom to top, varying from 14.3% to 20.2 %. The saturation in oil diminishes in the productive zones from top to bottom (from 77.8% to 60 %.). Therefore, one can prove a certain link between the porosity of the reservoirs and their saturation with oil. The yields of oil are considerable, reaching 20 m³/day.

10.4 Petrophysical Aspects, and Lithostratigraphy of Hassi R'Mel, Oued Noumer, and Haoud Berkaoui oil and gas fields

10.4.1 Scope

The Triassic Province of the Saharan Platform constitutes of four large oil and gas bearing systems within the sedimentary section; Hassi R' Mel, Oued Noumer, Ait Kheir, and farther to the southeast Haoud Berkaoui fields. The Oued Noumer area constitutes both the Oued Noumer and Ait Kheir oil and gasoline fields; each of the productive fields is controlled by its hydrodynamic system and probably by the regional oil and gas generating process.

The High structure of the Hassi R'Mel-M'Zab region is associated with the major uplift of the Pre-Cambrian basement which reached at depth of about 2,550ms; the Oued M'ya basin separates the Hassi R'Mel from El –Agreb Messaoud to the southeast, where the basement reached at depth of approximate 5,000ms.

The Paleozoic formations were eroded to different degrees from one structure to the other during the Hercynian period, and hence the Triassic section was unconformably deposited over different Paleozoic levels. **Fig 10.4** – A Subcrop map of the Hercynian Discordance shows lithological characteristics, depositional boundaries of the Triassic formation and Petroleum prospects of the Triassic Province of the Algerian Saharan Platform (Assaad, 1973)

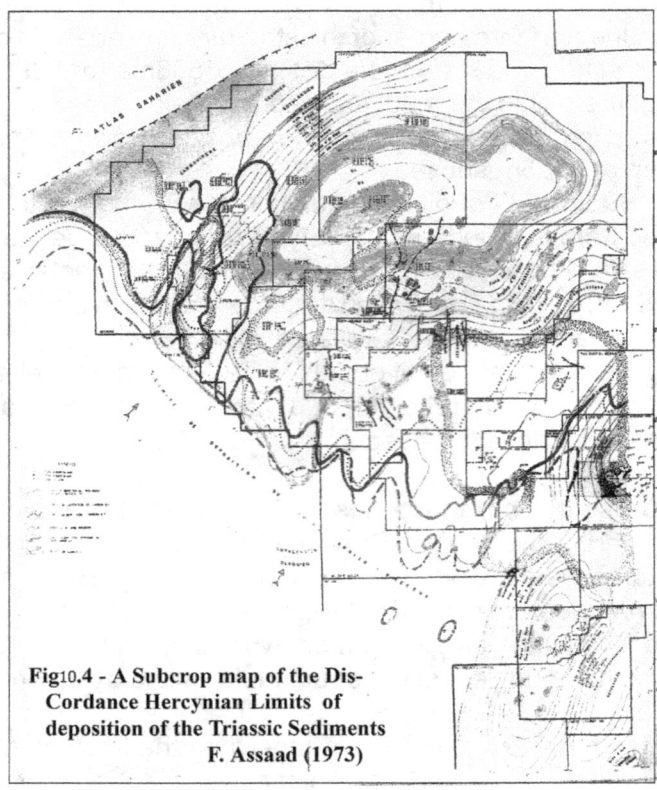

Fig10.4 - A Subcrop map of the Dis-Cordance Hercynian Limits of deposition of the Triassic Sediments F. Assaad (1973)

The Hassi R'Mel- M'Zab High was intensively eroded and the Paleozoic formation, represented by the Cambro-Ordovician, attained a thickness of 300-500ms. At the Oued Noumer area of the Oued M'ya Basin, the thickness of the Paleozoic section is nearly 1000ms, but in the most collapsed part of the depression, it is up to 1300-1400ms in thickness (Technoexport, 1971); the thickness of the Mesozoic deposits is increasing from Northwest to southeast; it amounts to approximately 2200ms at the Hassi R' Mel High, 2900ms in the Oued Noumer area, and 3750ms in the central part of the Oued M'ya depression. At the Oued Noumer area, the Hercynian erosion surface is about 2450ms subsea; and at DRT-1, is about 3531ms subsea. At OA-1, it is about 3800ms subsea and represents the center of the Oued M'ya Triassic basin.

At the Oued Noumer field, the Ordovician and Gothlandian deposits, overlain by the Mesozoic section, form a monocline with an angle of 2-3°, towards the eastern side.

10.4.2 The Hassi R' Mel Gas Field

10.4.2.1 Location

The Hassi R'Mel field is located approximately between latitudes 32° 50`- 33° 40` N and longitudes 03° 20`- 03° 40` E. It is about 400 km south of the capital, Alger and 280 km to Hassi Messaoud at the mideast of the Saharan Platform. It was discovered in the early 1957's, by the completion of HR-1 that had been drilled to a total depth of 2275ms by the "Societé Nationale de Recherche et d'Exploitation des Petroles en Algerie" (S. N. REPAL, 1961), in association with "Companie Francaise des Petrolier en Algerie" (C.F.P.A).

Hassi R' Mel Field was firstly awarded in January 1962 for a fifty years term, to the "Societe' d'Exploitation des Hydrocarbures d'Hassi R'Mel" (SEHR), covering an area of 3525 sq. kms. Since April 1971, the field was nationalized by the Algerian Government: "Societé Nationale De Transport et de Hydrocarbures, "Sonatrach". **Fig 10.5** – A geographic location map of the Hassi R'Mel field.

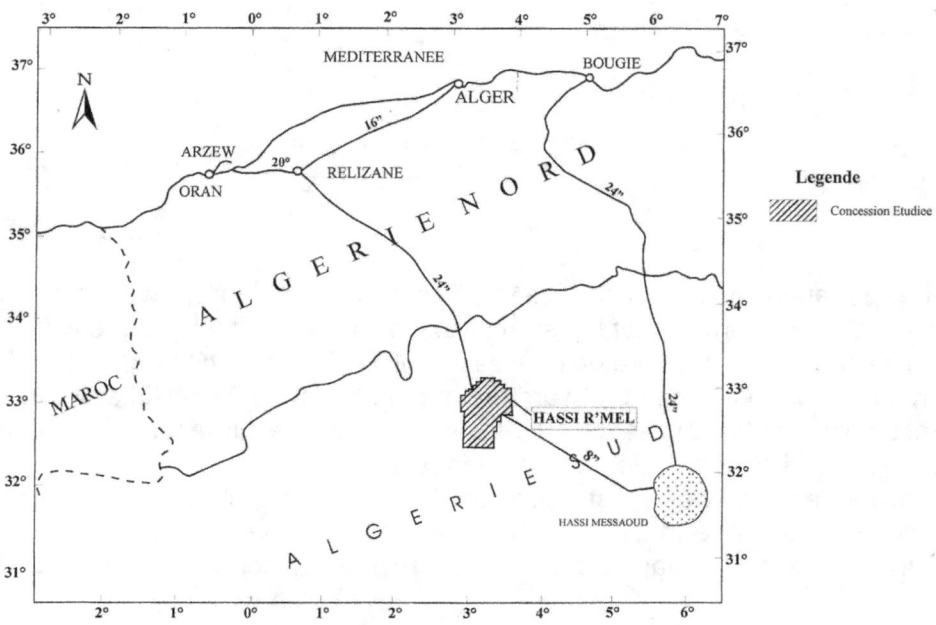

Fig10.5 - A Geographic **Location Map of Hassi R'Mel Field**

10.4.2.2 Structural Geology

The Triassic basin of the Hassi R' Mel dome-like structure, located at the northwestern extremity of the Algerian Saharan Plarform, covers an area of 3427.5 ms². It had been wrongly predicted by S.N. Repal (1961), as a large, arcuate, fairly symmetrical NNE-SSW anticline dissected by a series of parallel transverse faults into a system of tilted fault blocks (horst" and "graben" structures). The long axis is known by a series of normal faults10.6 here striking NW-SE; the arcuate distortion of the long axis may be due to either differential subsidence or a drag associated with lateral movement.

Fig 10.6 – An old isopaches contour map of the Triassic Argilo-sandstone "B-series"(old classification) at Hassi R'Mel field, prepared by S.N. Repal, 1972 as an anticline open structure

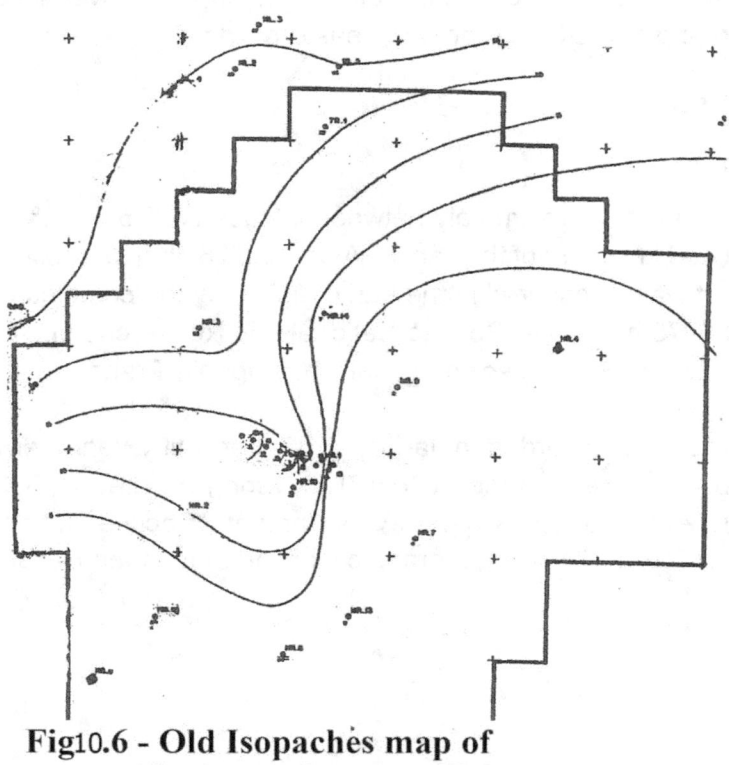

Fig10.6 - Old Isopachès map of Shaly sandstone unit B
SN REPAL 1972

Seismic re-interpretation on top of the detrital Triassic formation showed a complicated faulting system, especially in the northern part of the structure, which might affect the gas/water contact in some parts of the field. The enormous structure of the Hassi R'Mel is found at the intersection of two major axes of the substratum, one of which is the northward prolongation of the Hoggar-Idjerane-M'Zab High, with a slight westward nosing, whereas, the second is of E-W trend, at the Tilrhent and Djefara Highs, probably representing an unexposed NE extension of the Atlas relief (Technoexport, 1972).

The post-Hercynian erosion was most active on the Tinrhemt arch that resulted to a total removal of the Paleozoic sediments where the Triassic beds immediately overlie the rocks of the Pre-Cambrian basement. **Fig 10.7** – An isobaths map on top of the Hercynian discordance (Stoica, 1971).

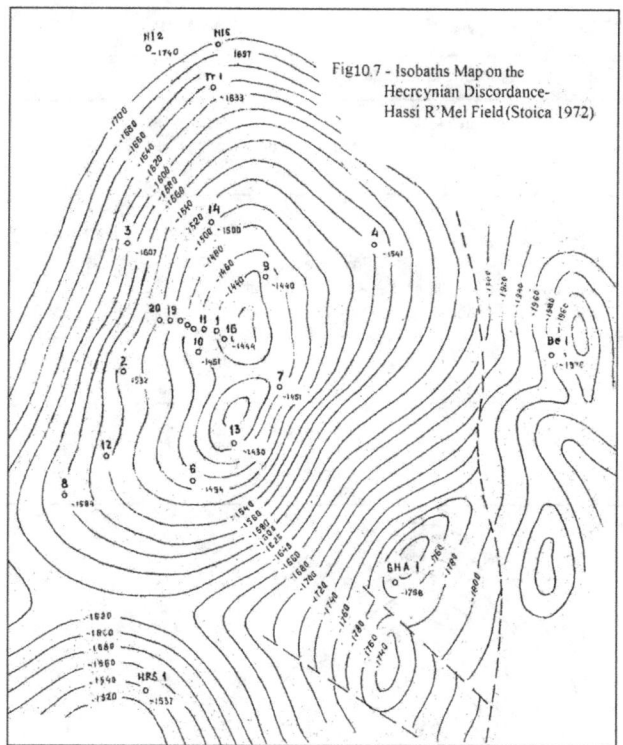

Fig 10.7 - Isobaths Map on the Hercynian Discordance- Hassi R'Mel Field (Stoica 1972)

An indication of a local unconformity, that occurred during the deposition of the Triassic sandstones at Hassi R'Mel structure, remained high during the deposition of the Lower sandstone series (SI-series), forming to the west, a small basin of piedmont series that were filled up with reworked materials from adjacent eroded reliefs of the local basement. The thickness of the lower series reaches up to 80ms.

10.4.2.3 Lithostratigraphy

Based on the geological and petrological data of the final well drilling reports at the Hassi R'Mel and the nearby areas, an electric log re-interpretation of the Triassic section was submitted to adopt a new regional nomenclature of the Triassic reservoirs, instead of the old one, by considering the base of Unit D_2 as one marker and the base of unit "3a" as another marker; the lithostratigraphic classification, based upon Gamma Ray/Neutron log, can be applied on the whole Triassic Province of Algeria (Stoica and Assaad, 1972):

(1) The Lower Series "Serie inferieur "S.I." - is a continental reworked series; its lithological characters reflect the local substratum and is very heterogeneous both horizontally and vertically. It immediately overlies various Paleozoic units on the Pre-Mesozoic (Hercynian) surface of erosion that affected to a certain extent the thickness and facies distribution within the series; the lower series is missing in the higher part of the structure (because of non-deposition).

Andesite eruptions are locally intercalating the lower series as contemporary of the sporadic shaly sandstone deposits in the southwestern part of the field. The fine grained size of the series is due to the fact that it originated from grain size sections of the Devonian and Gothlandian.

Table 10.2 - Old and recent Classifications of the Triassic Detritals (Stoica and Assaad, 1980). **Fig 10.8** - An isobaths map on top of the lower sandstone series (serie inferieur).

Table 10.2- Comparison between the former and the present classification of the Triassic formation

Former Classification		Present Classification			
Lower shaly series		3b-unit	Upper Shaly Sandstone (3)	Upper Trias	Upper Trias
T2-unit	(A-series)	3a-unit			
T1-unit	(B- series)	2c-unit	Equivalent of Argilo-carbonate (2)	Upper Interm	Intermediate Trias
		2b-unit		Middle Interm	
	(C- series)	2a-unit		Lower Interm	
Lower shaly sandstone (SI-series)		Lower Shaly sandstone (1)			Lower Trias

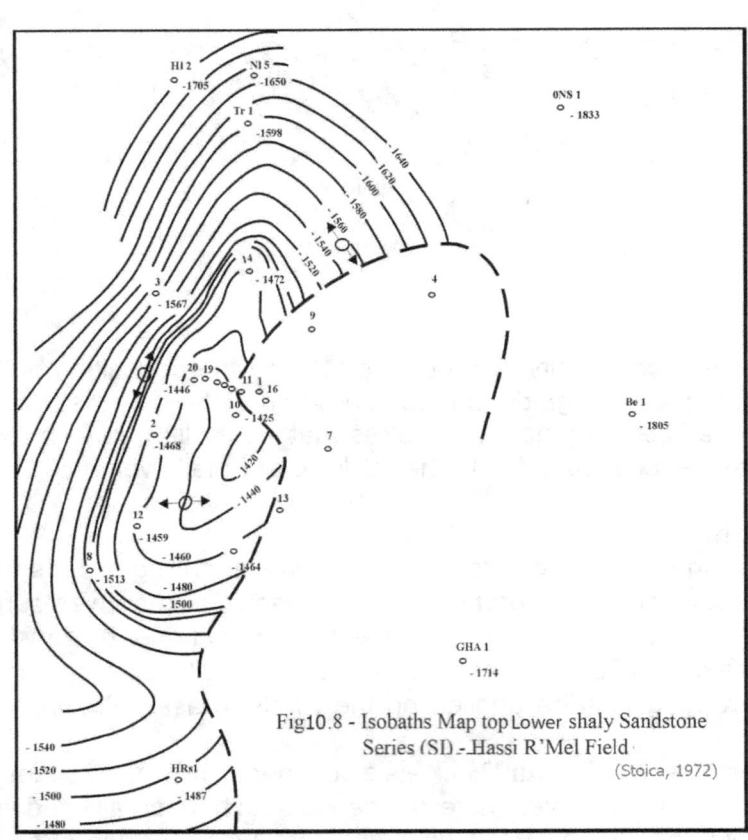

Fig10.8 - Isobaths Map top Lower shaly Sandstone Series (SI) - Hassi R'Mel Field

(Stoica, 1972)

The Intermediate series (Equivalent of the Argilo-Carbonate) constitutes of three units (2a, 2b, and 2c -units) or (the old B and C-Series combined of the T1-unit);

The intermediate series, a fluvial type of deposition, marked the beginning of flood periods that overlie uncoformably the shaly sandstone lower series; The source of the coarser deposits of the series together with the presence of rather fresh feldspars, cited at Morocco to the west or to the northwest at the Oranes High Plateaus, possibly resulted from the alteration of the granite massifs.

The intermidiate series are classified chronologically into three units:

(a) **The Lower-Intermediate unit (2a-unit) of** (the lower part of the old C-Series)- basically defined by well log data and grain size analysis (Sopefal 1971); it is the first massive sandstone layer that overlies the Lower series. It generally begins with conglomerates, followed upward by massive sandstones, intercalated with shale, quartzite pebbles, and anhydritic sandstone in most cases. Near the top, shales are intercalated by numerous breccia. A local unconformity are found at the base of the 2a-unit, which was missing further to the west of HR-2 well. At HR-3, there is a conglomerate bed of 2ms thick in the middle part of the 2a unit together with heavy minerals at the base. **Fig 10.9** – An isobaths map on top of 2a-unit

(b) **The middle Intermediate (2b-unit) of** (the upper part of old C-series and the lower part of the old B-series) consists of fine to medium grained, well sorted, argillaceous sandstone. Generally, the upper part is shaly and turns southwards to completely shales. At HR-4, HR-7, HR-9 and HR-13, the "2b"-unit is immediately overlying the Hercynian erosion surface. A low content of heavy minerals is encountered in the unit. The Hassi R'Mel High "Môl" completely emerged during the deposition of the "2b" - unit that covered the whole structure due to pinching out. **Fig 10.10** – An isobaths map on top of "2b"-unit.

(c) **The upper intermediate 2c-unit of** (upper level of the intermediate series); only exists at Be-1 well (Berriane); it is missing all over the Hassi R'Mel structure because of its regressive character, being affected by the unconformity that followed the Shaly carbonate deposition in other areas.

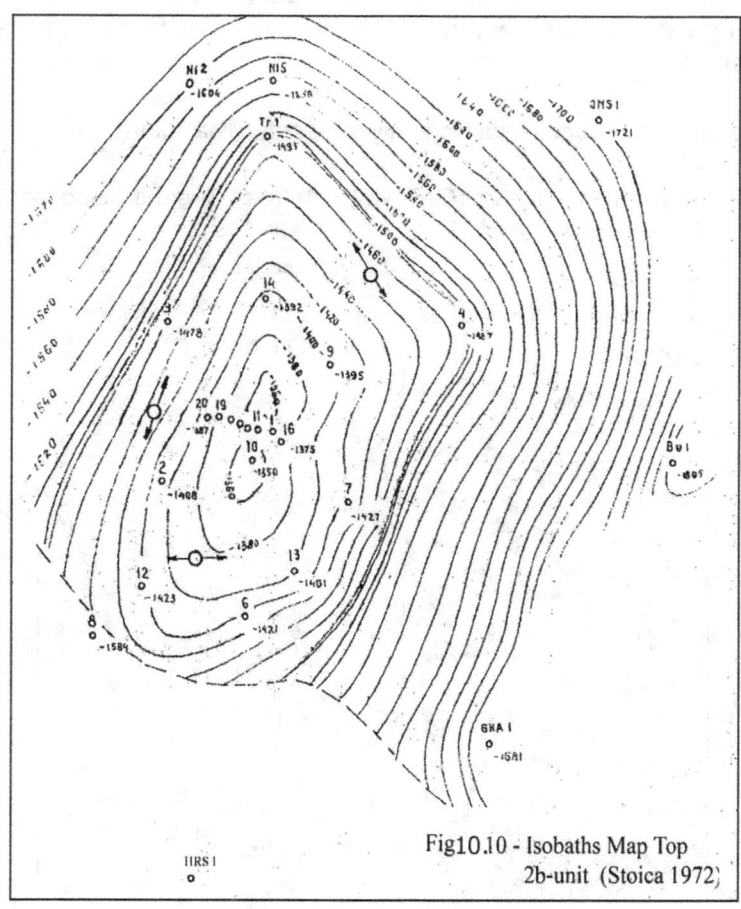

Fig10.10 - Isobaths Map Top 2b-unit (Stoica 1972)

3- The Upper Argilo-Sandstone series of 3a-unit

(a) **The 3a-unit (The old T2 or A-series)** –the main resevoir bed shows a sudden change of its aspect and probably its type of deposition as the finer sandstone grains constitute of the thickest zones whilest, the coarser sandstones are found in the thinner zones of the resrsrvoir **(see Table 10.2)**.

The "3a"-unit generally comprises of fine-grained sandstone with abundant feldspars, frequent sedimentary breccias and conglomerates (e.g. HR-3, HR-4 and HR-7), showing therefore, an unconformity and the beginning of a new transgressive cycle. The unit is intercalated by little shales, sometimes ferruginous towards the north and east of the Hassi R'Mel field. Heavy mineral content of "70gm/-100gm" of the rock unit from which zircon constitutes up to about 55%.

At HR-4, HR-7, HR-9, and HR-13, **the "3a"** sandstone unit comprises of clean sandstone, with a low content of heavy minerals, and overlies the Hercynian (Pre-Mesozoic) surface of erosion. The upper series which is the most transgressive and well developed of the Triassic formation, overlied by either "Shaly-carbonate" to the southeast of Hassi R'Mel field", or unconformably overlies the Lower shaly sandstone (SI- series), or the Paleozoic formation.

The northern High reliefs of the Talemzane ridge divide the basin into northern and southern parts and were considered as the origin of the feeding materials that might occur on a site where the coast line was submitted to frequent displacement, when the basin eventually became isolated from the open sea (Sonatrach, Technoexport, 1971).

Fig 10.11 – An isobaths on top of the upper Triassic shaly sandstone "3a"-unit.

Fig 10.12-Lithofacies limits of deposition of the 2a (C-series); "2b & 2c" (B-series) and 3a (A-series) sandstone units.

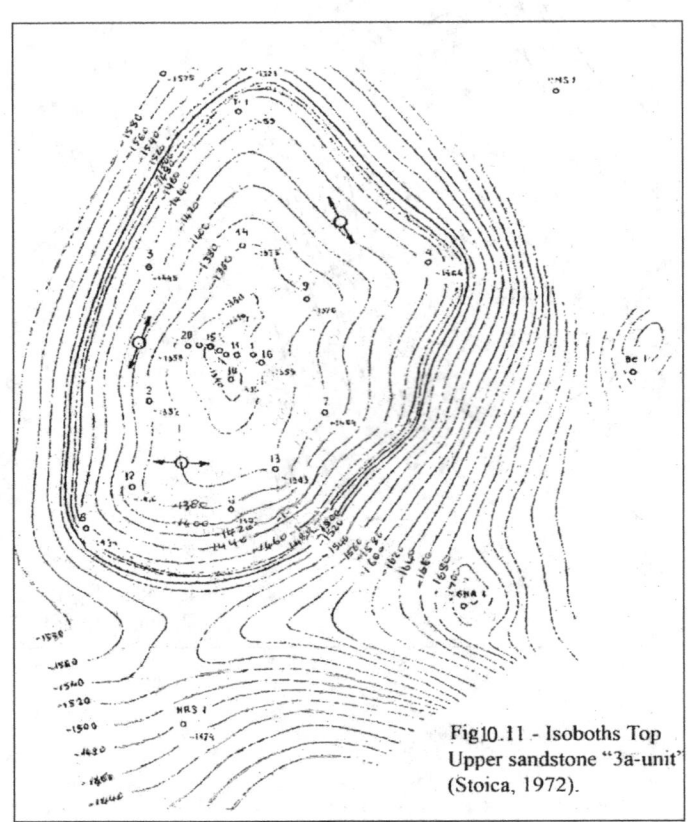

Fig10.11 - Isoboths Top Upper sandstone "3a-unit" (Stoica, 1972).

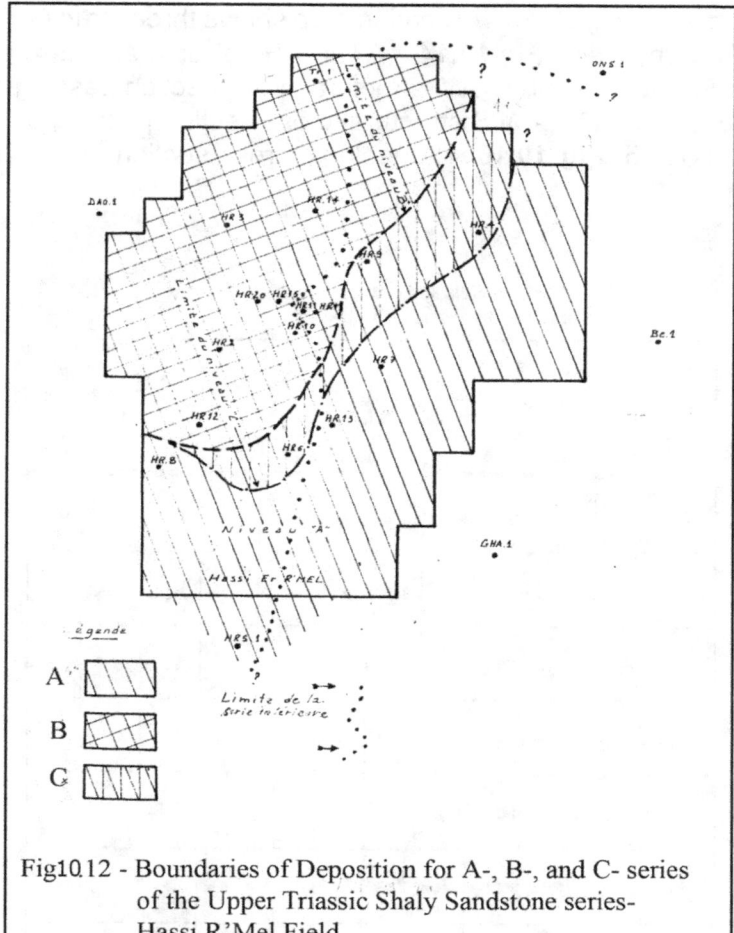

Fig 10.12 - Boundaries of Deposition for A-, B-, and C- series of the Upper Triassic Shaly Sandstone series- Hassi R'Mel Field

b- **The 3b- unit** (**likely equivalent to the old lower shaly bed**)- represents the final stages of the fluvial sequence and is likely the initial stage of basin deposits in which the Triassic sedimentation occurred between the m**arine incursions and the fluvial lagoon deposits of continental origin.**

In general, the overlying shaly lagoons were thus formed between the sea and the continent, together with the precipitation of carbonates and anhydrites. When the basin was isolated from the open sea, huge lagoons were established and the evaporitic cycle started with salt deposits.

10.4.2.4 Hydrocarbon Perspective of Hassi R'Mel Gas Field

The term gas and condensate (gasoline) field is practically used for a field whose hydrocarbons are in a gaseous state and partly pass to the liquid state as pressure decreases. If such conditions prevail, then in the course of the field's production, as pressure declines, a certain quantity of liquid components of the gas is separated due to retrograde condensation and accordingly it remains as a part of the reservoir. In such case the system of the reservoir exploitation has to be specifically designed, regarding the condensate content.

Hassi R'Mel Field, is productive of gas and gasoline at beds "3a", "2a", upper "2b", and "2c" (A and B-sandstone series) where generally, gas/water contact is at 1505ms subsea.

Structural isobathes maps were constructed for each sedimentary bed of the Triassic shaly sandstones of the Hassi R'Mel Triassic Reservoirs. The tested intervals of Gas/water contacts are shown on each cross section. **Table 10.3** - General well data and Petrophysical characteristics of Hassi R'Mel reservoirs; **Table 10.4** -Drill Stem Testing results of the Upper Triassic Reservoirs of Hassi R'Mel gas Field through perforation zones. **Fig 10.13** - A location map shows three structural cross sections passing by Hassi R'Mel field and nearby wells. **Fig 10.14** – A NW/SE - structural cross section passing by HR-7, HR-16, HR-1, and HR-3. **Fig 10.15** - A NE/SW -structural cross section passing by HR- 4, HR- 9, HR-1, HR-10, HR-12 and, HR-8. **Fig 10.16** - A N/S structural cross section passing by HRS- 1, HR- 6, HR-10, HR-11, HR-14, TR-1, and, NL-5; **Fig 10.16a-** A geologic cross section of the Triassic detritals among wells Hr 8, 6,3,4 & Hr 10.

Table 10.3- General well Data and Petrophysical characteriswtics of Hassi R'Mel Reservirs(A, B, & C)

Well Abbreviation	Z_s (m)	Z_t (m)	Final Depth (m)	Thickness (m)			Porosity % (ϕ)			Permeability (K)		
				A	B	C	A	B	C	A	B	C
Tr.1	767	771	2434	23.5	17.8	66.8	19.50	14	15	189	179	1451
HR.1	776	779	2275	16	12	32.5	16	21	21	919.2	928	1843
HR2	727	730	2308	13.6	9.5	14.5	22	7.5	17	500	85	2020
HR3	728	831	2352	36	26	16.6	13	19.3	/	1.5	382	/
HR4	718	721	2687	29.5	/	18	21,3	/	12	1855	/	25
HR.6 bis	762	765	2259	12.06	/	16.80	23-24.6	/	244	0.1-1680	/	162-4700
HR.7	762	765	2258	22.9	/	/	11.07	/	/	510	/	/
HR8	725	729	2336	19	/	/	16.7	/	/	1438	/	/
HR9	771	775	2217.0	19.5	7	29.5	20.5	3.7-9.5	17.07	250	0.1-1.9	315,7
NL5	771	775	2494.0	28	23	56	17.3	22	19	82.8-315.0	/	/

Table 10.4 – Drill Stem Testing Results of the upper Triassic Reservoirs of Hassi R'Mel Gas Field
Stoica (1972)

Well N°.	Nr. Test	Intervals Dept (top/base) origin (table) DP=0	Tested reservoir level/s	Results	Testing Hrs Interval top/ac. en678 DP=0	Perforations Results
HR-1	2C	2135 / -1354	3a+3b	☼ important flow of humid gas (un-measured)	2132 / -1351 2147 / -1365	3a
	4C	2135 / -1359 2149 / -1373	— " —	☼ m³/h + 0,350 m³/h ⊕	2157 / -1375 2166 / -1387	2b
	5C	2218 / -1442 2234 / -1456	AGI+Palz	☼ strong flow (un-measured)	2187 / -1408 2218 / -1439	3a
HR-2	1C	2087 / -1367 2140 / -1410	3a+2b	☼ strong flow of humid gas	3122 / -1352 3128 / -1408	3a
	2F	2170 / -1440 2187 / -1457	2a+2b	☼ " — "	2168 / -1428 2180 / -1450	2a+2b
HR-3	1F	2160 / -1429 2205 / -1474	3a+3b	☼ Humid gas		
	3F	2160 / -1429 2246 / -1515	3a+3b	☼ Gas (humid) + ⊕ 240		
	4F	2239 / -1508 2261 / -1530	2a	☼ 4.530 m³/h		
HR-4	1F	2164 / -1443 2208 / -1487	3a	☼ 4900 m³/h + 1300 m³/h ⊕		
	2F	2247 / -1526 2261 / -1540	2b	⊕ 0,350 m³/h		
HR-6 bis	2F	2159 / -1394 2291 / -1426	3a+2b	☼ 10 000 m³/h + ⊕		
	4F	2216 / -1451 2228 / -1463	2b	☼ 3000 m³/h + ⊕		
HR-7	1F	2154 / -1389 2196 / -1431	3a	☼ + ⊕	2168 / -1403 2198 / -1433	3a
HR-8	1F	2202 / -1474 2250 / -1521	3a	☼ + ⊕ ⊕	2226 / -1393 2228 / -1405	3a
HR-9	2F	2139 / -1364 2215 / -1440	3a+2b	☼ 503 000 m³/y		
NL.2	1F	2302 / -1580 2525 / -1542	3a	⊕ 15 m³/h, 30 l s/h		
	3F	2525 / -1742 2554 / -1776	AGI+Pz	⊕ 310 m³/h		
	4F	2540 / -1757 2555 / -1772	Pz	⊕ weak flow		
NL.5	1F	2300 / -1525 2321 / -1546	3a	⊕ 25 m³/h		

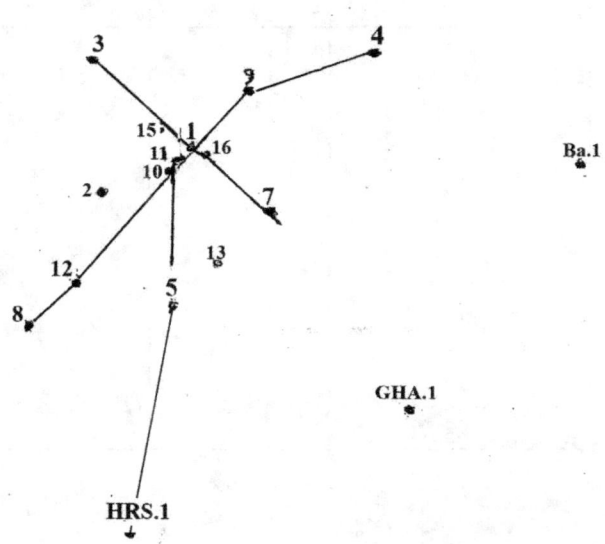

Fig10.13-Key map of well locations and selections of Three cross sections-Hassi R'Mel Field

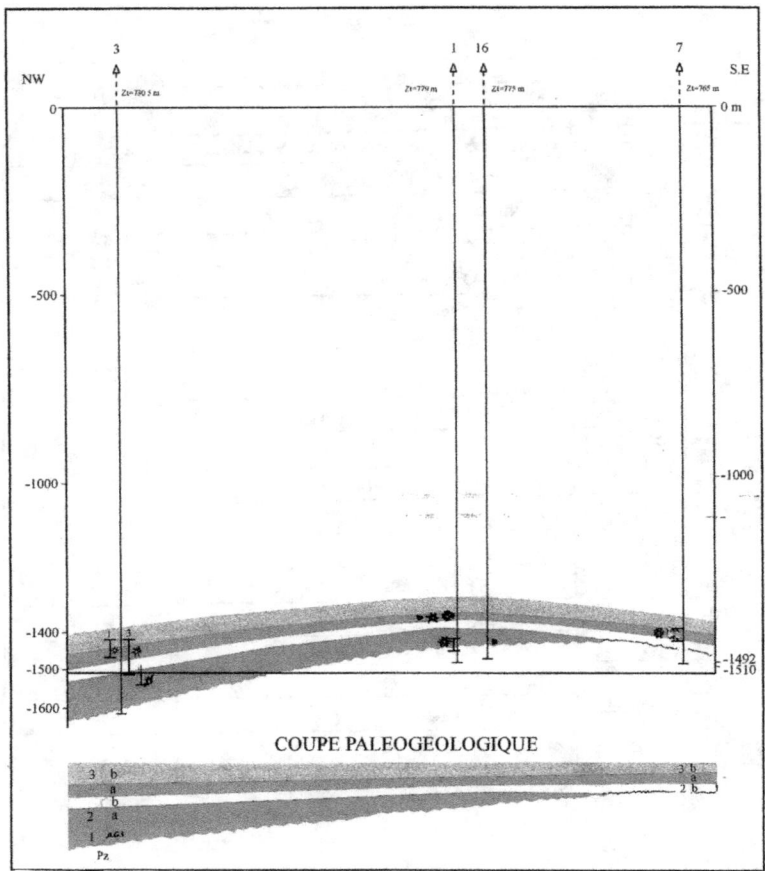

Fig10.14 - NW/SE paleogeologic and Structural Cross Sections - Hassi R'Mel field
(Stoica, 1972)

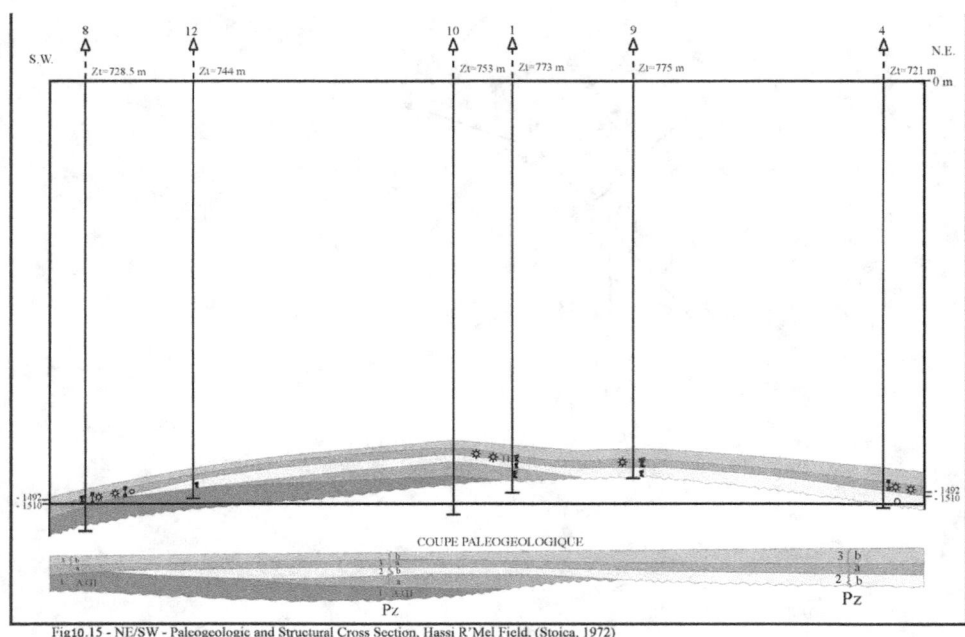

Fig10.15 - NE/SW - Paleogeologic and Structural Cross Section, Hassi R'Mel Field, (Stoica, 1972)

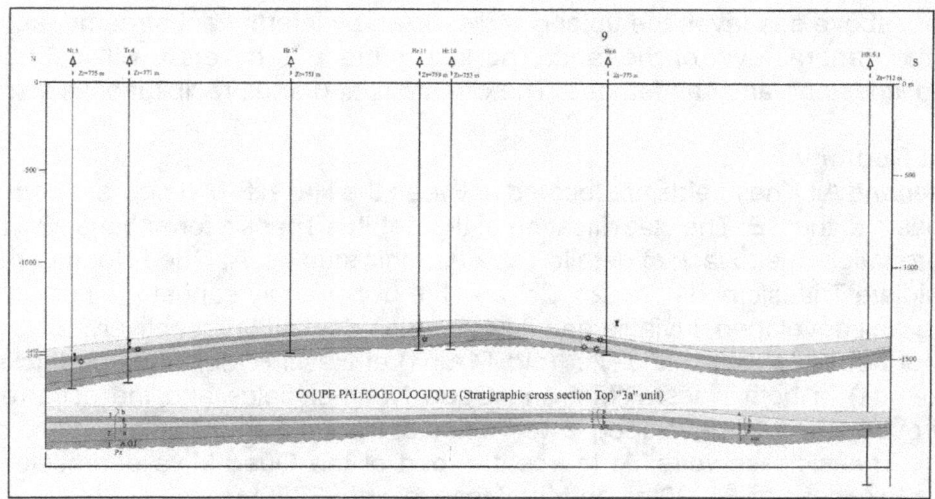

Fig 10.16 - NS Structural CrossSection - Hassi R'Mel Field
Stoica (1972)

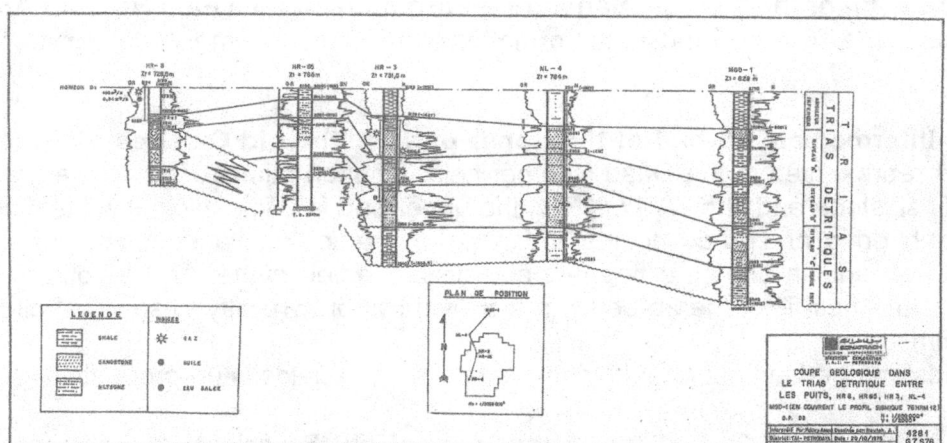

Fig 10.16a - A geologic Cross section of the Triassic detritals among wells of Hassi R'Mel Field

10.4.3 Oued Noumer and Ait Kheir Fields

10.4.3.1 Topography

The Oued Noumer and Ait Kheir Oil/condensate fields constitute the NW flank of the Oued M'ya basin and locate approximately between 32° 13′ 20″- 32° 26′ 30″ N and, 03° 56′ 28″ – 04° 02′ 37″ E; the Oued Noumer area is 70 sq. kms, SE of Hassi R'Mel field and 12.5 km NNE of Ait Kheir field. In the Oued Noumer field, only six wells were drilled in 1972, but only three wells are productive (ONR-1 bis, ONR-3, and ONR-5); whereas, in Ait Kheir field, all the first three drilled wells were productive (AT-1, AT-2 and AT-3). **Table 10.5** - General well data and data results from isobaths and isopachs maps on Oued Noumer and Ait Kheir wells.

At Oued M'ya basin, the pre-Triassic topography of the Gothlandian /Devonian includes valleys, and created a landscape with overlying deposits that were likely be the result of a very intensive ancient erosion of the surrounding high structures of Hassi R'Mel, Talemzane, and Hassi Messaoud.

The Oued Noumer field is characterized by broad, flat and rolling uplands, of low relief ~10 ms, and rising 200-500 ms. above sea level; the upland is dissected by intermittent streams, and numerous hills that rise above the general level of the land. The land surface is covered with pebbles and cobbles, mostly of silicified limestone and sandstones. The climate is arid and precipitation is less than 10 cm/yr.

10.4.3.2 General Geology

The Oued Noumer and Ait Kheir fields are located between the Hassi R'Mel high structure to the NW, and Haoud Berkoui basin to the SE. The classification of the detritus Triassic formations was mainly achieved by electric log correlation due to lack of detailed stratigraphic studies. At Oued Noumer field (Zs = 400ms MSL), the intermidiate Triassic beds "2a, 2b, 2c" and the upper shaly sandstone beds ("3a, 3b"); (or old T1+T2- units) are well developed fluviatile and fluvio-marine deposits respectively.

The thinning of beds "2a", "2b", and "2c" (old T1-unit) at Hassi R'Mel field, the thickening of 3a and 3b units (old T2- unit) in both Hassi R'Mel and Oued Noumer fields and the occurrence of eruptive phenomena that overly the well developed lower series at Haoud Berkoui field, caused some difficulties for correlating the Triassic reservoirs. At the central part of the Oued M'ya depression, the Paleozoic section attained a thickness of ~1000ms and the Mesozoic of ~3750ms.

The Oued Noumer reservoir is a closure of an assymmetric NNW-SSE anticline, located between a closure on a dome-like structure with a NNW-SSE fault on its eastern part.

10.4.3.3 Classification of the Triassic Sandstone Formation at the Oued Noumer Area

The Classification of the Triassic Sandstone Formation at the Oued Noumer Area is given in chronological sequence: :

(1) The Middle Intermediate 2b-bed of the upper part of "the old C-series and the lower part of old B-series" is represented in the Oued Noumer field, by alternations of gray, fine grained, quartizitic sandstone, shaly siltstone and shale. At ONN-1, the upper part is 24ms thick of white, fine grained, hard siliceous, fairly anhydritic, and grades to fine and medium- grained argillaceous and anhydritic sandstone with intercalation of reddish brown, silty shale and disseminated pyrite. At the bottom, it comprises of coarse grains. At Ait Kheir field, the 2b-bed is mainly reddish brown, silty to sandy shale; the total thickness is about 28ms.

The 2a- sandstone bed in both Oued Noumer and Ait Kheir fields is un-productive **(see Table 10.2)**.

(2) The Upper Intermediate 2c-bed (the upper part of old B-series) - At the Oued Noumer field, the sandstone of 2b and part of 2c-beds, of about 68ms, appears to be homogenous lithologically and petrographically. It consists of multicolored sandstone, fine grained, quartzitic, and poorly consolidated by siliceous kaolinitic cement. In the lower parts of the bed, it grades into medium to coarse sandstone with a thin conglomerate bed at the bottom. An intercalation of a very thin dolomitic-silty variegated shale bed (3m thick) is also encountered. At ONN-1, it is of about 38ms thick, beige in color, medium to fine-grained, very friable, subrounded, well sorted, intercalated by green and micaceous shale flakes, with few white pebbly and hard siliceous sandstone, sometimes argillaceous and anhydritic. **Fig 10.17** –Isopaches of bed "2b" (old B-series) of Oued Noumer oil Field. In Oued Noumer field, the sandstone 2b and 2c units.

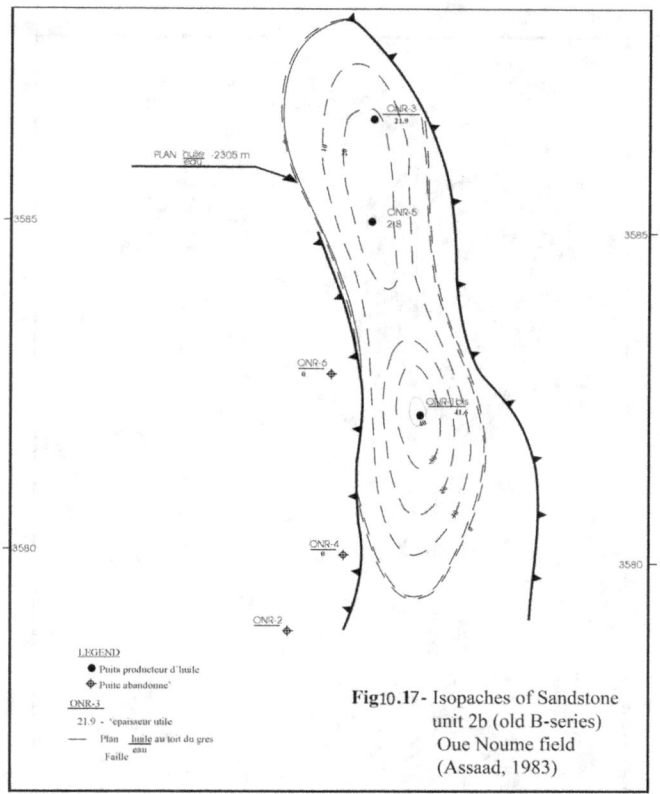

Fig 10.17 - Isopaches of Sandstone unit 2b (old B-series) Oue Noume field (Assaad, 1983)

In the Ait Kheir field, the sandstone, of the 2c-bed is beige in color, medium to coarse grained, friable, subrounded to subangular, well sorted, with intercalation of few centimeters of green, micaceous shale flakes. At AT-3, to the NW of the field, the sandstone turns to brownish gray and maroon in color, very fine to medium-grained, subangular to subrounded, poorly sorted, hard to semi-friable and micaceous; at the bottom, there are intercalations of gray, hard and micaceous shale with hard, argillaceous siltstone.

(3) The Upper Triassic argilo-sandstone of 3a -bed (~A-series) constitutes of sandstone with intercalation of siltstone and sandy shale, that are frequently wedging out. The sandstone is fine to medium-grained, with oblique and cross- stratification in some locations. It is poorly consolidated by chlorite-illite cement, grades into shaly siltstone and variegated shales with varying silt content.

At ONN-1, to the NE of the Oued Noumer field, the 3a -bed, of 78ms thick, is dark brown to black siltstone, turns downward to fine and medium grained sandstone, subangular to subrounded, friable with argillaceous siliceous cement, sometimes anhydritic, micaceous with few intervals of reddish brown shale. In Ait Kheir field, the 3a-bed is siltstone, dark brown to black gradually turns to beige in color, and fine to medium, well sorted, subangular to subrounded, friable sandstone, with an argillaceous and micaceous cement and inclusions of green shales and thin intervals of bituminous sandstone.

10.4.3.4 Characteristics of the Reservoir Rocks

In Oued Noumer field, the reservoir rocks are intercalated by rather clean shales with good isolating properties; the intercalating beds (between 3a and upper part of 2c-beds), grade to shaly sandstone and siltstone away from the productive limits (e.g. at ONR-2); in other words, the Sandstone series of "3a" of the upper shaly sandstone, and the Intermediate Triassic bed, might be in hydrodynamic connection and represent one single hydrodynamic reservoir within the productive areas. **Fig 10.18** – An isobaths map on top of "A" sandstone series or 3a-unit" at Oued Noumer field. **Fig 10.19** – A revised isobaths map on top of "B" series, or "upper 2b/2c beds" in both fields.

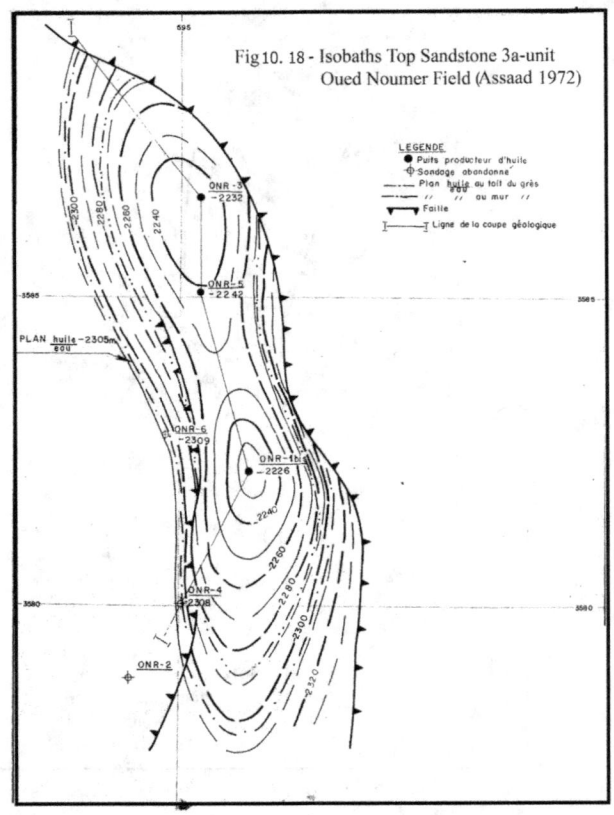

Fig 10.18 - Isobaths Top Sandstone 3a-unit Oued Noumer Field (Assaad 1972)

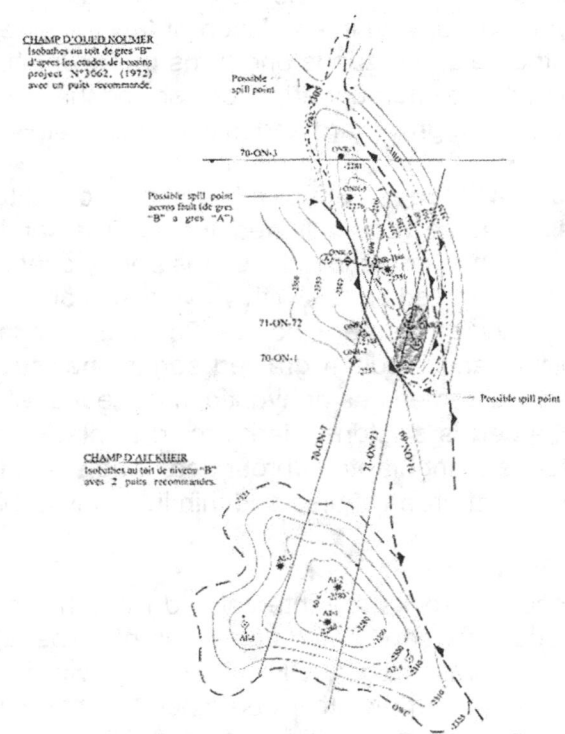

Fig 10.19 - Isobaths Map Top "B"/(2b-2c) unit of Oued Noumer Field

Detailed studies of seismic sections, predicted a possible spill point across the western and eastern faults of NNW/SSE trend that give the horst-structure of Oued Noumer field.

Fig 10.20 – An Isopaches map of "A" series, or "3a-unit" at Oued Noumer Field.

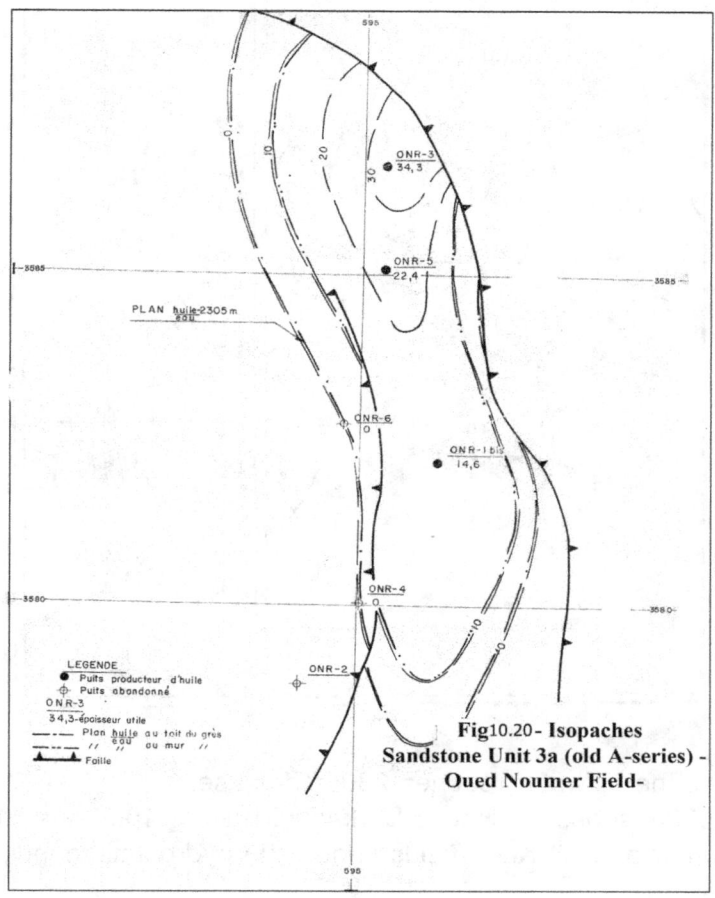

Fig10.20- Isopaches Sandstone Unit 3a (old A-series) - Oued Noumer Field-

At Oued Noumer field, the intermediate Triassic shaly sandstone beds:"2b, and 2c" (old T1-unit) are oil producing, where oil/water is at 2305m subsea; the upper 3a-unit is gasoline producing, where Gasoline/oil is at ~2255m subsea).

According to seismic re-interpretation, the isochrone map of the upper part of the sandstone "B-series", shows a structural closure of NNW-SSE faults on the eastern flank of the horst structure of the Oued Noumer field. **Fig 10.21ab** - Isochrone map on top of the Intermediate series of the Triassic shaly sandstone series (top of "B" series) with an explanatory structural cross section in both fields.

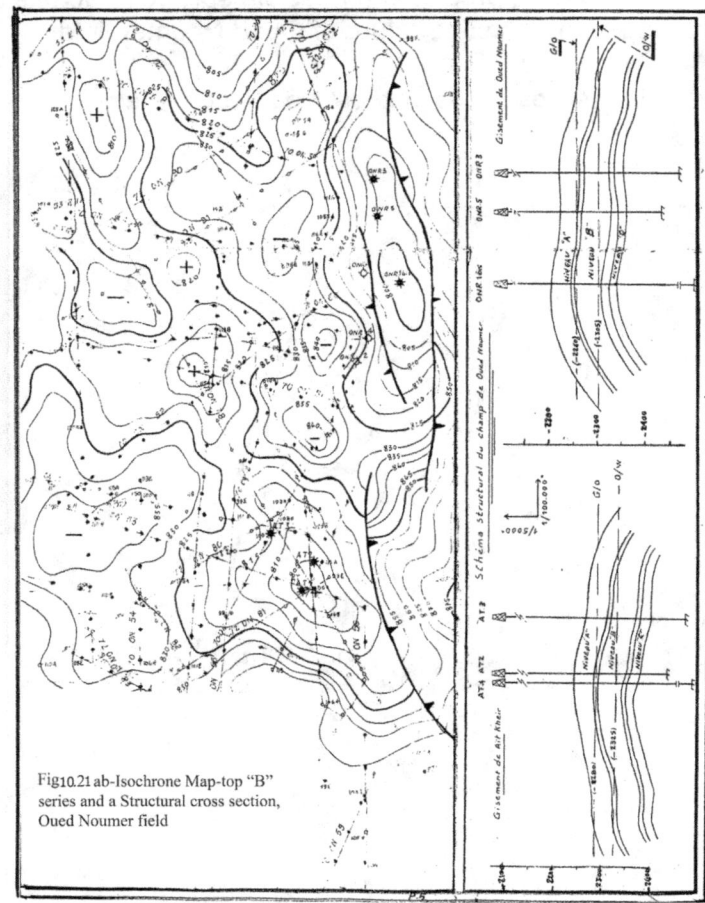

Fig 10.21 ab-Isochrone Map-top "B" series and a Structural cross section, Oued Noumer field

At Oued Noumer field, the oil/water reaches 2305ms subsea and at Ait Kheir, the oil/water ranges from 2325m subsea to 2500m subsea at Haoud Berkaoui field. **Fig 10.22** – A structural geological cross section of the Oued Noumer and Ait Kheir fields shows the hydrostatic conditions of the reservoirs in both fields.

Due to various geological criteria of evaluation of oil and gas potentialities, the Oued Noumer field is a favorable structure for oil and gasoline exploration; the GOR values range from $1200m^3/m^3$, for the upper "A" sandstone series, considered then as a condensate bearing bed, to $470m^3/m^3$ in average (450-$480m^3/m^3$), for the lower "B" series, considered as oil bearing, even though, the relation of GOR versus depth did not show either abrupt or gentle change between the two beds "A" and "B".(Personal communication, SN Repal, 1982).

The gasoline/oil in Ait Kheir field might be arbitrarily considered at depth of 2280ms subsea, while in Oued Noumer field, it could be higher for 20m, i.e. at 2260m subsea. A local fault of 20m of displacement is presumably situated between the two reservoirs.

The productive wells, DRT-1 and HKA-1 are located near the central part of the Triassic basin whereas; ONN-1 well is located on its northwestern part. The productive sandstone (B-series) shows higher average porosity (20.4%) than that of the upper "A-series" (14.7%). In general, the average porosity of both series is estimated as 17.5% and the average permeability is estimated as 790md and the total net pay sandstone of the Triassic detritals is in average of ~30m, and is considered as a good granular reservoir. The Lower series "Serie Inferieur" is oil productive at DRT-1 where the oil/water is at 3491ms subsea.

10.4.4 Haoud Berkaoui and Haniet El-Mokta Oil Fields

Haoud Berkaoui and Haniet El-Mokta Oil Fields discovered in 1968, as oil productive in the Lower reservoir series "Serie Inferieur, (SI-series)" to the far SE of the Oued M'ya basin, and are considered as reworked from local substratum, of a fluvial or rather a fluvio-deltaic origin (IFP, 1967). The lower reservoir series are well developed, overlain by eruptive magmatic rocks where O/W is at depth 2500ms subsea. (Zs = 200m; MSL).

The Triassic beds are unconformably overlying the Hercynian discordance, which extends upward to the Liassic age. A marker dolomitic bed "D2" is considered as a useful, easily recognized datum, (near the producing zones), wide spread over the whole basin of Oued M'ya, and of no unexpected facies change. However, Horizon "B", and the Discordance Hercynian are also good seismic and safe markers for preliminary isobaths-mapping.

The isopaches and iso-percentage sandstone maps, of the above fields, show that the "2b/2c" of the intermediate series (or B- and C- series of old unit T1), thickened to the East and NE where shale intercalation increased and might be dolomitic and/or anhydritic. e.g. at ONN-1, and HBK-1; there is an inverse relation between the maximum and minimum thickness of the above series with that of the iso-percentage sandstone maps. In general, the lower "B" and "C" units of the T1-series filled up the lows to the East and NE where the Argillaceous material increases; the upper series "A" of the lower part of the old T2-unit, deposited as a continuous mantle of more shallow marine environment to the west and northwest.

10.5- Application of Gussow Theory on the Triassic Reservoirs. The Efolution of the Hydrocarbon Migration -Methodology

10.5.1 – Discussion

The application of the Gussow Theory of petroleum hydrocarbon migration, depends on differential entrapment of petroleum, which can be applied within the Triassic reservoir rocks from the northwest at Hassi R'Mel, to the southeast at Oued Noumer, and farther to the east at Haoud Berkaoui field.

Hassi R'Mel had attained much lower structure probably during the Barrimean from the northwest, to the southeast, where relatively higher structures of both the Oued Noumer, Ait Kheir and Houed Berkoui, presuming that the three fields were originally connected; the following is a postulated evolution of the hydrocarbon migration through time:

a- Top Barremian - The primary hydrocarbon migration probably took place, followed by the secondary migration of petroleum hydrocarbons during the Middle Alpine orogeny up to Top Barremian, through the Hercynian discordance due to high temperature and pressure, resulting in the initial separation of the petroleum hydrocarbon from the water environment **(Fig 10.23a)**. Water was then squeezed out and expelled from the shale during their compaction by the load of the overburden on the reservoir rocks and a fraction of the petroleum hydrocarbons (and/or asphalt) was entrained in the water of the reservoir rocks "or carrier beds" where it was carried until trapped.

When oil was saturated with gas in great depths and pressure was released at Hassi R'Mel structure, oil and gasoline migrated updip till it reached its spill point at the down dip trap.

Gas and gasoline (condensate) continued to be trapped in Hassi R'Mel structure together with oil at the bottom of the fold, and when reached above the spill point, gasoline and oil moved updip, and accumulated on top of the Oued Noumer and Ait Kheir structures. As compaction continued, petroleum hydrocarbon was then barred to form petroleum pools where oil and gas spilled out to the higher structure, concentrate and accumulate at Haoud Berkaoui basin (at Haniet El-Mokta well, HKA-1).

Petroleum Hydrocarbons moved therefore, from the structurally Low area at Hassi R'Mel, (Northwest of the Oued M'ya basin) to the structurally higher areas to the Southeast.

b. Top Maestrichtian - Faulting and Tilting, the basic requirement for hydrothermal activities, are more essential in the hydrocarbon migration processes other than deep burial that provide pressure and temperature differentials. In the absence of any unbalanced forces, petroleum hydrocarbon became deeply buried for a long time. Local fluid potential or temperature gradients might cause local movements and would require regional disturbance (e.g. regional folding, tilting, and mountain building).

c. Top Mio-Pliocene – A Tectonic Transitional phase- might be postulated during that period due to the continuous structural adjustment towards the present structural configuration; vertical migration of fluids might occur when erosion began, due to the release of pressure of the gas and gasoline (condensate) traps in the Triassic reservoirs of Hassi R'Mel and Oued Noumer areas; as digenesis of the source rocks advanced, the upward vertical permeability lessened until finally the lateral permeability became greater and therefore, the fluids moved out laterally along the bedding planes through the Hercynian discordance.

The Saharan Platform attained the present structure during Late Alpine movements, and the structures became reversed, where Hassi R'Mel field became uplifted into a dome-like structure, associated with anticlines at the nearby Oued Noumer and Haoud Berkoui fields systematically at lower much levels.

Both the Oued Noumer and Ait Kheir fields of the Oued M'ya basin seem to be a transitional zone of hydrocarbon accumulation, before the orogenic movements took place through Late Alpine orogony.

d. In Recent Time, Due to continuous erosion, vertical migration of fluids might take place in the Triassic Reservoirs due to the release of pressure of gas and gasoline trap at Hassi R'Mel field because of density factor; the thickness of the overburden of post Triassic sediments in Hassi R'Mel field were approximately of 2200ms, whereas, in both Oued Noumer and Haoud Berkoui Fields of 2700m (**Fig 10.23b**).

References

Assaad, F., (1983) An approach to "Halokinematics" and Interplate Tectonics (North- Central Algeria), Jr. Petroleum Geology, Vol.6, No. 1, July 1983.

------------, (1981) A further Geologic Study on the Triassic Formations of North- Central Algeria with Special Emphasis on Halokinesis, Journal of Petroleum Geology, 4,2, pp. 163-176.

Aliev, M.M. et al, Technoexport, Sonatrach (1971) – Geological Structures and Estimation of Oil and Gas in the Algerian Atlas and the Algerian Sahara, Algiers.

------------, (1972) Contribution to the study of the Triassic formations of the sersou-Megress Region (High Plateau) and the area of Daia M'Zab (Saharan Platform), No. 84 (B3), Eighth Arab Petroleum Congress, Algeria, 1972.

SN RÉPAL (1961) Les séries Permo-Triassiques dans le Nord Sahara. Études Pétrographique du cycle Détritique. Etudes du Cycle Salifére (texte et planches).

Sonatrach (1970) – Exploration Directorate, Agha, Alger (Internal Report).

Stoica, I., and Assaad , F.A. (19\82) Geological Studies on the Triassic Reservoir of Condensate and Gas at the Hasssi R'Mel Field (Internal Report, Sonatrach).

Sopefal, (1971) – Trias du M'Zab et de L'Oued M'ya, Etude de Sunthese, Algerie, (Rapport Int.) Sonatrach.

Hobson, G.D (1973) Modern Petroleum Technology, Fig 8/p20. Reader in oil Technology Ltd, Essex, G.

Levorsen, A.I (1954) Geology of Petroleum, W.H. Freeman (540, 541,574), edited by Frederick A. F. Berry, Univ. California, Berkeley, San Francisco and Company, USA.

SNPA (1964) Étude Micro stratigraphique et Sédimentologique des séries du Trias á Tertiaire recontrées par les Sondages de l'Hydraulique des Chotts Rharbi et Chergui

Chapter (11)

An Approach to Halokinematics and Interplate Tectonics- Development of The Triassic Salt domes in North Algeria –its Stratigraphic Relation with that of the Gulf of Mexico. The Economic Aspects of Salt structures

11.1 Introduction

The World Lithosphere is composed of six large and several small tectonic plates that have been subjected to movement in the course of the Earth's geological history since the Pangea era.**(Fig 11.1)**. According to plate tectonics theory, the upper layer,"or the crust", of the earth is broken into major plates; e.g. the Arabian, African, Eurasian, Somalian, Mexican plates, etc., such plates intersect at one of three boundary types as, spreading, subduction, or transform (lateral) zones. At subduction and transform boundaries, the plates generally move past each other in abrupt lurches, known as earthquakes. Motion at plate boundaries is suggested to concentrate along a primary plate boundary fault, which can lead to an additional set of secondary faults which then accommodates secondary stresses associated with the primary faults (**Komatina 2004**).

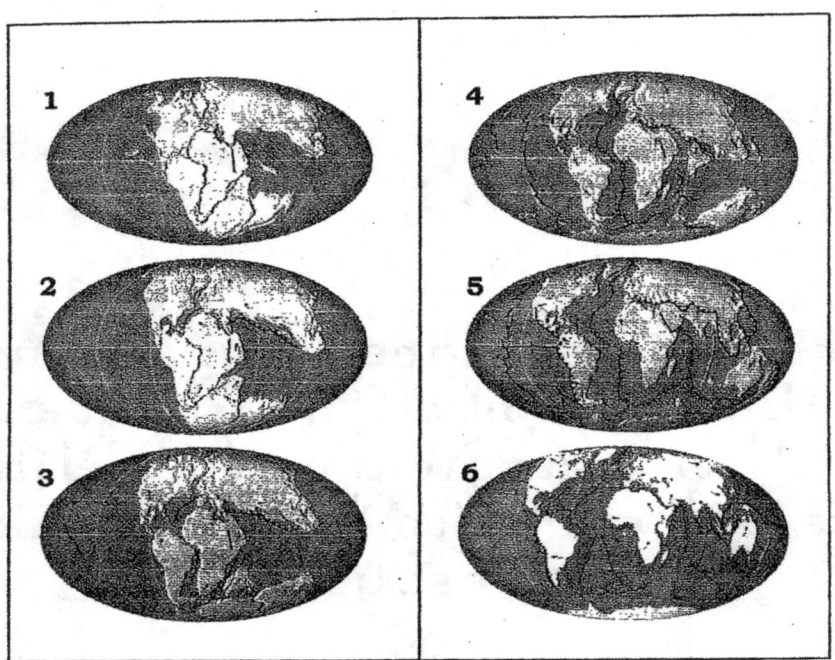

Fig 11.1 - Movement of Continents from Pangea Period to Nowadays
(Komatina, 2004)

The formidable strength of the crust, allows enormous plates to remain intact, whereas, the motion of the crust occurs at their boundaries. The crustal plates ride atop the mantle, in which the ongoing process of convection is accommodated by gradual flow. Earthquakes cannot occur in the mantle because rocks are too hot and plastic to develop faults, whereas the upper layers of the crust are brittle, and breakable under some circumstances.

Apart from the occasional terrifying and catastrophic earthquakes, plate boundary zones which often offer far more than their share of geographical amenities, run directly or nearly directly along coastlines. Even when coastlines are not along active plate boundaries, such areas can still be horribly vulnerable to Tsunami damage from quakes.

Hough and Billham (2006), stated that the process of subduction, builds mountains not far inland, whereas, oceanic crust sinks along coastlines, creating a narrow strip of hospitable coastline along which populations congregate. The Atlantic coast of North America, separates the North American continental crust from the Atlantic oceanic crust, by a passive boundary as North America drifted away from Africa and Europe by a recent generated oceanic crust along the Mid-Atlantic Ridge.

A mixed zone of strike-slip and extensional faulting through northern Mexico, took place together with a long subduction offshore zone in most of Mexico; a long subduction zone runs along the entire coast of Chile and a continuous belt of active plate boundaries, including subduction zones, surround the Caribbean (Hough, and Bilham, 2006).

11.2 Paleostructural History

During the Carboniferous/Jurassic era, Pangea (Super continent) gave rise to (1) Laurasia that constituted the Canadian and Eurasia shields in the northern Hemisphere and, (2) Gondwana Proto-continent in the Southern Hemisphere. Later, seafloor spread and resulted into the formation of continents **(see Fig 8.3- Klett 1997-**Paleogeographic Map of Gondwana Land**).**

Pomerol (1975) in his "Stratigraphy and Paleostratigraphy of the Mesozoic Era" stated that the break-up of Gondwana land took place within the Upper Triassic, and has continued in the present due to sea-floor spreading.

The breakup of the Gondwana Proto-Continent took place most probably from Late Triassic (200M yrs) to Early Liassic times (updated by Assaad, 1983), due to indirect forces generated by Plate Tectonics movements; Inter-continental Tectonics movements took place during the Early Alpine orogeny that caused the separation of Gulf of Mexico (South America) from Africa; The Triassic/Liassic sedimentary basin in the Gulf of Mexico seems to attain the same stratigraphic settings as that of the Arabian Maghreb, NW of Africa. **(see Fig 9.1** -A geologic Map in North Algeria, showing the four main structures of Algeria with locations of the related wells**)**.

Two intracontinental rift zones might be possibly formed within the two main Atlassic flexures of the Arabian Maghreb, south of the Tellian Atlas (e.g. at Bourlier area, NE of the High Plateau "BO-1 well"), and south of the Saharan Atlas, and became the possible source of basalt flows, where a binary subduction complex exists due to tectonic distention forces:

a) The descending of the oceanic crust that resulted to the formation of the Tellian Atlas.
b) The descending of the northern rim of the Saharan Platform that resulted into the formation of the Atlassic plate.

The above events might cause the formation of both subduction zones at the southern Tellian and Atlassic flexures; it might also explain the repetition of the Upper Jurassic deposits (Malm) at Ain Mahdi well (AMI-1) at intervals 1540--1750ms" (210ms thick) and "4160-5020m" (860me thick), due to reverse faulting, where the total depth of the well reached Dogger formation at depth 5020m. **Fig 11.2** – A Sketch Diagram showing the two subduction zones at the southern Tellian and Saharan Atlassic flexure belts.

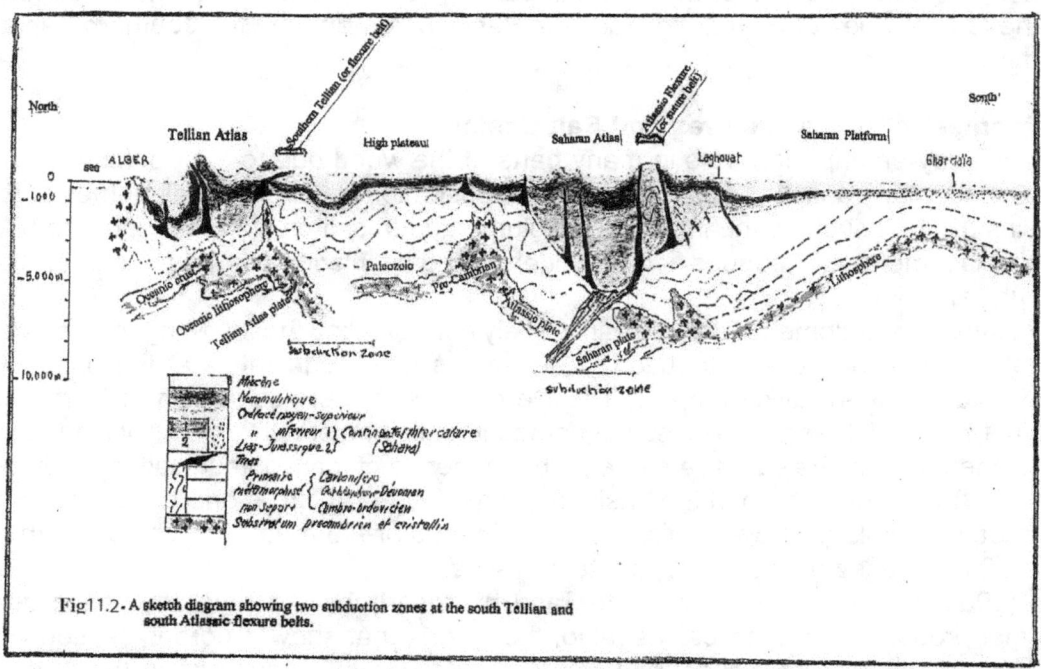

Fig 11.2 - A sketch diagram showing two subduction zones at the south Tellian and south Atlassic flexure belts.

11.3 Evaporite Rocks of the Eastern and Western hemispheres

11.3.1 Scope
Evaporite deposits have concerned man before the dawn of the recorded history; salt domes have been the subject to scientific investigations over a Century and have increased in scope and intensity during the last Century. Several scientific references were presented by European and Ametican geologists to better study and interpret the stratigraphical and structural evolution of salt deposits and its movements.

Salt rocks constitute the cap rock of the upper detrital Triassic reservoirs in NE zone of Algeria, whereas the eruptive rocks mainly basalt and spillites, constitute the cap rock of the lowermost series of the lowermost series to the SE; east. The absence of the Cretaceous formations to the SW of the Pre-Atlas zone may maily attributed to the presence of salt domes.

The major salt basins of the Eastern Hemisphere are in the Miocene Red Sea basin of Yemen; the Triassic and Permian basins, in South and Northwest Germany (Trusheim, 1960), respectively; and the Cambrian salt basin of the Persian Gulf Province of Iran.

Salt basins of the Western Hemisphere are those of the Triassic-Jurassic in South America and Gulf of Mexico, as well as the Triassic-Liassic salt basin of the Mediterranean zone of Morocco, Algeria, Tunisia and possibly to the Western Sahara of Libya.

The major salt basins of the world, e.g. Gulf regions, Mediterranean zones, North Germany, and Northwest Africa, reflect major negative elements that have been probably related to the reactivation of basement structure. The present continental margins that include many of the major coastal salt basins, their bounding orogenic belts and cratonic elements are positioned along ancestral zones of orogenesis. Many coastal salt basins must be viewed as segments of much larger marine basins and should not be considered in isolation.

Hydrocarbon prospects occur at salt dome structures of the upper Jurassic rocks of Louisiana, East Texas, the elongate salt anticlines of the paradox basin in southeastern Utah of the United States; the Miocene of the Gulf of Suez of Egypt, the Red Sea Basin in Yemen, and the southern Iranian Region of Central Foothills Province.

11.3.2 Development of Salt structures and Salt Basins
Salt movements play an important rôle in many parts of the world due to their outflow from older rocks deep in the ground. Many assumptions have been proposed for the origin of salt domes and the mechanism of their installation; such as the increase in volume due to salt crystallization; or the phenomena of dissolution, displacement and redeposition by underground circulating connate water, or due to indirect volcanic activity, etc.

The localization of salt domes within a basin is likely controlled by the following four factors, in relation to the general structure and the distribution of sediments: (1) Irregularities at the base of the salt as a result of structural pattern at depth, e.g. The German Zschestein salt basin in North Germany; (2) Irregularities at the top of the salt as a result of compact faults of a small throw or of lithological irregularities; (3) Heterogeneities in the salt layers due to the presence of local potash salt, anhydrite or dolomite accumulations; (4) General pattern of the basin, e.g. one degree of dip, is enough to cause the first slide of salt that creates irregularities over which dome structures are formed. In such cases, the halokinesis would begin in the deeper zones and propagate in waves..

Hedberg (1964) considered the evaporites and hydrocarbons are frequently associated because both are normal products of closed basins; also, the basins that show a porous section with overlying evaporites, include some of the richest oil-bearing provinces in the world as in the Algerian Saharan Triassic basin and that of Gulf of Mexico.

The occurrence of salt-domes could be probably shown by air photography, since their circular cross sections affect the drainage patterns of the surface beds. Precise photogeologic studies and reflection seismic surveys may reveal the presence of salt dome structures nearby areas of halokinetic piercing diapirs, as well as in other areas away from the direct effects of regional tectonics and may probably lead to the discovery of stratigraphic petroleum traps in the Pre-Atlas zone.

It is worthy mentioning that Salt migration processes in shallower regions can be clearly reconstructed by using reflection seismic surveys, whereas, in some areas, salt could partly escape through fractured zones, and a larger part was left behind at depth.

11.3.3 Types of Salt Movements and the controlling factors- North Algeria

There are two theories of salt movements, Halotectonics (by compressive tectonic forces) and Halokinesis (the autonomous and hydrostatic salt movement); both theories were applied on Zechstein salt masses in North Germany by Trusheim (1960), and since then, they were in a widespread use by European geologists, as well as by the author in 1972 (The Eight Arab Petroleum Congress, Algeria, 1972).

Salt movements and the diapirism of salt bodies penetrating younger beds are characterized by two phenomena: "halokinesis", and "halotectonics"; both explain the mechanism of salt migration

11.3.3.1 Halokinesis and Triassic/Liassic Salt Dome Structures

Salt movements primarily considered the effect of gravity and pressure of sediments over the deep salt beds, which were firstly submitted by Lachman and Arhenius (1912) and more detailed by Barton (1933), whereas Nettleton (1936) related salt movements to the geostatic hypothesis that depends on the following two main properties:

a) **Salt's plasticity**: The plasticity is clearly demonstrated by the observation of salt beds in mines where salt domes are exploited; such beds are always warped, crumbled, and folded up in very packed folds; on some piercing salt domes that reach the surface, real outflows could be observed, forming salt glaciers (e.g. Iran). The plasticity threshold for a normal salt rock, estimated approximately as 200 kg./sq. cm., which corresponds to about 500 ms thick of overburden, whereas the plasticity of the fresh sediments is very low. The geologic study of salt bearing basins shows that halokinesis inaugurates when a salt layer of at least 300ms thick and is overburden by 1000ms of sediments; the process of flow is thus initiated and forms salt pillows, salt domes and/or salt diapirs. It was noticed that potash salt are more plastic than sodium salt and would be the first to move, acting as a "halokinesis Vanguard" (Trusheim 1960); however, the presence of water would inhibit the movement (Nettleton 1936).

b) **Density contrast between salt and overlying sediments** deep in the earth, is likely the prime mover for the salt's ascension according to the Archimedis hypothesis. The salt's average density is 2.19 gm/cc and remains constant with depth. In controversy, the overburden sediments get compacted and density increases from under 2.0gm/cc for freshly deposited shale up to 2.6gm/cc for shale buried under several thousands of meters. **Fig 11.3** - A diagram showing the density contrast with depth between salt and overlying sediments. The point of contrast of density between salt and sediment, became zero at depth of 600ms, where the density of sediment equals that of salt (2.19gm/cc), and increases with a negative value with increase of depth of burial, up to a density of sediment of 2.6 gm/cc.

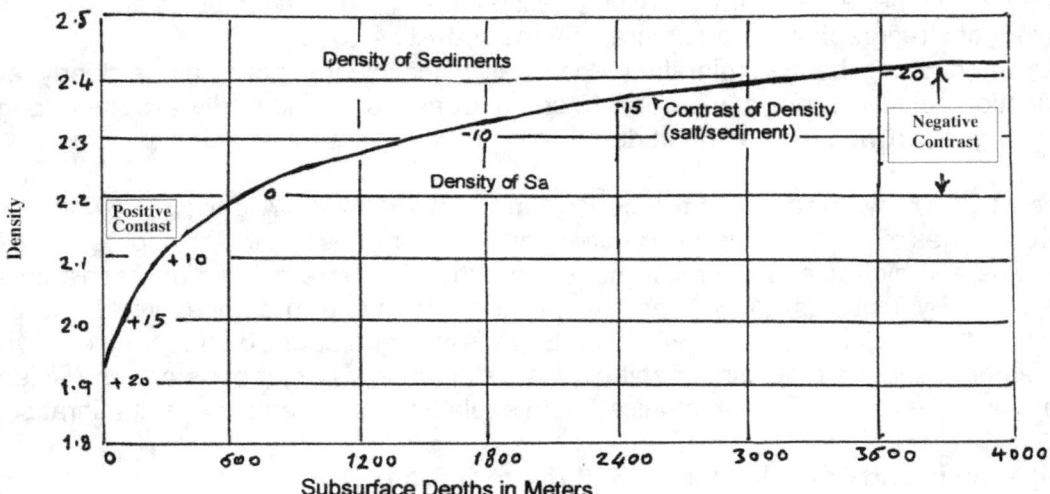

Fig 11.3 – Density Contrast with Depth, between Salt and overlying sediments (Assaad, 1983, adapted from Levorsen, 1954)

In other words, the density contrast changes from positive to negative as the salt layer is covered with more than 600ms (1800 ft) of sediments. Once the thickness of burial increases, the plasticity threshold is reached and the salt turns from elastic to plastic in an unstable status, tending to rise upward if the dip of the formation is at least one degree, causing the first slide and leading to halokinesis. The term Halokinesis was used by Trusheim (1960), to designate the formation of salt dome structures as a result of the autonomous movement of the Triassic salt under the indirect influence of gravity; Trusheim discussed the mechanism of the Permian salt migration in north Germany basin, due to halokinesis; the salt structures were classified into salt pillows, salt stocks, salt walls, and extrusions along fissures, which were accompanied by primary, secondary and third order peripheral sinks. **Fig 11.4** – Evolution of the Permian salt domes in North Germany (Zeshestein formation).

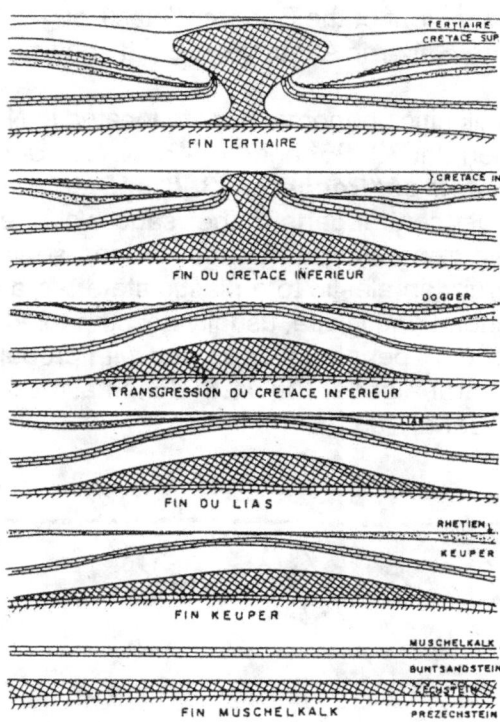

Fig 11.4 - Evolution of Dome structures in North Germany

The mechanism of Development of salt domes and Diapirs- The salt accumulation related to the pillow stage, is associated with primary peripheral sinks, developed as an attribute of the salt pillow; the primary peripheral sinks are characterized by the thinning of the overlying sediments within the rim of the salt dome structure, where the periphery subsides as the pillow rises and the resulting depression fills with sediments; when the supplying salt begins to shrink, a gradual destruction of the pillow is accompanied by shearing cracks or fissures, followed by a subsidence of the adjacent overlying sediments that thickened towards the dome, and resulting to a secondary peripheral sink of possible stratigraphic petroleum prospects. The lateral migration of salt dome to the ENE trend (Saharan Atlas trend), is reflected by the progressive shifting of the primary peripheral sink with time; **(See Appendix 11-A1a & 11-A1b)**.

Salt Crests of Diapir Folds – Guillmot (1964), made a distinction between salt dome structures and salt crests of diapir folds. Salt domes consist of a column of saliferous formation that has intruded into more recent sediments, with the salt stamping out, in more or less considerable thickness of stratigraphically younger beds. The deformation of salt crests of diapir folds, on the other hand, is then due to the displacement of salt masses without any significant influence of tangential tectonic movements. In such a case, the salt crest of diapir folds, only plays the rôle of a disharmonic plastic mass which emphasized and complicated deformations that occurred due to folding; the diapirism occurrence is not always due to salt, as diapirs are sometimes associated with argillaceous core.

11.3.3.2 Salt Domes and Piercing Diapirs

Salt movements affected greatly the present structures of the Triassic Province of Algeria. The pronounced disharmony between the subdued and flat pre-salt Triassic rocks and the strongly disturbed post salt rocks is due exclusively to the mobility of the Triasssic/Liassic salt over which the structural features (anticlines, synclines, faults, over thrusts, etc), are largely independent of the underlying basement; The mobility of the Triassic- Liassic salt deposits played the most important rôle during the Alpanian orogeny in the post-Liassic structural development of the Saharan Atlas trough, constituting the deepest Jurassic basin in Algeria, whereas, in the pre-Saharan Atlassic region, it shows different locations of diapiric structures of saliferous shale.

Salt domes formed due to halokinitic phenomenon, are located in N.E Algeria, within the Pre-Atlassic zone and covered an area between Lat 33° 25´ and 34° 45´ N. Salt domes extend from Mecheria in the West at Aflou; to the east at the ***Diapir of "Rocher de Sel" of Djelfa;*** salt domes probably exist in linear trends along margins or in the most central parts of the "saucer-shaped" basins on ENE/WSW-striking tectonic lineaments of the Atlassic trend where the load of deep seated sedimentary trough has been sufficient to cross the boundary from an elastic to a plastic state. As a calm tectonic area, the development of salt domes between Mecheria and Djelfa, usually accompanied by shearing cracks and fissures, associated with primary and secondary peripheral sinks and with probable oil traps. **Fig 11.5-** A Location of salt dome structures in NE Central Algeria. **Fig 11.6**–Schematic Diagrams showing evolution of salt dome structures in NE Algeria (Assaad, 1983).

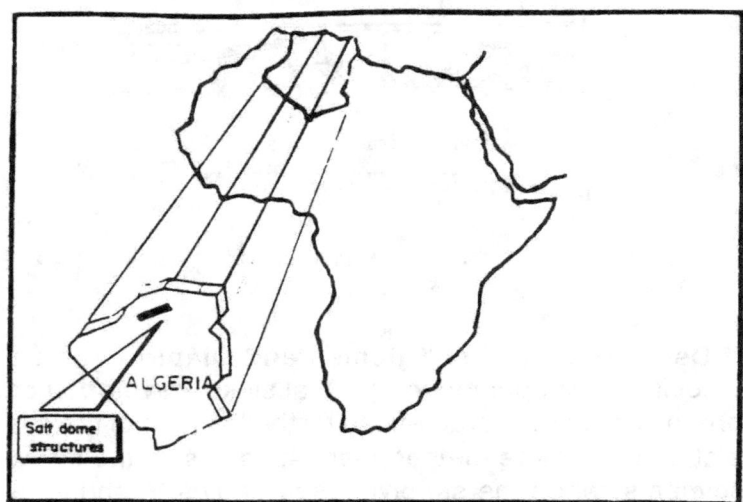

Fig11.5 - Location of salt dome structures in the north-central zone of Algeria.

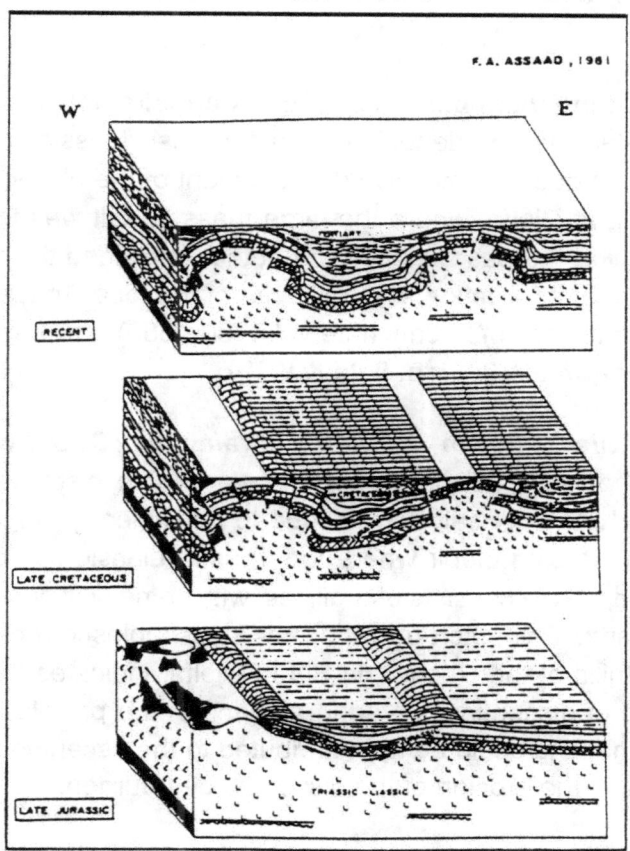

Fig.1.6-Schematic diagrams showing evolution of salt dome structures (north-central Algeria).

The Piercing Diapirs – The diapir stage started when the swelling of the salt pillow, increased under continuous supply of the mother salt rock. The extended sedimentary roof over the swelling salt pillow is gradually forced to break along shearing cracks leading to the formation of a piercing diapir that is characterized as the stage of predominantly vertical salt migration for a period of time and can rise until the gravity equilibrium of the rock system surrounding the salt is reached. If the overburden sediments are very thick, and the salt replenishment is limited, then the "diapir" remains stationary at a certain stratigraphic level for any length of time. Controversially, if the overburden is thin, the underlying salt can find its way to the surface and flows out as diapirs. In North Germany, such outflows seemed to take place predominantly under water.

Age of Salt Diapirs of the High Plateaus in Algeria- The large Menna diapir of the Aurés, appeared in the Pre-Miocene time, didn't show any salt because of dissolution; **The El-Qutaya diapir**, the Algerian biggest diapir, west of the Aurés rim structure, appeared in Post-Pliocene and continued to grow, but sustained its thickness due to dissolution of salt, which balanced its upward movement.

In most cases, the absence of salt at the surface of the diapirs is likely, in relation with their age, since the salt of the more ancient (Pre-Miocene) diapirs disappeared by dissolution, and therefore, the general absence of salt at the surface of the diapirs at the Sersou region should not be interpreted as an original absence. Reasoning from the superficial absence of salt, one might think that all the diapirs of the Sersou region were established prior to Miocene time. However, the absence of dated Neogene terrains in the vicinity of Djelfa Salt rock prevents us from confirming a more recent establishment.

11.3.3.3 Disclosure of several Issues related to the Discovery of a new Triassic basin in NE Algeria

a) The Mis-interpretation of the disappearance of salt deposits at Cedraia area (CED-1 well)
Caratini (1964) mis-interpreted the non-deposition of the Triassic/Liassic salt at Cedraia area (CED-1), in NE of the High Plateaus, as due to a massive displacement of the plastic layer towards the outlets at the diapir of "Rocher de Sel" at Djelfa, where the large mass of salt was formed as a result of vertical rise together with lateral flow at the base; though the author explained the disappearance of salt at the Cedraia area, due to the fact of its location at the NE rim of the local Triassic/Liassic basin, where only anhydrites and shales of a thickness of 270ms (interval 4380-4650), has been encountered at the end of the evaporitic cycle **(see section 9.4.2/ Figs. 9.9a & 9.9b).**

b) *The "Rocher de sel" of Djelfa, 60 km South of Cedraia area (CED-1 well)*
it is interpreted by the author as a normal outlet of the massive displacement of plastic deposits in the ENE/WSW trend of the "chain" of salt domes (due to halokinesis), originated from Mecheria in the west; it is an example of a piercing diapir where the Triassic/Liassic salt outcrops as a massive, fine crystalline, white to pink, and mainly consists of argillites, with some well developed by-pyramidal crystals of some centimeters of quartz. The outcrop of the pre-Triassic plastic rocks (Permian puddings) cited at the diapir's northern foothills of the Rocher de Sel of Djelfa, indicates its emergence to the surface, which reveals the presence of discontinuities and fracturing in the pre-plastic Triassic substratum. The absence of Cretaceous formations could be then attributed to the ascending salt domes from the SW of the pre-Atlas zone followed by the erosion of the dissected overburden.

c) Age of Salt Diapirs in NE Algeria- The large menaa diapir of the Aurès, appeared in the Pre-Miocene time, didn't show any salt because of dissolution; the El-Qutaya diapir, theAlgerian biggest diapi, west of the Aurès rim structure, appeared in Post-Pliocene and continued to grow but sustained ite thicness due to dissolution of salt, which balanced its upward movement; the absence of salt at the surface of the diapirs is likely in most cases , in relation with their age, since the salt of the more anient (Pre-Miocene) diapirs disappeared by dossolutionand therefore, the general absence of saslt at the surface of the diapirs at Sersou region should not be interpreted as an original absence; resoning from the superficial absence of salt, one might think that all the diapirs of Sersou region were established prior to Miopliocene time. However, the absence of dated Neogene terrains in the vicinity of Dgelfa salt rock prevernts us from confirming a more recent esytablshment.

11.4 Halotectonics and Salt Outcrops of the High Plateaus

The Triassic-Liassic deposits outcrop of different ages of emplacement, located in an area between latitude 34° 15′ t and 34° 55′ E., are represented by saliferous shales and other associated rocks, mainly produced as a result of compressive tectonic forces. The term diapir is a Romanian term used for salt-dome type, and should not be considered as a halotectonic phenomenon; however, every conceivable transition between both the Halokinesis and the Halotectonic is expected.

At the northern part of Dj. Nador of the High Plateaus, the Triassic and Liassic outcrops are found at many locations, as gypsiferous red shales with a number of bipyramidal quartz crystals. e.g. North of Dj. Ben En Nsour and Dj. Es Safeh. **Fig 11.7a-** Triassic outcrops at Dj.Nador (Assaad, 1972).

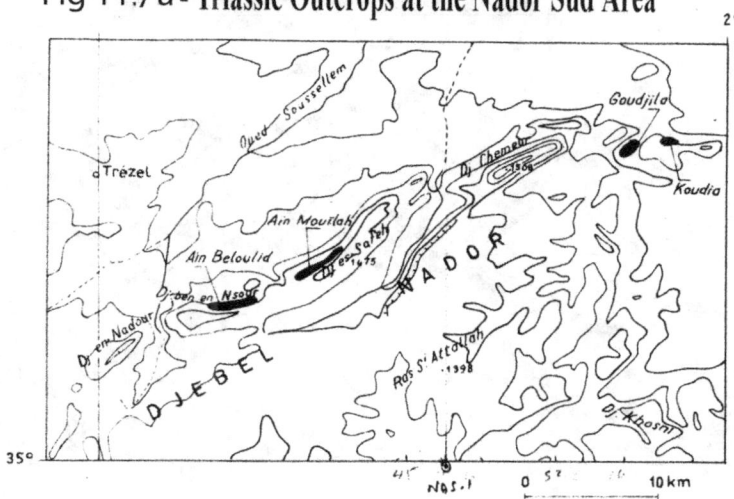

Fig 11.7a - Triassic Outcrops at the Nador Sud Area

At the Region of Sersou (ex-Chellala Reibel) The Triassic/Liassic deposits outcrop on the surface as saliferous shales which are concentrated along major faults or outcropping on top of faulted anticlines, e.g. the horsts of Ghar Rouban and Tiffrit **to the NW of the High Plateaus**, whereas to its west part, the outcrops are represented by volcano-sedimentary formations, which mainly comprise of brownish red conglomerates, green and red shale, basalt, and rarely dolomites and marls; the Triassic outcrops are recorded in six areas: At Djebel Zerga, Draa ez Zoubiat, Zerguine, A. Fedoul, Keudia and Djouabi, where at Dj. Zerga, the Triassic outcrops are massive, mainly of yellow to gray Cargneules, frequently present on the surface as isolated blocks of few meters at a maximum; such rocks are cavernous, sometimes include gravels of bluish, calcarenous dolomites that resulted from differentiated dissolution (De-dolomitization) of the original rock; Sandstone beds, are sometimes abundant on the NE peripheral part of **Djouabi**, as yellow, poorly sorted, cemented with calcareous shales, with no stratification, and form a part of the diapir. Other sandstone beds at Djouabi that form blocks of less than one cubic meter, are encountered in the plastic shale deposits, and might be produced by fluvio-aeolian sedimentation (of Barremian or Albian Age), due to lack of diagenesis of cementing material (Caratini, 1964); to the north of **Dj. Koudia:** Outcrops as intercalations of red marls and important quantity of quartz grains; *at Zerguine "diapirs"*, some sources of mineral water, which are partially present, and circulating at depth through the extrusive evaporites, became mineralized and warmed up by an exothermic chemical reactions and therefore, such thermomineral waters can't be discharged as "vadose"; **to the north of Dj.Zerga and SW of the Zerguine extrusion** (near Ain el- Morr), quartzites outcrop as black rounded hard blocks, (difficult to define its Age). In the vicinity of Zerguine "diapir", and on its top, there are four thermomineral sources of the same chemical composition and can be classified as: "Chloro-sulphurized" type, characterized by their abundance of dry residues and of high sulfate content, due to the dissolution of gypsum. **Fig 11.7b** – Geographic Location of the Triassic outcrops at Sersou (ex-Chellala-Reibell).

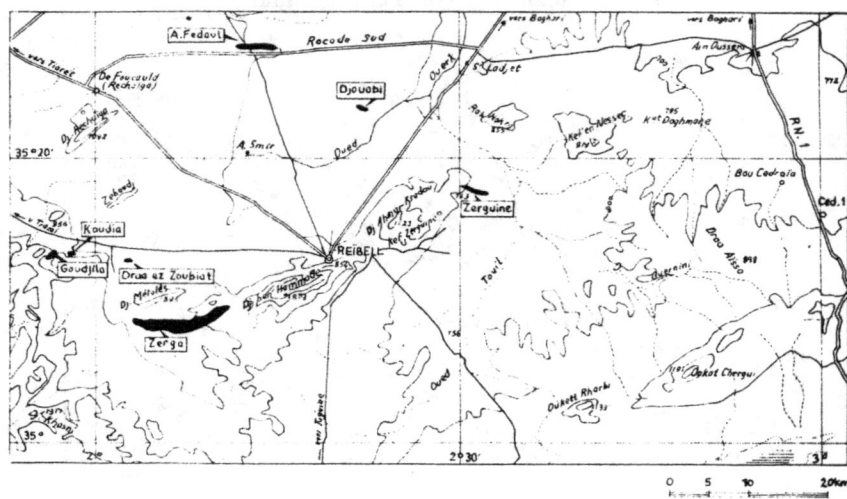

Fig 11.7 b - Triassic Locations of outcops at Sersou rRegion "ex-Chellala- Reibel)

11.5 Halotectonics and the salt outcrops at the Saharan Atlas

The Triassic/Liassic deposits outcrop along the crests of faulted anticlines, e.g. at Djebel Bou Lerfad area, and comprise of red and violet shale forming a chain of hills, capped by doloritic basalt, green and red shaly siltstone beds and salt rock, with Liassic reefal limestone on top. In the northern portion of the Saharan Atlas, at Djebel Melah, the Triassic/Liassic outcrop consists of slightly sandy argillites, reddish green shale, and gypsum with an apparent thickness of 54ms.

In the Saharan Atlas, the outcrop of salt structures at Djebel Bou-Lerfad are found in the Miocene continental formations, therefore concluding that they continued to develop until Miocene time. The Khoudiat El-Adjouan area is characterized by Nummulitic (Upper Eocene) continental deposition and is rich in autochthonous quartz and gypsum of Triassic/Liassic origin whereas, salt is missing due to dissolution.

11.6 Plate Tectonics and Halokinematics

Plate tectonics dominates geological thinking; the study of paleotectonic events on the Epihercynian and African platforms, the wide distribution of spilites in the lowermost series of the Triassic detritals and the discovery of flysch facies in the Tellian Atlas (at BO-.1, to the far North) may support the existance of two interplates, the Saharan and the Atlassic plates. However, detailed geological and geophysical studies would be necessary to confirm such theory.

Salt movements, are explained due to halokinesis in terms of plate kinematics. The term Kinematics is the study of movements exclusive of the influence of mass and force; "halokinematics" was proposed by the author (Assaad, 1983) to define disturbances of salt flow caused by continuous indirect geologic events which might be a consequence of plate kinematics; it indicates the causative effect of plate tectonics on salt kinematics; in other words, it qualifies salt tectonics more precisely according to their cause; although the causes are always speculative, it is assumed that salt movements due to halokinesis may be a consequence of plate kinematics, besides the lunar and tidal forces mentioned by Trusheim (1960); however, the daily disturbance of the earth's cust by such forces may have had a definite influence on the maintenance of salt-flow in recent times.

Theoretically, it is possible that salt deposits in NE of the Atlas trough, were formed immediately after the major Early-Alpine orogeny at the margins of probable structural basins, thus leading to the first creation of salt-dome structures due to halokinesis. The stresses generated in times of the regional unrest during the Middle and Late Alpine orogeny, could have accelerated the flowing movement of salt. However, the author believed that the disturbance of the earth's solid crust due to plate tectonic movements, is likely to have had some indirect influence on the initiation and maintenance of the salt flow that led to halokinetic structures.

During the Mesogene, the climate of North Africa was tropical with warm seawater that caused the deposition of carbonates. The successive southward migrations of the Equator were accompanied by the deposition of evaporitic sediments, and many of neritic basins became lagoon due to strong variations in salinity. The Alpine Orogeny had its early phases in Jurassic and Cretaceous times and reached its maximum in the Cenozoic.

11.6.1 Salt Movements and Plate Kinematics

Great progress in the elucidation of Mediterranean geology has been possible because of the interpretation of observations in the light of plate-tectonics concepts. The Lower Triassic fluvio-deltaic deposits and the principal Upper Trias-Lias evaporites may be explained in terms of plate tectonics, while the successive southerly migrations of the Equator during Late Triassic-Early Liassic times led to the formation of several lagoons. **Fig 11.8-** A composite paleogeographic map from the Tethys Ocean to the Mediterranean Sea showing a Plate tectonics model of the evolution of the Western Alpine System (B.Biju Duval et Montadert, L., 1979).

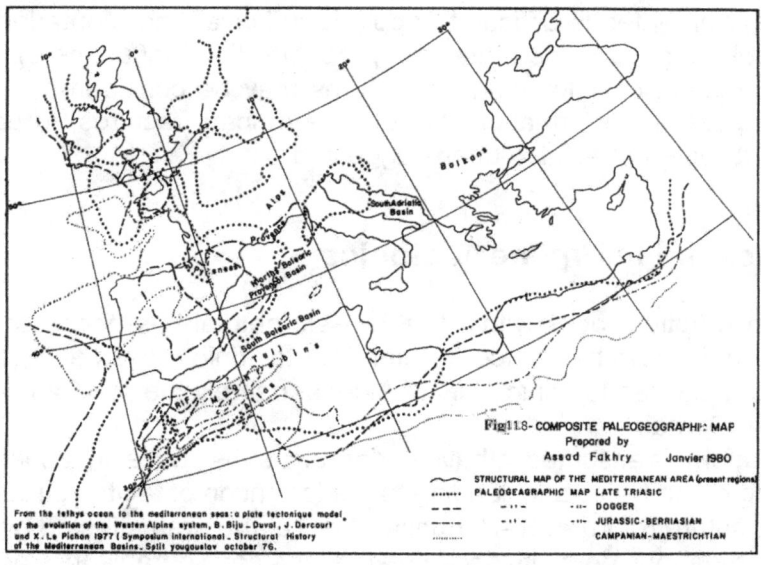

A systematic attempt to reconstruct the development of salt dome structures in NE Central Algeria within the framework of plate kinematics:

(a) **Triassic- Jurassic:** A wide Tethys Ocean between Europe and Africa was present in particular during Liassic and Jurassic. That was a period of extension, which manifested itself in the formation of continental margins and affected the whole area. Primary salt-dome structures might have been inaugurated in ENE/WSW "Atlassic" trend at the end of that period, but in general, that occurred immediately after the major Early Alpine orogeny, as the prevailing stresses might have been converted into the initial salt movements by the indirect effects of plate kinematics. The thickness of inhomogeneities within the salt or in the underlying or overlying beds, the size of the basin and the rate of subsidence, etc., all played a part. It is postulated that the development of salt structures normally began with salt pillow stages.

(b) Jurassic- Upper Cretaceous: All data pointed towards the birth and evolution of a spreading "ocean" and its passive continental margins. In general, Africa moved left-laterally with respect to Europe from the Triassic to Upper Cretaceous. The initial Tethys had been completely consumed in Maestrichtian times, whereas the Mesogean Ocean attained its maximum size.

Salt pillows continued to be formed as salt accumulations, where the migration took place, from the margins inwards. The absolute rate of salt flow was estimated by Trusheim (1960) as 0.3mm/year on average; this rate is much lower than that of the lateral movement of Africa, estimated as 2.4cm/yr.

(c) Upper Cretaceous -Yepresian: Africa ceased to move left-laterally and started a right lateral movement of limited extent, which continued until the Ypresian. The mass displacement of salt might have caused primary peripheral sinks, which are typical attributes of salt-pillow formations. During the Cretaceous period, accumulating sediments tended to thin and exhibit facies changes across ascending domes, whereas fault systems probably played a part in eroding the Cretaceous overburden **(see Fig 11.6)**. If the rate of surface upwarp was sufficiently high, no Cretaceous formations were deposited above the salt domes and a local disconformity was produced. The absence of Cretaceous formations to the SW of the Pre-Atlas zone might therefore be attributable to the presence of salt structures.

(d) Post-Ypresian – Present: Africa rotated about a trigometric pole of rotation in the region of Morocco, resulting in north-south compression to the east. That was a time of collision, during which the concept of plates is difficult to apply (Biju-Duval and Montadert, 1977). Halotectonic emplacement took place and therefore, the probable direct influence of plate tectonics movements on the formation of halotectonic structures may be considered in the High Plateaus of North Algeria. A possible deformation due to intra-African fractures, in particular the so-called "South Atlas" fault system, has also been suggested.

11.7 Eruptive Rocks, and Interplate Tectonics

A general review of the eruptive rocks within the Triassic detritals, is necessary, as they were most probably deposited in the Triassic basin due to interplate tectonic movements, as is evident from the paleostructural history of the three tectonic units of Algeria, namely the pre-Cambrian, the Epihercynian platforms, and the Tellian Domain.

The Epihercynian platform is assumed to have undergone subsidence, in respect to the Pre-Cambrian Platform at the bigenning of the Paleozoic, leading to the formation of faulted structures, which extended along the south Atlassic suture (of Lutetian-Bortgalian Age) at the area affected by the downward movement (Assaad, 1972). Terrigenous deposits accumulated for a considerable thickness (1,200m at AMI-1), over the large Paleozoic "geosyncline" in the Atlas trough (Atlas Flexure), south of the Saharan Atlas Mountains.

According to the concept of plate tectonics and due to forces generated by its movements, the fragmentation of the Gondwana proto-Continent took place from Late Triassic to Early Liassic times and resulted in the formation of rift zones which were the possible source of basalt flows together with their associated tectonic events; a rift zone might be predicted within the oceanic crust, whereas the Triassic/Liassic salt structures in the African and the Mexican plates are very much correlatable. However, the subduction zones have been active from the Mesozoic to the present time in the Mediterranean Alpine area.

Petrographical and mineralogical studies of the eruptive rocks of the Triassic Province in NE zone of Algeria have led to the conclusion that they are mainly composed of basalts and spilites, not as wrongly identified before as andesites and dolerites (personal commnication, 1975). The flow of eruptive rocks has been attributed to the presence of rift-zones and their associated tectonic events.

Intracontinental plate tectonics may have gradually developed when convergent and divergent movements began in Late Ordovician times, as evidenced by magmatic intrusions in the surrounding sediments of north Algeria.

A rift zone was presumably formed as a result of tectonic distensional forces, and therefore a subduction complex is assumed to have been formed in the southern Atlassic flexure zone where the 'Saharan Plate" descended below the "Atlassic plate" and thick Mesozoic sediments have been deposited.

It is interesting to mention that at Bourlier graben (BO-1), far to the north of the High Plateaus, thick terrigenous subcontinental sediments of flysch facies, of Late Paleozoic Age, were deposited due to the Hercynian orogeny, and attained a thickness of ~2,000ms.

In the Tellian domain, "geosynclinal" phases of evolution prevailed from the Jurassic to the Upper Cretaceous, and the orogenic phases took place from Oligocene to Pliocene times. The troughs of the early "geosynclinal" phase were formed during the early Alpine stage, whereas during the Late orogenic phase, the folded Tellian mountain chain, characterized by Miocene Nappes, was bounded to the south by a marginal trough filled with Oligocene and Miocene molaasse complexes. The latter was separated from the Epihercynian platform by folded structures (or flexures), where many outcrops of Triassic evaporites have been found along the Tellian Atlas faults (Technoexport, 1971).

Fig 11-9- Schematic Diagrams showing the evolution of the Intracontinental plates in North Cenrtral Algeria; **Fig 11.10** – A generalized N/S Schematic Cross section showing the Tectonic forces and Structural features of NE Algeria (Assaad, 2014).

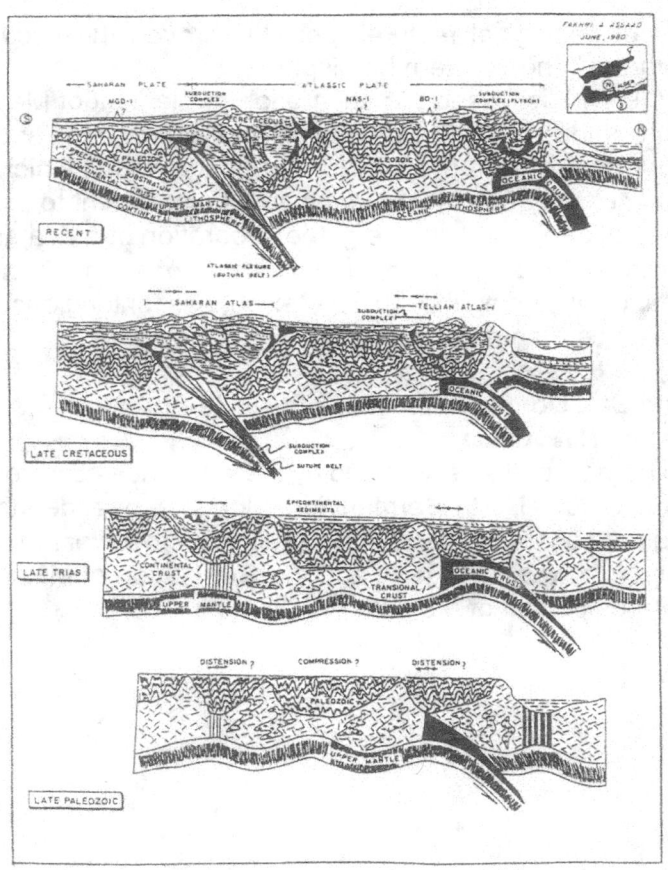

Fig 11.9 – SCHEMATIC DIAGRAMS SHOWING EVOLUTION OF intracontinental PLATES
(NORTH CENTRAL ALGERIA)

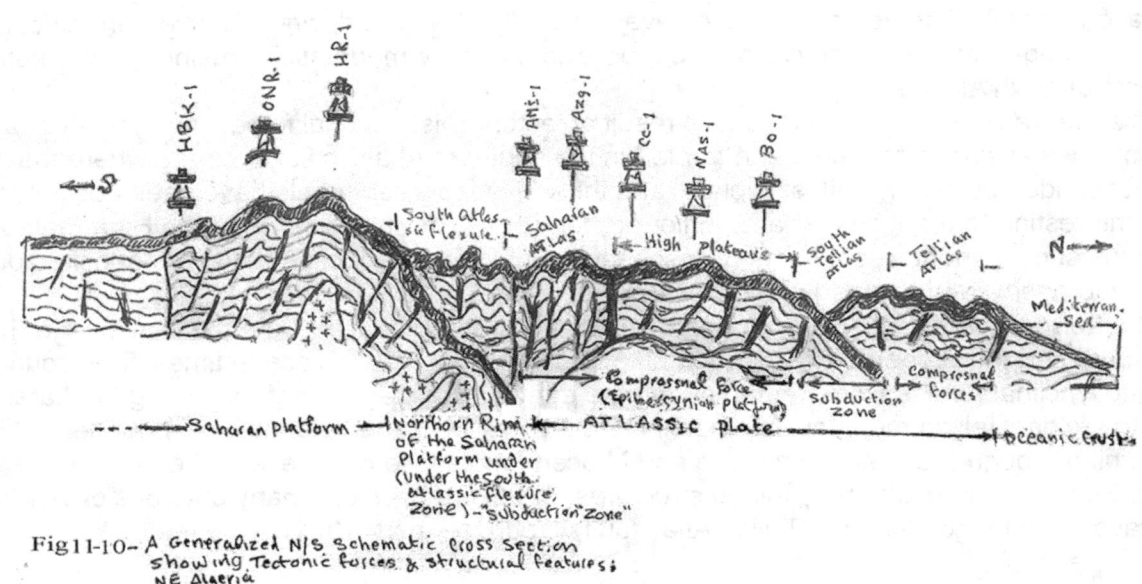

Fig 11-10 – A Generalized N/S Schematic Cross Section showing Tectonic forces & structural features; NE Algeria

11.8 Evaporitic Deposits in NW Africa and the Gulf of Mexico

In Northwest Algeria, several geological features possibly proved the occurrence of the African and Mexican intercontinental plates of the Western hemispheres.

Hogens, M.J.O 'Brien (1994), discussed the separation of the Gulf of Mexico and Africa, and stated that the Sea floor spreading implies that the continents have been split and spread apart as a result of upwelling of new crustal material from the earth's mid-ocean ridges which are known to exist in the Atlantic, Pacific and Indian oceans. Such movement would be sufficient to produce the phenomenon of continental drift in the course of geological time, e.g. the separation of Africa and South America, and the associated growth of the Atlantic Ocean. The striking similarity of shape of the west coast of Africa and the east coast of South America, including the Gulf of Mexico, has for long encouraged the belief that they were once united and have since then moved apart.

11.8.1 The Triassic/Liassic Salt Dome Structures of the Gulf of Mexico and its Correlation with that of the Arabian Maghreb, NW Africa

The Permian period was extremely arid as evidenced by the worldwide occurrence of evaporates, e.g. The Permian Zechstein Salt structures of North Germany are elongate meandering salt ridges in the deepest part of the basin. Salt domes are characterized by the structural growth of many salt stocks, which can be correlated with those of the Northwest of Algeria and the Gulf region of Mexico. **Fig. 11.11a** –A location of Prominent structural features of Gulf of Mexico.

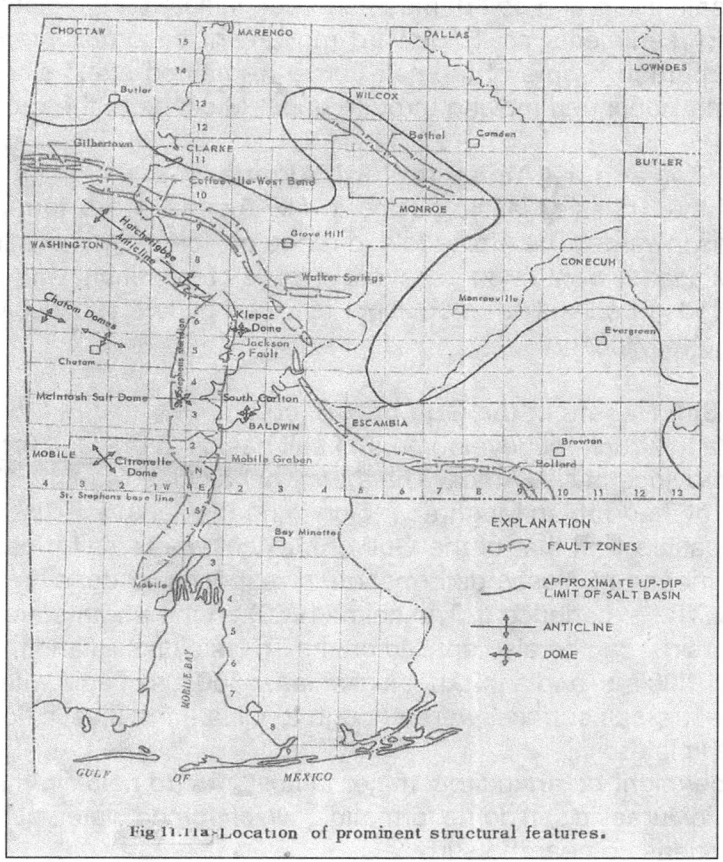

Fig 11.11a - Location of prominent structural features.

A plenty of data have been collected on the evaporitic rocks of the Jurassic age in the coastal province of Mexico and southwest of USA. The Gulf Coast basin of the southern United States shows five salt dome basins of maximum deposition. Fig 11.11b – A diagramatic map outlines the major structural features of the southern United States and Gulf Region (Halbouty, 1979).

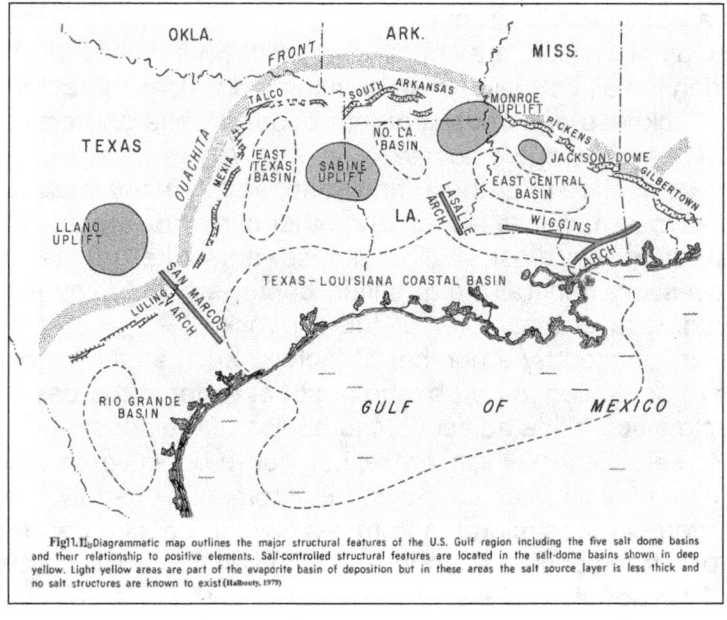

Fig11.11b Diagrammatic map outlines the major structural features of the U.S. Gulf region including the five salt dome basins and their relationship to positive elements. Salt-controlled structural features are located in the salt-dome basins shown in deep yellow. Light yellow areas are part of the evaporite basin of deposition but in these areas the salt source layer is less thick and no salt structures are known to exist (Halbouty, 1979)

The flow of salt into salt pillars and salt domes had been initiated immediately after the burial of the Mesozoic and the Tertiary sediments as its upward movement occurred intermittently, depending on the thickness of the overburden. Some of the salt domes remained static since the end of Oligocene whereas, others apparently continued moving through upper Miocene to the recent time.

11.8.2 Salt Dome structures and the Alpine in North Algeria
Salt dome structures of the Triassic/Liassic basins in NE Algeria, were formed in Post Early Alpine Orogeny from the margins inwards. The Alpine folded range of North Algeria which occurred through the lower Jurassic and the Miocene, went through several stages of evolution, with abrupt tectonic changes that covered lengthy periods of subsiding or uplifting, followed by two successive orogenic movements from the upper Paleogene to Miocene.

11.8.3 Development of Salt Basins in the Gulf of Mexico
The depositional history and structural development of salt basins at the Gulf region of Mexico coincides with that of the Triassic province of NW Africa. The theory of halokinesis of the Triassic salt structures in terms of halokinematics, at the Arabian Maghreb region in northwest Africa, (including Morocco, Algeria, and Tunis), can be compatible with that of the Gulf of Mexico, where salt masses flowed into elliptical dome-like structures by means of plastic deformation in response to density differences between salt and overlying sediments. The experience of Trusheim (1960) on the parameters used for salt flow at the German Zechstein salt basins, can be also applied on the Triassic/Liassic and the Gulf of Mexico basins; the overburden of about 1,000 ms. and a thickness of at least 300 ms of salt, with very few degrees of dip of the basement or of the pre-salt surface, were enough to initiate the plastic flow process. A rift zone is expected in between both plates.

The volcanic influence might be an indirect major factor " due to halokinematics " in the initiation or continuation of the Triassic/Jurassic salt dome formation, which forced water, salt, sulfur, and oil and gas, to pierce upward and condensate near the surface.

11.8.4 Configuration and Composition of Salt Dome Structures in the Gulf of Mexico
Salt domes of the Gulf of Mexico are circular to broadly elliptical, whereas in USA, they are circular or modified circular slopes. The tops of individual salt stocks range in diameter from one-half mile to four miles or more; the average diameter of domes in the Gulf region is two miles. Some salt stocks are vertical at one depth and inclined at others. The age and mode of deposition of the Gulf salt ranges between Permian and Triassic to Middle Jurassic age.

Salt structures in the Gulf of Mexico, are coarse crystalline with individual crystals ranging from ¼ to ½-inch in diameter. Layering in salt deposits, is the dominant physical characteristic. The layers average one inch to ten inches in thickness and consist of interbedded white and gray to black salt bands with minor impurities, mainly of anhydrite and dolomite.

The cap rock of salt is composed of granular anhydrite, and in many instances is the only consistent of the salt dome material. Gypsum, calcite, sulfur, and other minerals, are actually hydrothermal solution products of anhydrite alteration. On most shallow domes, anhydrite grades upward into gypsum and calcite with or without accessory minerals (e.g. sulfur, barite, etc). A marly limestone and/or dolomitic limestone overlie the anhydrite layer that develops the cap rock.

Brine crystallization is influenced by a number of factors, such as temperature, and solubility of the salts in the brine, which are controlled in part by the depth of water in the basin, its bathymetry, and the geo-morphological characteristics of the adjacent land bodies of the basin.

A large percentage of salt domes which are either non-productive or have produced only minor amounts of oil or gas, can be explained as Late-occurring domes. Generally, the vertical growth of a salt dome occurred after the hydrocarbons migrated from nearby source, and trapped; On the other hand, all known types of hydrocarbon traps occur in connection with salt dome structures, show different distinct types of oil traps (**Fig 11.12** Halbouty; 1979):

(1) Simple dome anticline
(2) Graben fault trap over dome
(3) Porous cap rock (limestone or dolomite)
(4) Flank sand pinchout and sand lens
(5) Trap beneath overhang
(6) Trap uplifted and buttressed against salt plug
(7) Unconformity
(8) Fault trap downthrown away from dome
(9) Fault trap downthrown toward dome

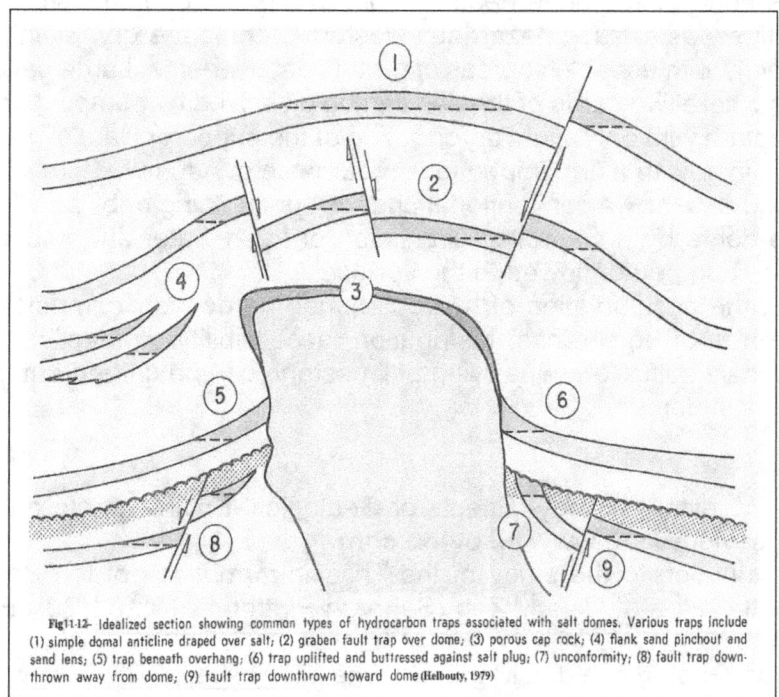

Fig.11-12- Idealized section showing common types of hydrocarbon traps associated with salt domes. Various traps include (1) simple domal anticline draped over salt; (2) graben fault trap over dome; (3) porous cap rock; (4) flank sand pinchout and sand lens; (5) trap beneath overhang; (6) trap uplifted and buttressed against salt plug; (7) unconformity; (8) fault trap downthrown away from dome; (9) fault trap downthrown toward dome (Halbouty, 1979)

11.9 *Separation of Gulf of Mexico and Africa:*

Sea floor spreading implies that the continents have been split and spread apart as a result of upwelling of new crustal material from the earth's mid-ocean ridges which are known to exist in the Atlantic, Pacific and Indian oceans. Such movement would be sufficient to produce the phenomenon of continental drift in the course of geological time, e.g. the separation of Africa and South America, and the associated growth of the Atlantic Ocean. Africa and Gulf of Mexico; the striking similarity of shape of the west coast of Africa and the east coast of South America, including the Gulf of Mexico, has for long encouraged the belief that they were once united and have since then moved apart;

11.10 Is Oil Migration related to the partial subduction along the Tellian Flexure at the Western Coastal Zone of Algeria

Oceanic basins were initiated by continental separations that begin with intracontinental rifting, e.g. between the Tellian Atlas and the Epihercynian platform.

The immense petroleum accumulations of the Triassic province of Algeria at the southern Tellian flexure belt, may owe their origin due to partial subduction, formed from crustal collisions where the rifted margin sediment prisms are drawn down against and beneath the suture belt. During such partial subduction, fluid hydrocarbons may be driven updip away from the subduction zones to accumulate in reservoirs along adjacent platform margins and within foreland fold-thrust belts (Halbouty, 1979).

11.11 Economical Aspects of Salt Structures

The tectonic outline in areas of thick salt deposits, should be carefully studied to define the different types of salt dome structures buried deep underneath the ground for the eventual storage or disposal of high level solidified radioactive waste; such hazardous waste materials are dry, mainly impervious to water, and not associated directly with useable sources of groundwater wastes. Large caverns can be mined out at depths up to 300 ms where two thirds of the salt can be dissolved by pumping hot water, and removed from the salt dome structure with only slight deformation of the support pillars by pumping hot water; then the cavern can be then lined with a firm impervious material, e.g. Asphalt.

Salt can be extracted either by a conventional underground mining or by solution mining in which the salt is dissolved in the dome by a controlled circulation of fresh water and subsequent evaporation of saturated brines that are pumped or flowed to the surface.

Taking into account the accumulation of hydrocarbons, the economic importance of salt structures can be described in terms of occurrence, distribution, accessibility, productivity and other factors. In genearal, the mining of salt, sulfur, and quarrying of limestone can be carried out from the cap rock and/ or from salt pillows and salt domes.

References

Komatina, M.M (2004) Medical Geology- Effects of Geological Environments on Human Health, ISBN 0-444 –51615-8, Elsevier, USA; www.elsevier. com

Assaad, F. (1972) Contribution to the study of the Triassic formations of the Sersou-Megress Region (High Plateau) and the area of Daia M'Zab (Saharan Platform), No. 84(B3), Eighth Arab Petroleum Congress, Algeria, 1972.

Caratini, C. (1964) Etude Geologique des Monts Du Chellala Reibell, SPHP.

Trusheim (1960) Mechanism of salt Migration in Northern Germany. Bull., AAPG., 44, (91), 1519-1540, 1960.

Biju-Duval, B. and Montadert, L. (1977) Geological evolution from the Tethys Ocean to the Mediterranean from Mesozoic to the present (Symposium International – Structural History of the Mediterranean basins- Split, Yugoslavia. pp 13-17.

Halbouty, M.T (2nd Ed., 1967, 1979) Salt Domes – Gulf Region, United States and Mexico. Figs: 1.3 & 6.1. Gulf Publishing CO, P. O. Box 2608., Houston, Texas 77001.

Nettleton, L.L. (1936) History of Concepts of Gulf Coast Salt-Dome Formations, Bull., AAPG, V.19, no.12, pp. 2373-2383.

Barton, D. C. (1933) Mechanics of Formation of Salt Domes of Texas and Louisiana, Bull., in AAPG- Oil Fields, pp. 97-99, AAPG (1936).

Hough, S, E and Bilham, R.G. (2006), After the Earth Quakes-Elastic Rebound on an Urban Planet, ISBN 0-19-517913-7, p 8, Oxford University Press Inc. Mexico,, Figs 1.3 & 6.1; Gulf Publishing CO., Box 2608, Houston, Texas 77001.

Pomerol, C.H. (1975) Manuel du Stratigraphique et Paleogeographique; Era Mesozoique.

Guillmot, J (1964) Cours de Geologie du Petrole

Levorsen, A.I. (1954) Geology of Petroleum, W.H. Freeman (540,541, 574), edited by Frederick, A. F. Berry, Univ.California, Berkeley, San Fransisco and Company, USA

Assaad, F., (1983) An approach to "Halokinematics" and Interplate Tectonics (North- Central Algeria), Jr. Petroleum Geology, Vol.6, No. 1, July 1983.

------------, (1981) A further Geologic Study on the Triassic Formations of North- Central Algeria with Special Emphasis on Halokinesis, Journal of Petroleum Geology, 4,2, pp. 163-176.

By: Hogins, M., M, J.O ' Brien (1994) Salt sill deformation and its implications for subsalt exploration; Leading Edge, v. Aug. 1994, pp.849-851.

Cited References (UAB Libraries):

Margaret A.McCoy and JudithA. Salerno (Iune 2010), Assessing the effects of the Gulf f Mexico oil spill on Human health:a summry of the June 2010 - rapporteurs Institute of Medecine of the National Academies (2010)/ UAB-Library:#TD427 L44 2010 GC1221 A87 2010eb.

Konrad, John and Shroder, Tom (2011) – Fire on the Horizon: The unfold storyof the Gulf oil Desaster; UAB Library # HD7269-P42 M615 (2011)

Priest, Tylor (2007) Offshore Imperative:Shell Oil's search for Petroleum in postwar America UAB library # HD9569.s55 P74 2007

Carvenar, Bob (2010) High Stakes, high risks, and the story behind the deepwater well blowout, UAB # TD427.P4 C38 2010

Hudec, Michael R., et al (2011) – The Salt Mine: a digital Atlas of SaltTectonicsView all formats and Lanhgauges: a Database/WorldCat-Publisher AAPG. (2011).

Appendix 11.A1- *More about Evporite Rocks (Special Readings- Aa a Reviewer of the Env. ` Geol Jr, Springer; Heidelberg, Germany, 2006:*

- Warren, J.K., (Springer, Berlin, Heodelberg, 2006)- stated that the worldwide, bedded and allochthonous evaporites are highly effective seals; even though evaporites constitute less than 2% of the world's sedimentary rocks, one half of the world's largest oilfields are sealed by evaporates, the other half are sealed by shales.

  Salts such as halite, gypsum, and trona have been long affairs of man, both as a preservative and as industrial feedstock. About 660 BC, the ancient Egyptians had been using natrun, an impure mixture of trona and other sodium salts, for several thousands years, both as an important part of the mummification process and for salting food. In the present, halite is a major feedstock to the chemical manufacturing industries, as are other varieties of salt. Salts are also important in trapping, sealing and perhaps generation of hydrocarbons and various metal deposits.

- Hodgkins and O'Brien, 1994;Ratcliff and Weber, 1997, stated that salt tectonics is a general term which encompasses notions of lateral and vertical salt flow, trans-stratal salt movement, salt pillowing and diapirism; it refers to tectonic deformation involving halite or other evaporites as a substratum or a soutrce layer. The authors emphasize that concepts of salt tectonics have undergone a major revision in the last fifteen years, due mainly to advances in 3D-seismic acquision and processing, essentially 3D prestack depth imaging, and the realization that salt tectonics is best modeled by considering salt as a pressurized fluid layer overlain by brittle sediment.

- Over time, a decimeter-to-a meter thick salt beds can stack one atop the other to form evaporite units, that can be hundreds of meters thick. When conditions are suitable, huge thicknesses of shoal water vaporite can accunulate in very short time scale.

- Solubilities vary from one type of salt to the other; halite is much more soluble than gypsum; its saturation concentration for halite is ~350,000ppm at 25° C. *(See Fig 11.A1)*

Chapter (12)

An Estimate of Petroleum Reserves of the Triassic reservoirs at Hassi R'Mel-M'Zab High and Oued M'ya Basin, NE of the Algerian Saharan Platform

12.1 Scope

An estimate of oil and gas reserves of the Triassic Deposits, carried out by the author, at Ouad M'ya basin and Hassi R'Mel-M'zab High of the Algerian Saharan Platform, and was instructed by the Algerian Ministry of Energy, which supervised both the Exploration and Production Directorates of the National Algerian Company "Sonatrach".

Personal efforts to review different publications of research studies on the Triassic reservoirs, were accomplished besides the available French publications and documents from the Archieve of both the Exploration and Production Directortaes, for a better idea on the Stratigraphical, structural and geological maps, besides the regional tectonic history and sedimentation of the Triassic Province.

It was noticed that the old French contractors during the French colonialism, used to submit different and confusing geological terms and nomenclatures, and had carried out their work separatedly, without communications among each other; a fact that led the author to consider such issue as the first priority to unify the related terms. *(See Chapter 10).*

12.2 Introduction

The Hassi R'Mel- M'zab includes Hassi er R'Mel field, whereas, the Oued M'ya basin comprises both Oued Noumer area and Haoud Berkaoui field. The Oued Noumer area is located on the southeast of the Hassi R'Mel- M'Zab structure and on the western side of the Oued M'ya depression **(see Fig 10.1).**

The classification of the detrital Triassic formation mainly depends on electric well log correlation and petrographic characteristics. The upper T2- unit of the Triassic detritus includes the "A"-sandstone reservoir at its lower part (or 3a-unit), which is overlain by the shaly bed (or 3b-unit); the T2-unit is underlain by the T1-unit of the lower Triassic detritus that comprises "B" (or 2a &2b) and "C" (or 2c) sandstone beds (Stoica, 1981). The Lower Series "Serie Inferieur; SI", is the lowermost series of the Triassic detritus primarily discovered at the Haoud Berkoui oil field further to the SE of the Oued M'ya basin, and is considered of fluvio-deltaic origin, **(see Table 10.2, IFP 1967).**

12.3 Petroleum Hydrocarbon Reserves of Oued Noumer Field

12.3.1 Discussion

The Oued Noumer area comprises both the Oued Noumer and Ait Kheir fields; the Oued Noumer field is 70km SE of Hassi R'Mel field, and is 12.5km NNE from Ait Kheir field.

According to seismic reinterpretation on top of the intermediate Triassic shaly sandstone (B-; or parts of 2a +2b- reservoirs), the isobaths & isochrone maps show the Oued Noumer field as a closure of a more or less symmetric NNW/SSE anticline, located between two faults of the same trend forming a horst-like structure. Ait Kheir field generally represents a closure of a dome-like structure of a NNW-SSE fault; the Oued Noumer field comprises three productive wells: ONR-1bis, ONR-3 and ONR-5, are three non-productive wells: ONR-2, ONR-4 and ONR-6, whereas at Ait Kheir field, there are three productive wells: AT-1, AT-2, and **AT-3 (see Table 10.5).** The main objectives are the upper Triassic shaly sandstone: (T2 or "A- unit"), underlain by (T1 or "B + C"-unit). The following two sandstone units are separated by few meters of shale intercalations:

a) The T2-unit - mainly constitutes of sandstone in the lower part (3a) with intercalation of siltstone and sandy shale that are frequently wedging out. The sandstone is fine to medium-grained, with oblique and cross- stratification in some locations. It is poorly consolidated by chlorite-illite cement and grades into shaly siltstone and variegated shales with a varying silt content.

At the wildcat well ONN-1, northeast of Oued Noumer field, the "A"- sandstone unit of 78m thick, is dark brown to black siltstone, turns downward to fine, medium grained sandstone, subangular to subrounded, friable with argillaceous siliceous cement, sometimes anhydritic, micaceous and few iintercalations of reddish brown shale.

In Ait Kheir field, the sandstone of "A"-unit is dark brown to black and gradually turns to beige in color, fine to medium, well sorted, subangular to subrounded, friable, with an argillaceous and micaceous cement with inclusions of green shales and thin intervals of bituminous sandstone. To the NW (at AT-3), the sandstone becomes brownish to reddish gray, argillaceous and sometimes micaceous sandstones, with intercalation of shale and light brown, very fine argillaceous sandstone, grading to siltstone with clay patches.

b) T1-Unit ("B" and "C" sandstone beds) - In the Oued Noumer field, the "B" sandstone is lithologically and petrographically homogenous. It consists of multicolored sandstone, fine grained, quartzitic, poorly consolidated by a siliceous kaolinitic cement. In the lower parts of the unit, it grades into medium to coarse sandstone with an intercalation of a very thin bed of dolomitic-silty variegated shale and a thin conglomerate bed at the base, (3ms thick).

In Ait Kheir field, the T1-unit is mainly sandstone, beige in color, medium to coarse grained, friable, subrounded to subangular, well sorted, with intercalation of few centimeters of green, micaceous shale flakes; at AT-3, the sandstone turns to brownish gray and maroon in color, very fine to medium, subangular to subrounded, poorly sorted, compact to semi friable and micaceous. Intercalations of gray, micaceous shale and hard argillaceous siltstone, are encountered at the base. The total net pay sandstone thickness is about 63m.in both fields.

In the Oued Noumer field, the C- sandstone series is represented by alternations of gray, fine grained, quartzitic sandstone, shaly siltstone and shale. At ONN-1, the "C" sandstone series is 24m. thick, white, fine grained, compact, grades to siliceous, fairly anhydritic, fine to medium grained sandstone with intercalation of reddish brown, silty shale and disseminated pyrite.

In Ait Kheir field, the C-sandstone series is reddish brown silty to sandy shale, and attains a total thickness of about 28m.

The reservoir rocks, in both the Oued Noumer and Ait Kheir fields, determined by Gamma-Sonic logs, are defined by isobaths maps (Figs 12.1 and 12.2a);

Fig 12.2b- A reviewed isobaths map on top of "B" sandstone in both fields, used a detailed study on seismic sections, and predicts a possible split point across the western amd eastern faults of NNW/SSE trend that gave the horst structure of Oued Noumer field. Figs. 12.3 and 12.4 are both isopaches maps of the Triassic reservoir series "A" and "B" respectively.

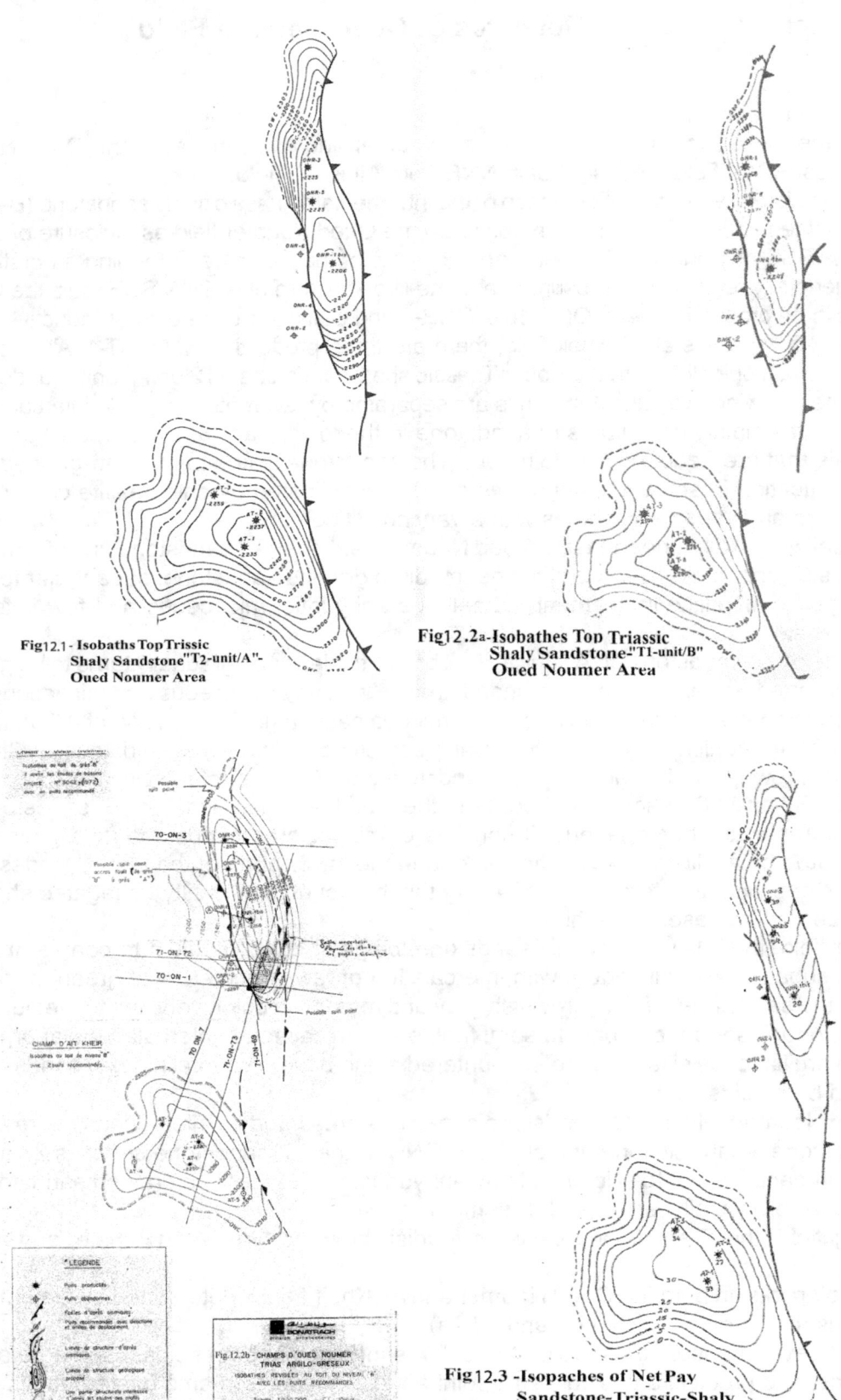

Fig.12.1 - Isobaths Top Trissic Shaly Sandstone "T2-unit/A" - Oued Noumer Area

Fig.12.2a - Isobathes Top Triassic Shaly Sandstone "T1-unit/B" Oued Noumer Area

Fig.12.3 - Isopaches of Net Pay Sandstone - Triassic-Shaly Sand stone "T2-unit/A" - Oued Nouer Area

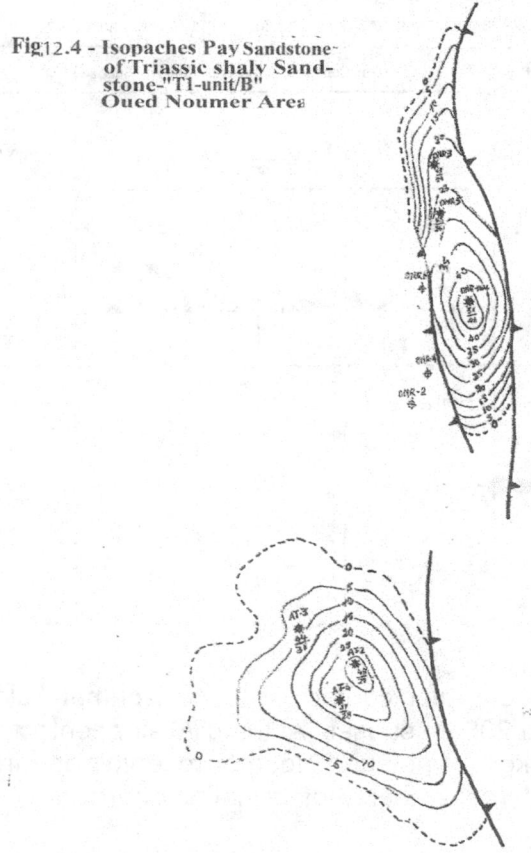

Fig.12.4 - Isopaches Pay Sandstone of Triassic shaly Sandstone-"T1-unit/B" Oued Noumer Area

c) **The lower series** "Serie Inferieur- or Si-series" are reworked sandstone series.

12.3.2 Reservoior Characteristics

From the paleotectonic point of view, Oued Noumer structure is mostly developed by the Alpine movements, and throughout the entire Mesozoic era and remained as a positive structure, favorable for hydrocarbon accumulation.

In the Oued Noumer field, the productive rocks are intercalated by rather clean shales with good isolating properties; the intercalating beds between (A) and (B) series, grade to shaly sandstone and siltstone away from the productive limits (e.g. at ONR-2), where the sandstone series "A", "B" and "C" might be connected and represent one single hydrodynamic reservoir. The C- sandstone bed is unproductive in both Oued Noumer and Ait Kheir reservoirs.

Table 12.1- shows different positions of oil/water contact in different wells according to electric well log correlation, lab tests, and core analysis of the residual hydrocarbon contents; differences of positions of oil/water contact among reservoirs of each well can be attributed to the different methods applied for its determination; e.g. lack of a definite control of the measurements of the drill pipe, of the cable used for logging operations, and /or due to incorrect measurements of surface elevations (Zs).

Table 12.1 - DIFFERENT POSITIONS OF OIL/WATER CONTACT BASED ON DIFFERENT ANALYSES OF THE RESERVOIR ROCK

Productive Well	Oil/Water Contact Position (Meters Subsea)			
	based on log data	based on core analysis	based on test data	For reserve evaluation purposes
			Technoexport 1971	
ONR-1 bis	-2302	-2306	-2302	
ONR-3	-2306	-2307	-2306 to -2308	-2305
ONR-5	-2303 to -2306 (zone d' interface)	-2306,5	-2301 to -2309,4	
AT-1	-2325 to -2329 (zone d' interface)	-	-	
AT-2	-2332,5	-	-	-2325
AT-3	-2334 to -2335.5 (zone d'interface)			

For reserve evaluation, the oil/water contact at Oued Noumer field is defined at 2305m subsea, whereas, at Ait Kheir field, it is at 2325m subsea. At the oil/water contact in both reservoirs, the formation pressure ranges from 408-410 kg/cm² at Oued Noumer reservoir and from 410-412 kg/cm² at Ait Kheir reservoir; both indicate a considerable reserve of formation energy.

12.3.3 Hydrocarbon Potentialities

The Triassic "B" sandstone of the Oued Noumer area is well developed; the porosity log shows a higher average porosity (20.4%) than that of the above "A- sandstone"(14.7%). In general, the average porosity of both series was estimated as 17.5%; differences in porosity data were noticed according to the use of different tools. Table 12.2 - shows the porosity of "A" and "B" reservoirs at both Oued Noumer and Ait Kheir fields. The total pay sandstone of the Triassic formations is in average of ~30m.

Table 12.2 - Determination of porosity percent of the Triassic Shaly Sandstone of "A" and "B" series - Oued Noumer and Ait Kheir Fields

Productive wells of Oued Noumer and Ait Kheir Fields	Porosity of Sandstone "A" series %							
	V/O TECHNOEXPORT (1971)			SN REPAL (Person. Comm) (1973)	V/O TECHNOEXPORT (1971)			SN REPAL (Person. Comm.) (1973)
	Core Samples ϕ Car.	Electri Log ϕ D.	Porosity of Produc Layer		Core Samples ϕ Car.	Electric Log ϕ D.	Porosity of Produc Layer ϕ V	
ONR-1 bis	18,3	19,0	19,0	13,7	22,1	22,3	22,3	22,6
ONR-3	16,4	16,2	16,2	16,1	19,5	18,5	18,5	15,8
ONR-5	13,0	15,8	13,0	14,2	20,9	22,3	20,9	21,0
Average Porosity per series	—	—	15,5	14,8	—	—	20,9	21,4
	SONATRACH 1973				SONATRACH 1973			
AT-1	17,1			18,3	22,4			20,5
AT-2	20,5			15,0	19,9			18,2
AT-3	10			11,0	18,6			
Average Porosity per series	15,9			14,7	20,3			19,3

At Oued Noumer field, the average permeability of "A-series", determined from core samples, ranges from 144md to 185md, and for "B-series," from 1200md to 1364md; whereas, at Ait Kheir field, the "A" series ranges from 15md to 84md and for "B-series", from 980md to 1034md.

Table 12.3- shows permeability values for both reservoirs at Oued Noumer and Ait Kheir Fields.

Table 12.3- Determination of porosity percent of the Triassic Shaly Sandstone of "A" and "B" series - Oued Noumer and Ait Kheir Fields

Productive wells of Oued Noumer and Ait Kheir Fields	Porosity of Sandstone "A" series %							
	V/O TECHNOEXPORT (1971)			SN REPAL (Person. Comm.) (1973)	V/O TECHNOEXPORT (1971)			SN REPAL (Person. Comm.) (1973)
	Core Samples ø Car.	Electri Log ø D.	Porosity of Produc Layer		Core Samples ø Car.	Electric Log ø D.	Porosity of Produc Layer ø V	
ONR-1 bis	18,5	19,0	19,0	13,7	22,1	22,3	22,3	22,6
ONR-3	16,4	16,2	16,2	16,1	19,5	18,5	15,5	15,8
ONR-5	13,0	15,8	13,0	14,2	20,9	22,3	20,9	21,0
Average Porosity per series	—	—	15,5	14,8	—	—	20,9	21,4
	SONATRACH 1973				SONATRACH 1973			
AT-1	17,1			18,3	22,4			20,5
AT-2	20,5			15,0	19,9			18,2
AT-3	10			11,0	18,6			
Average Porosity per series	15,9			14,7	20,3			19,3

In general, the average permeability is estimated for both "A" and "B" in both fields of Oued Noumer area, as 790md, and is considered good granular reservoirs.

The productive wells, DRT-1 and HKA-1 are located nearby the central part of the Triassic basin where the Lower series "Serie Inferieur" is oil productive at DRT-1 at much deeper depths.

The Oued Noumer area seems to be a transitional zone of hydrocarbon accumulation between the lower oil productive Triassic series **(SI-series),** at DRT-1 and KHA-1, and the upper gas productive Triassic reservoir at Hassi R'Mel-M'Zab. However, the Gussow theory of differential entrapment of petroleum among the said fields, had been applied to explain the regional migration of hydrocarbons of the Triassic Province in the Algerian Sahara **(see Ch.10-Section 10.4)**.

The gasoline/oil contact in the Ait Kheir reservoir might be arbitrarily considered at 2280ms subsea, whereas, at Oued Noumer reservoir, it could be 20ms higher, i.e. at 2260ms subsea. A local fault of 20ms of displacement can be presumably situated between the two reservoirs to explain the two different contacts, or probably due to the possible split point across the western and eastern faults of NNW/SSE, trend, shown by the seismic section, and gave the horst structure of the Oued Noumer field.

According to various geological criteria used for the evaluation of oil and gas potentialities, the Oued Noumer area is a favorable structure for oil and gasoline exploration; GOR values range, from $1200m^3/m^3$, for the upper "A" sandstone to, $470m^3/m^3$ in average ($450-480m^3/m^3$) for the lower "B" sandstone, considering the former reservoir as gasoline bearing rock, and the latter as oil bearing bed; the relation of GOR versus depth did not show either abrupt or gentle change between the two rocks (Personal communication).

A review of the geological and geophysical data shows, that some wells drilled in the surrounding areas of the Oued Noumer region are located at unfavorable geological conditions, e.g. wells Baa-1, Ga-1, Be-1, and ONN-1, or not drilled on optimal structural conditions; also, the negative results of the tests in wells EHA-1, AF-2, AF-4, can be explained by lack of favorable traps to the southwestern border of the Oued M'ya Triassic basins.

12.3.4 Parameters used for Measurement of Reserves

The original reserves of hydrocarbons in-place of the Oued Noumer and Ait Kheir fields for both reservoirs "A" and "B" are approximately estimated volumetrically, depending on the structural closure on top of related isobaths, taking in consideration the available isochrone map on top of "B".

The measurement of gas condensate (or gasoline) reserve is carried out from top of "A" to the condensate/oil contact, whereas, oil reserve is estimated from top of "B" to the oil/water contact. **Table 12.4** –Petroleum factors used for the estimation of gasoline and oil reserve of both the "A" and "B" reservoir rocks at the Oued Noumer area respectively.

Table 12.4 - Permeability Measurements of Well core samples at Oued Noumer Fields

Puits	Niveau	V/O Technoexport (1971)		Sonatrach (1973)	SN. REPAL 1973
		Moy. Arithm. (Ka)	Moy. Médiane (Km.)	Moyenne Médiane (Km.)	Average per Reservoir
ONR. 1 Bis	A	254,60	290	185	Oued Noumer reservoir 800 md.
	B	1.382,20	916	1.364	
ONR. 3	A	201,9	120	150	
	B	1.467,7	1.060	1.287	
ONR. 5	A	150,9	1,3	144	
	B	975,4	660	–	
AT. 1	A	–	–	34	Ait Kheir reservoir 750 md.
	B	–	–	1.034	
AT. 2	A	–	–	35,8	
	B	–	–	979	
AT. 3	A	–	–	15	
	B	–	–	630	

The following formulas are used for the estimation of condensate and oil reserves knowing that the quantities of the condensate (gasoline) are at normal pressure and temperature (NTP);

{First} Estimate of Oil and gasoline reserves in- place:

$$Qoil_{place} = F \cdot h_u \cdot \emptyset \cdot \check{S} \cdot \breve{G} / F.V.F$$

Where,
   F = area (m²)
   h_u = pay sandstone (m)
   $\emptyset$ = average porosity (%)
   $\check{S}$ = average saturation of hydrocarbon (%)
   $\breve{G}$ = specific mass (gm/cm²)
   F.V.F. = a Formation Volume Factor that indicates the quantity of hydrocarbon obtained on the surface (atmospheric pressure at 15 °C) in one meter cube of fluid under bottom conditions (408 kg/cm²; of 87°C);

{Second} Estimate of Gas Reserve:

$$Qgas_{\text{in-place}} = F \cdot h_u \cdot \emptyset \cdot \check{S} \cdot T \cdot P_o / P.Z.$$

Where:

$$T = \frac{T_{273} + \text{Atm. Temp } (15.5°C)}{T_{273} + \text{Formation Temp. } (87°C)}$$

P_o = initial formation pressure
P = Atmospheric pressure (1,033 kg/cm²)
Z = Gas deviation factor (1,086)

The average porosity ($\emptyset$) is determined for an approximate estimation of reserves in the Oued Noumer area, taking the pay sandstone thickness (hu):

$$\emptyset = \sum \frac{hu_1 \cdot \emptyset_1 + hu_2 \cdot \emptyset_2 + hu_3 \cdot \emptyset_3}{\sum hu_1 + hu_2 + hu_3}$$

The average saturation (Š) of hydrocarbons is determined in the same way; the net volume of the reservoir rock has been determined planimetrically from isopach maps. Hydrocarbon reserves of both Oued Noumer and Ait Kheir fields can be estimated from related charts.

12.3.5 An Approximate Estimate of Reserves of Oued Noumer and Ait Kheir fields

The study of petrophysical properties is based on the data analysis and interpretation of well log data. In Oued Noumer and Ait Kheir reservoirs, measurements were carried out according to available graphics and equations given by Dresser Atlas and Schlumberger.

In Oued Noumer field, the horizontal permeability of beds, ranges from insignificant values near (0.05md) up to 1000md in "A" and up to 6500md in "B" sandstone reservoirs. The maximum permeability is observed in "B" sandstone at ONR-1 bis (Technoexport (1971).

1. The Oued Noumer Field:

$Q_{in-place}$ for gas (A-series) = $8 \times 10^9 m^3$
$Q_{in-place}$ for cond (A-series) = 0.95×10^6 tons
$Q_{in-place}$ for oil (B-series) = 26.0×10^6 tons

2. The Ait Kheir Field

$Q_{in-place}$ for gas (A-series) = $6 \times 10^9 m^3$
$Q_{in-place}$ for cond (A-series) = 4×10^6 tons
$Q_{in-place}$ for oil (B-series) = 20.3×10^6 tons

Noticing that:

GOR: Gas = 1200 m3/m3 & Oil = 470 m3/m3
GOR Condensate (Gasoline) = 500-800 m3/m3
Formation Tempretaure = 87°C
Formation Pressure = 400 kg/cm²

12.4 Petroleum Hydrocarbon reserves of Hassi R'Mel Gas Field

12.4.1 Historical and Petrographical Aspect

Hassi R"Mel field is located on a dome-like structure; the dip of the sandstones from the crest of the structure is consistently less than one degree. It was discovered in the early 1957's, by the completion of HR-1 well which was drilled in the northern part of the field in 1958; the field was explored by drilling eight additional wells during the period of May 1957 to May 1960, in order to provide an estimate of hydrocarbon reserves, contained within the Upper Triassic sandstone section of the field (T2-unit "A-series"or 3a-bed); HR-10, drilled in October 1965; Hr-11, completed in November 1968; wells of HR-12 to HR-18, completed in 1971, and wells from HR-19 to HR-22 were completed together with HRS-1 in the beginning of 1972.

HR-4 well is the deepest well of the Hassi R'Mel field, where the depth of the basement (from ground surface), was reached at 2562.7ms; drilling was continued till the total depth reached at 2685.4ms. **Fig 12.5-** A Geographic Location Map of Hassi R'Mel Field.

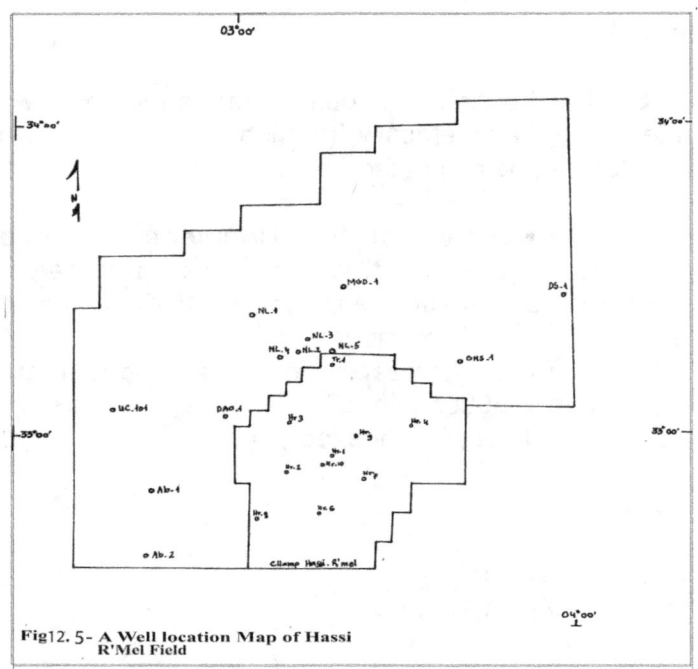

Fig12.5- A Well location Map of Hassi R'Mel Field

The distinct Triassic sandstone reservoirs: "A", "B", and "C"-series are clean sands, irregularly cemented by clay or dolomite with anhydrite streaks. The Triassic reservoir beds are intercalated by impermeable shale beds of at least 5ms in thickness, except in some locations where "B" and/or "C"-series were not well developed (e.g. HR-.4, HR-6 bis, HR-7 and HR-8).

No gas or oil shows were noted overlying the Triassic reservoirs, whereas the hydrocarbon productivity was limited to the upper Triassic detrital section (T2-unit); however oil shows was detected in the Ordovician formation.

The Senonian or Turonian sediments of the upper Cretaceous, outcrop in the Hassi R'Mel field; the Lower Cretaceous attained 600ms; the Jurassic 800ms, and the Triassic sediments that overlie the Cambro-Ordovician quartzites attained 600ms in thickness. **Fig 12.6 -** A subcrop map of Hercynian Discordance surface.

12.4.2 Sedimentology

The fluviatile type deposition, of the Lower series (Serie Inferieur, SI), marks the beginning of the T-1 unit, of different grain size, as the lower series is finer in grain size that might result from a south or southwest source of the fine grained Devonian or Gothlandian formations. The source of the T-1/C series, possibly came from the west (Morocco) or from the northwest (Oranese High Plateaus), where massive granites outcrop and supply a rather fresh feldspars; or might result from the erosion of the Pre-Atlassic flexure that divide the Saharan Atlas from the High Plateaus, or generally from the Atlassic Hercynian folded area (Technoexport, 1971).

The T-1/B and T-1/C series are closely associated, and are reasonably similar in time and space. The shaly character of the T-1/B series is possibly due to the accumulation during flood periods that resulted in the establishment of vast basins and became natural sites for shale deposition; thus the T-1/B unit attained its continental basin environment of deposition.

The stratigraphic level of the upper T-2 unit, is characterized by a basin type of deposition that suddenly changes its aspect and probably its type of deposition; finer grains are found in the thickest areas and the coarser grains in thin zones at pinchout areas. However, the T-2 unit is comparable and forms an identical sequence to the T-1/B unit. Therefore, the T-2 unit is the final stage of the fluviatile sequence and is likely the initial stage of the basin deposits. The Triassic sedimentation occurred between the marine incursions and the fluvio-lagoon deposits of continental origin.

The basins, created between the seas and the continent where carbonates and anhydrates precipitated, were then isolated from the open sea; the evaporitic cycle then started where great lagoons were established leading to the formation of salt deposits.

12.4.3 Reservoir Chracteristics

The reservoir characteristics of the three main sandstone reservoirs at the Hassi R'Mel field:

- **The "A" sandstone** - The determination of the gas /water contact of the "A" reservoir, was based on the assumption of the presence of a common water level in all parts of the field at 1500m subsea. The report of DeGolyer and Macnaughton, (Jul 1965) considered the G/W contact at 1500m subsea, and the Core Lab report, defined it at 1505m subsea (Oct 1965). However, data was insufficient to indicate any tilt in the water level or whether minor faulting had affected the contact. The net thickness of the "A"-reservoir at HR-8 well, is 16.4m thick. A drill stem test in HR-8 at a depth of 1501.5m subsea, recovered small quantities of gas and gasoline within the reservoir. The DST #3 for well HR-3, indicates the presence of gas and gasoline and water. At Tilrhemt (TR-1), the first DST was performed in July 1965 and yielded gas at the rate of 164,000m^3/day at 1491ms subsea, therefore, indicating that the gas/water contact should be below that point; a second DST was carried out at depth 1501.5ms subsea, which might represent the gas/water contact. **Fig 12.7** - is an isobaths map on top of "A" sandstone series (Unit 3a) of the Hassi R'Mel field.

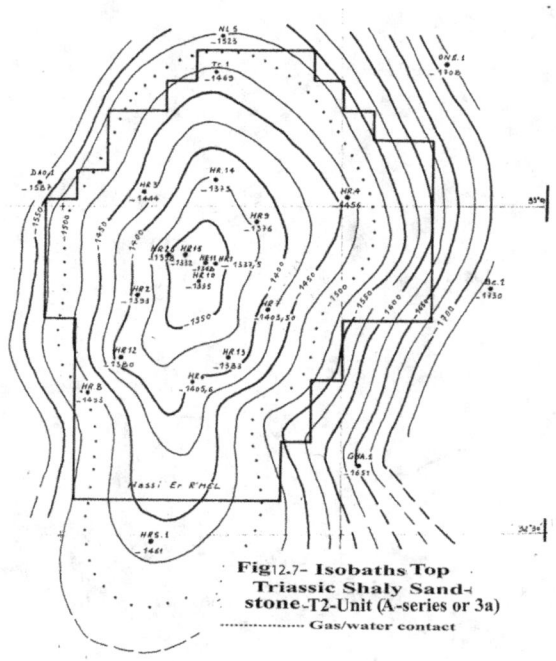

Fig 12.7- Isobaths Top Triassic Shaly Sandstone T2-Unit (A-series or 3a)
----------- Gas/water contact

- **The T-1/B unit**– At The Tinrhemt well (TR-1), the salt water-bearing zone is found at a depth 1504m subsea, whereas the electric well logs suggested a possible gas/water contact of HR-3, at a little lower elevation of 1504.5m subsea. **Fig 12.8** – An isobaths map on top of "B-series" (2a, 2b) sandstone.

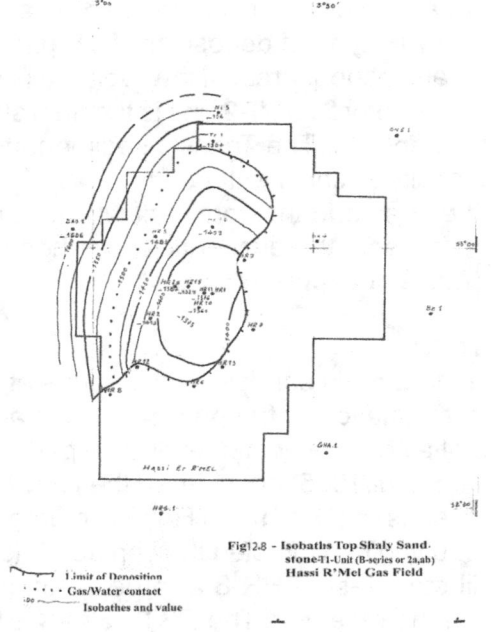

Fig12.8 - Isobaths Top Shaly Sandstone-T1-Unit (B-series or 2a,ab) Hassi R'Mel Gas Field

- **The T-1/C unit** –At HR-3, the salt water-bearing zone is found at a depth of 1520ms subsea, whereas, at HR-6, it was tested as a gas-bearing zone at a depth of 1462.4ms subsea, Thus, the effective lower limit of the gas accumulation was not precisely determined. **Fig 12.9**– is an isobaths map on top of "C" sandstone. The "B" and "C" sandstone series normally contain salt water at the Hassi R'Mel field. Generally, Gasoline at Hassi R'Mel may be restricted on the southern area where units T-1/B and T-1/C were already pinched out.

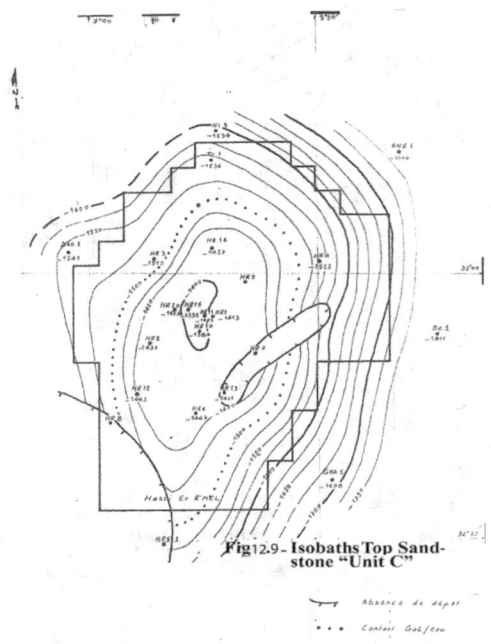

Fig12.9 - Isobaths Top Sandstone "Unit C"

- **The lower series** (SI) should be regarded as a piedmont series formed by reworking of the local substratum, filled with materials from the erosion of a nearby relief. The series includes important eruptive flows mainly of "andesites".

At the edge of the dome-like structure of the Hassi R'Mel field (Assaad, 1983) at HR-3, HR-4, HR-8, and TR-1-wells, were extremely poor gas producers being located in the proximity of the gas/water contact. The rest of Hassi R'Mel wells, that penetrated the Triassic section at higher structural levels, are excellent gas producers.

To the south of the field, an exploration well (HRS-1) was proposed by the author and a montage report was accomplished in 1972 (Internal report: Bulletin Sonatrach, Assaad, 1972); the "A" sandstone was the objective reservoir that yield gas at the rate of 192 m³/day to 104m³/day and gasoline at the rate of 3500 liter/hr. The bottom hole pressure was 312 kg/sq.cm, approximately at the gas/water contact of 1504ms subsea.

12.4.4 An Approximate Estimate of Gas Reserves

Gas and gasoline reserves of the Hassi R'Mel field were estimated by the author in Sept.1975, on the assumption that the Triassic sandstone forms a closed dome like-structure; the original gas in-place in each of the three main sandstone reservoirs (A, B, and C), has been approximately estimated volumetrically.

Figs 12.10, 12.11, and 12.12- show the sand Isopaches maps for "A", "B", and "C" sandstone series respectively, which were prepared and planimetered to obtain the total net productive sand volumes; the "A" zone is continued over all the entire system; the "B" zone is only developed to the NW at wells of TR-1, HR-1, HR-2, HR-3, HR-9, HR-10, HR-12, HR-14 and, HR-20. The "C" zone is not represented to the east at HR-7 and HR-13-wells (Stoica and Assaad, 1972).

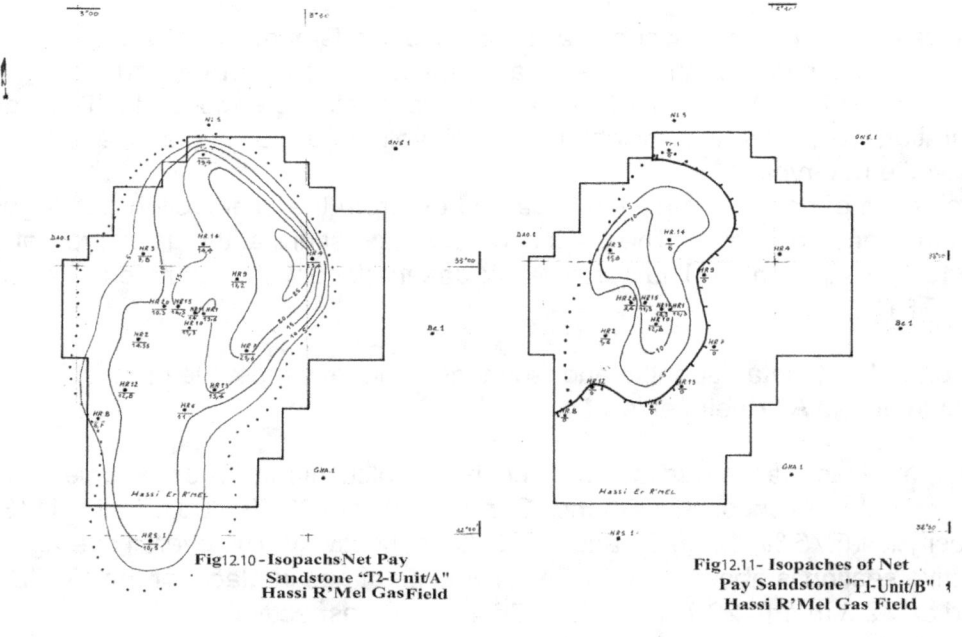

Fig12.10 - Isopachs Net Pay Sandstone "T2-Unit/A" Hassi R'Mel GasField

Fig12.11- Isopaches of Net Pay Sandstone "T1-Unit/B" Hassi R'Mel Gas Field

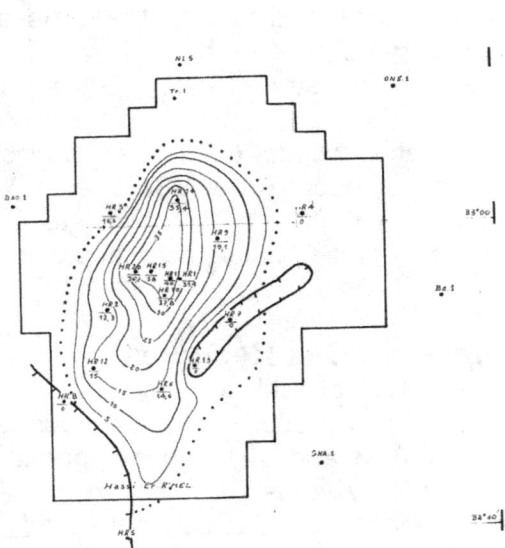

Fig 12.12 - Isopaches of Net pay Triassic shaly Sandstone "T1-Unit/C"

The Triassic reservoirs can be described as a massive sandstones, slightly cemented, and of medium to coarse grained with occasional shale intercalations. Calculations from electric well logs, indicate that water saturation ranges from 12% to 26%; the average connate water content is 20%; average porosity and water saturation were integrated to obtain the unit volume of gas in-place, and to estimate the gas in place of the whole reservoir.

The determination of gas accumulation is carried out from the intersection of the gas/water contact up to the top of the sandstone zone. The original reservoir pressure at the gas/water (at 1503m subsea) was equal to 310-312 kgm/cm.sq. The quantities of gas-in-place were estimated at normal temperature and pressure (NTP).

Table 12.5 – General well data, Isobaths and Isopaches data of sandstone units "A", "B" and "C", for both Oued Noumer and Ait Kheir fields

Table 12.6 – summarizes the various factors used in the calculations of gas reserves. The average permeability of the three zones: "A", "B" and "C" ranges from 435.78, 303.07, and 1143md, and the average porosity is 15.75 %, 14.36 %, and 17.63 %, respectively. The overall average porosity for the entire productive section is about 16%; the initial volume of gas in-place for the "A", "B" and "C" zones are estimated as 32.8 m^3/m^3, 29.7 m^3/m^3, and 36.4 m^3/m^3, respectively.

Table 12.5 - Petroleum Factors for Reserves of Oued Noumer and Ait Kheir oil, Gasoline and Gas Fields

interprété par Fakhry A. ASSAD

Caractéristiques des Fluides du Gisement	Unité de Mesure	Champ d'Aït Kheir			Champ d'Oued Noumer		
		Grès "A"	Grès "B"		Grès "A"	Grès "B"	
		gaz et condensat	cond.& huile	condens. & huile	gaz et condens.	cond.& huile	condens. & huile
Superficie prouvée	km2	25,50	48,10	50,50	20,40	13,10	23,60
Hauteur moyenne utile de grès	m	13,00	2,20	9,70	11,59	1,80	23,95
Volume utile de roches magasins	$m^3 \cdot 10^6$	331,04	105,82	489,85	236,45	23,58	565,25
Porosité moyenne	%		14,70	19,30		14,80	21,40
Saturation moyenne en hydrocarbures	%		67,70	79,50		78,60	85,00
Volume des roches saturées en hydroc.	$m^3 \cdot 10^6$	32,62	10,53	75,16	27,51	2,74	102,81
Facteur de gaz à condensat	m^3/m^3	-	-	-	660	460	-
Quantité de condensat correspondant à 1 m3 de gaz produit	cm^3/m^3	-	-	-	1515	2174	-
Facteur de volume (408 kg/cm2, 87° C)	-	-	2,445	2,43	2,77	2,445	2,43
Densité du gaz aux conditions de surface (air = 1)	g/cm^2	0,67	-	-	0,67	-	- *
Dens. du condensat, conditions de surf.	$g/cm2$	-	0,72	-	-	0,72	- *
Dens. de l'huile, cond.de surf. (eau)	$g/cm2$	-	-	0,80	-	-	0,80 *
Pression ds le gisement / Tempér.	$kg/cm2$ / °C	410/87	-	-	408/87	-	-
1/F.V.F.	m^3/m^3	-	0,409	0,412	0,36	0,409	0,412
V.∅.S Hydrocarb./PVF	$10^6 m^3$	-	4,21	30,97	-	1,10	42,15
Coefficient de compression du gaz Z (à l'air)	-	-	1,06	-	-	1,06	-
Réserves de gaz & condensat en place	$10^9 m^3$	9,74	-	-	8,18	-	-
Réserves de condensat & huile en place	10^6 T.	-	3,03	24,77	-	0,79	33,72
Totaux	10^6 T.	-		27,80	-		34,51

* Considering the petrophysical characteristics of Ait Kheir Field as that of Oued Noumer Field

Table 12.6 - A summary of Reservoir factors of gas and gasoline Field- Hassi R'Mel Field

Concession Nature de Fluide	Hassi R'Mel Gaz + eau			Septembre 1972	
Characteristics of reservoir Fluids	Unité de Mesure	Trias "A"	"B"	"C"	Total	moyen
Area top reservoir	Km^2	3427.5	1125.0	1625.0	3427.5	
Average Net pay sandstone	m	11.11	5.41	13.7		19.69
Volume Net Pay Sandstone	$m^3 \times 10^6$	38079.53	6790.75	22262.50	67187.98	
Average Porosity	%	15.75	19.56	17.65		16
Average Water Saturation	%	20.0	20.0	20.0		20.0
Reservoir Temperature	°C	90.0	90.0	90.0		90.0
Subsea top reservoir (Prof. toit)	m	-1503	-1503	-1503		-1503
Original pressure of reservoir (Pi)	kg/cm^2	310	310	310		310
Reservoir Pressure	kg/cm^2	58.5	58.5	58.5		58.5
Compressibility Factor (original) (Zi)		0.949	0.949	0.949		0.949
Final Compressibility Factor (Zt)		0.921	0.921	0.921		0.921
Original gas in place	m/m.mch	32.1	29.70	56.1	32.72	
Recoverable gas	" "	26.76	23.59	29.52	26.7	
Recoverable gas in place	" "	6.04	6.11	6.78	6.02	
Gas in place (Gp)	$m^3 \times 10^6$	1825846.96	196661.82	1022302.1	2206898.92	
Gas recoverable (Gr)	" "	997642.7	160058.95	652172.22	1796160.92	

By applying the above factors to the volume of net reservoir rock in the three sands, it yields a total of approximately 2.21×10^{12} m³ of gas-in place; a sum of about 29.4 % more than that estimated by Core Lab, in 1965, as the total volume of gasoline in-place had been estimated approximately of 1.7×10^{12} m³.

The calculation of recoverable gas was based on an abandonment pressure of 50kg/sq.cm at the well head. The equivalent abandonment pressure at bottom hole conditions was 58.5 kg/sq.cm and the recoverable gas reserves had been calculated as approximately of 1.8×10^{12} m³.

Assuming a cumulative production in the first ten years from 1960 –1970, to be equal to 0.0175×10^{12} m³ of gas; the remaining recoverable reserves of gas should be approximately equal to 1.7787×10^{12} m³.

The production data during the two years of 1963 and 1964 indicated that shrinkage of 4% by volume occurs at the center of separation because of the extraction of gasoline. In addition, the field usage and losses have mounted to 6% of the produced gas (DeGolyer and MacNaughton, 1965).

The author considered the shrinkage factor of 8 % because of plant and field losses, which resulted to a volume of 1.6364×10^{12} m³ of commercial gas; if total losses can be reduced to 5 %, a volume of approximately 1.6898×10^{12} m³, can be obtained of remaining and commercial gas reserves.

References

Assaad, F., (1983) An approach to "Halokinematics" and Interplate Tectonics (North- Central Algeria), Jr. Petroleum Geology, Vol.6, No. 1, July 1983.

--------------, (1981) A further Geologic Study on the Triassic Formations of North- Central Algeria with Special Emphasis on Halokinesis, Journal of Petroleum Geology, 4,2, pp. 163-176.

-------------, (1972) Contribution to the study of the Triassic formations of the sersou-Megress Region (High Plateau) and the area of Daia M'Zab (Saharan Platform), No. 84 (B3), Eighth Arab Petroleum Congress, Algeria, 1972. sondages Algero-Tunisians

SN RÉPAL (1961) Les séries Permo-Triassiques dans le Nord Sahara. Études Pétrographique du cycle Détritique. Etudes du Cycle Salifére (texte et planches).

SNPA (1964) Étude Microstratigraphique et Sédimentologique des séries du Trias á Tertiaire recontrées par les Sondages de l'Hydraulique des Chotts Rharbi et Chergui.

Core Lab (1965) Estimaion de Gas et Condensat, Hassi R'Mel, Core Lab., Alger, Algiers.

Sonatrach; Aliev, M., Ait Loussine et al, (1971) - Geological Structures and estimation of oil and gas in the Sahara of Algeria: Spain, Altamira-Rotopress, S.A., 265p

Stoica, I., and Assaad, F.A. (1972) Geological Studies on the Triassic Reservoir of Condensate and Gas at the Hasssi R'Mel Field (Internal Report, Sonatrach).

Appendix 12. A1

English Translation

For Table 12-5 - Petroleum Factors for reservoirs of Oued Noumer and Ait kheir Oil, Gasoline and Gas fields

Superficie prouvèe = Area top reservoir
Hauteur moyenne utile de grès = Average net pay sandstone
Volume utile de roches magazins = Volume net pay sandstone reservoir
Porositè moyenne = Average porosity
Saturation moyenne en hydrocarbures = Average hydrocarbon Saturation
Volumes des roches saturèes en hydroc. = Volume of saturated hydrocarbon bearing reservoir rock
Facteur de gaz à condensat = Factor of gasoline
Quantitè de condensat correspondant a $1m^3$ de gaz produit = Quantity of gasoline related to $1m^3$ of gas produced
Facteur de volume ($408kg/cm^2$, 87 °C) = Volume factor ($408kg/cm^2$, 87 °C)
Densitè du gaz aux conditions de surface (air =1atmos.) = Gas density at surface (air =1atmos.)
Densitè du condensat aux conditions de surface = Gasoline density at surface
Densite' de l'huile, cond. du surf. (eau) = Oil density at surface conditions (water)
Pression {de la gisement} = Pressure/Temperature (of the reservoir rock}
Temperature=Temp.
1/ F.V.F = 1/F.V.F
V.O.S Hydrocarbures 1/FVF = V.O.S Hydrocarbons 1/FVF
Coefficient de compression du gaz Z (à l'air) = Compressibility coefficient of gas Z (in the air)
Réserves de gaz & condensat en place = Reserves in place of gas and gasoline
Réserves de condensat & huile en place = Reserves in place of gasoline &oil
Toteaux = Totals

Other translated French geologic expressions in some figures:

tronquent = truncating
selon le degre' = according to the degree
par la suite = later on
introduire = to introduce
nayant souvant qu'un = but, mostly
tout a' fait = only
remittent en cause = calling into question
jusqua' ici = up till now
en effet= as a matter of fact
recueillis = collected
sont identiques = are identified

Presque = nearly
Sauf sur les bords du basin= except on the edges of the basin
Pourrait indiquer =might indicate
Peu accentue' = little dipping or gentle relief
Sou unite' = subunit
Quelques règles = some rules
La fermature initiale =initial shut in pressure

Appendix 12.A2- A Possible Gravemetric method for Hydrocarbon Prospects

The problem of new discoveries of oil and gas prospects in Algeria remained unsolved. Since the giant discovery of both Hassi R'Mel and Hassi Messaoud fields in 1956, many other smaller Triassic fields were found at considerable expenses as the seismic survey of a multiple closure remained the main method for geophysical prospecting of promising structures, and deep drilling might take place before equally covering the necessary geological studies of the encountered areas, regardless of the required number of seismic lines and sections. Though it happened that some areas were surveyed many times through shooting tests by different seismic crews of different companies where other areas were left virtually unexplored. The gravimetric methods are much less expensive than that of the seismic and the volume of the seismic work, can then be reduced, and therefore inrease its effectiveness by using either a scale of 1:100,000 or 1: 50.000, thus allowing better locations of seismic profiles.

The gravimetric application may accordingly play an important role in exploring structural prospects within the Triassic province as it depends on the differences in rock densities, e.g. the difference in densities between the basement and the Cambro-Ordovician was found to be 0.1-0.2 gr/cc, whereas, that between the Upper Principal saliferous formation and the overlying Jurassic was equal to 0.2-0.3 gr/cc.

Fig 12.A1- A Graphic presentation of Gravimetric anamolies in a productive and non-productive reservoir.

Appendix 12.A3

a) List of most Types of Petroleum Conversions, Measurements, and Geological Symbols

The following Abbreviations, conversions, Geologic Strutures, and AGI-sheets are given below:

One mile	= 1.6Km
One PSI	= 70.3067 cm^2
One Sq. mile	= 640 acres
One sq. Km	= 247.104 acres
One cubic ft.	= 7.489 gallons
One Ft.	= 3314 cm
One meter	= 3.14 ft
One meter	= 39,370 in
One inch	= 2.25 cm
One foot	= 12 in
M	= Thousands
MM	= Billion barrels
BBO	= Billion Barrels of Oil

Natural gasoline:

                    In HPT   = gallons/ thousands
                    In MCF   = thousands cubic ft

b) The following Images (JPG) comprises:

(1) Conversions of different scales
(2) Geological Stuructures
(3) AGI measurement Conversions
(4) AGI Gemological Weigts & Measures
(5) AGI Geologic Symbols
(6) AGI Oil Conversion Data
(7) AGI Data sheets 52.2 ^& 51.1

c) *Glossary* -General Stratigraphic and Structural Terms

- **Flysch**—Widespread deposit of sandstones, marls, shales, and clays, found in the north or south portions of the Alps.
- **Spilite**—A basaltic rock with albitic or sodic feldspars. The albitic feldspar is usually accompanied by autometamorphic minerals, or minerals characteristic of low-grade greenstones such as chlorite, calcite, epidote, silica or quartz. "**NB**: Albitization is the transformation process of basalt to spilites under low grade metamorphism."
- **Orogenic Movement**—Mountain making- denoting the process of formation of mountain ranges by folds, faults, upthrusts, and overthrusts, affecting comparatively narrow belts and lifting them into great ridges.
- **Epeirogeny and Epeirogenic**—Designating the broad movements of uplift and subsidence which affect the whole or large parts of continents and the oceanic basins
- **Epeirogenic Movements**—Movements of the earth's crust which produce and maintain the continental plateaus and the broad depressions which are covered by the sea.
- **Fluvial Deposits**— pertaining to rivers or produced by a fresh water river.
- **Alluvial Deposits**— unconsolidated materials of recent times (deltaic), made up of alluvium.
- **Alluvial Fan**—A gradually sloping mass of alluvium that widens out like a fan from the place where a stream slows down little by little as it enters a plain.
- **Delta**—An alluvial deposit, usually triangular, at the mouth of a river.
- **Deltaic Deposits**—Sedimentary deposits laid down in a river delta.
- **Nappe**—A large body of rock that has moved forward more than one mil from its original position, either by overthrusting or by recumbent folding.
- **Nappes**—Faulted overthrusted folds.
- **Lacustrine**—having to do with a lake or lakes; found or formed in lakes
- **Terrace**—A geologic formation; any of a series of flat platforms of earth with sloping sides, rising one above the other, as a hillside; or a raised, flat mound of earth with sloping sides.
- **Diapir**—A dome formation in which the rigid top layers have been split open by pressure from an underlying plastic core.
- **Geosyncline**—A very large, trough-like depression in the earth's crust containing masses of sedimentary and volcanic rocks.
- **Geosynclinal Prism**—The load of sediments which accumulates in the down-warped area of a geosyncline.
- **Geosyncline Trough**—A region subsiding over a long time, containing sedimentary and volcanic rocks of great thicknesses, primarily interpreted as an evidence of subsidence.

- **Orogeny**—The formation of mountains through structural disturbance of the earth's crust, especially by folding and faulting.
- **Foreland**—The relatively stable area, lying in shallow water, represented by the continental platform; or theresistant block towards which the geosynclinal sediments move when compressed; in structure, it is the region in front of a series of overthrust sheets.
- Sedimentary Basins—A wide, depressed area in which the rock layers all incline toward a center.
- **Appalachian Mountains**—A mountain system in eastern North America; "known for its tea plant leaves used in pioneer times; Appalachian tea plants."
- Island Arcs—A group of islands having a curving arc-like pattern. Most the island arcs lie near the continental masses, but inasmuch as they rise from the deep ocean floors, they are not a part of the continents proper.
- **Rif**—A Mountain range along the northeastern coast of Morocco, extending from the Strait of Gibraltar to the Algerian border; highest peak 8,000 ft (2,440 m).
- **Rift**—A large fault along which movement is mainly lateral, or simply an opening caused by splitting with a lateral movement (a fissure).
- **Mesogeosyncline**—A geosyncline between two continents and receiving clastics from both of them.
- **Deltaic Deposit**—A sediment deposit laid down in a well developed delta; can be characterized by cross bedding, mixture of sand and clay, and remains of brackish matter.
- **Great Rift Valley**—Depression of southwestern Asia and East Africa, extending from the Jordan River across Ethiopia and Somalia to the lakes region of East Africa.
- **Ridge**—A relatively narrow elevation which is prominent because of the steep angle at which it rises.
- **Ophiolite**—A basic igneous rock belonging to an early phase of the development of a geosyncline, and subsequently altered into rocks rich in serpentine, chlorite, epidote, and albite.
- **Andesite**—A Volcanic rock (with Mafic constituents, as plagioclase)
- **Pangea**—The hypothetical single landmass that split apart about 200M yrs ago, and formed Godwana and Laurasia.

www.ingramcontent.com/pod-product-compliance
Lightning Source LLC
Chambersburg PA
CBHW081142180526

45170CB00006B/1898